Practical Applications of Model Predictive Control and Other Advanced Control Methods in the Built Environment

Practical Applications of Model Predictive Control and Other Advanced Control Methods in the Built Environment

Editor

Etienne Saloux

Basel • Beijing • Wuhan • Barcelona • Belgrade • Novi Sad • Cluj • Manchester

Editor
Etienne Saloux
CanmetENERGY, Natural
Resources Canada
Varennes
Canada

Editorial Office
MDPI
St. Alban-Anlage 66
4052 Basel, Switzerland

This is a reprint of articles from the Special Issue published online in the open access journal *Buildings* (ISSN 2075-5309) (available at: https://www.mdpi.com/journal/buildings/special_issues/Model_Control).

For citation purposes, cite each article independently as indicated on the article page online and as indicated below:

Lastname, A.A.; Lastname, B.B. Article Title. *Journal Name* **Year**, *Volume Number*, Page Range.

ISBN 978-3-7258-0473-3 (Hbk)
ISBN 978-3-7258-0474-0 (PDF)
doi.org/10.3390/books978-3-7258-0474-0

Contents

About the Editor . . . . **vii**

Preface . . . . **ix**

Etienne Saloux
Practical Applications of Model Predictive Control and Other Advanced Control Methods in
the Built Environment: An Overview of the Special Issue
Reprinted from: *Buildings* **2024**, *14*, 534, doi:10.3390/buildings14020534 . . . . **1**

Etienne Saloux, Kun Zhang and José A. Candanedo
A Critical Perspective on Current Research Trends in Building Operation: Pressing Challenges
and Promising Opportunities
Reprinted from: *Buildings* **2023**, *13*, 2566, doi:10.3390/buildings13102566 . . . . **6**

Martin Gabriel and Thomas Auer
LSTM Deep Learning Models for Virtual Sensing of Indoor Air Pollutants: A Feasible
Alternative to Physical Sensors
Reprinted from: *Buildings* **2023**, *13*, 1684, doi:10.3390/buildings13071684 . . . . **35**

Pooria Norouzi, Sirine Maalej and Rodrigo Mora
Applicability of Deep Learning Algorithms for Predicting Indoor Temperatures: Towards the
Development of Digital Twin HVAC Systems
Reprinted from: *Buildings* **2023**, *13*, 1542, doi:10.3390/buildings13061542 . . . . **56**

Kashif Irshad, Md. Hasan Zahir, Mahaboob Sharief Shaik and Amjad Ali
Buildings' Heating and Cooling Load Prediction for Hot Arid Climates: A Novel Intelligent
Data-Driven Approach
Reprinted from: *Buildings* **2022**, *12*, 1677, doi:10.3390/buildings12101677 . . . . **80**

Chris Price, Deokgeun Park and Bryan P. Rasmussen
Cascaded Control for Building HVAC Systems in Practice
Reprinted from: *Buildings* **2022**, *12*, 1814, doi:10.3390/buildings12111814 . . . . **98**

Navid Morovat, Andreas K. Athienitis, José Agustín Candanedo and Benoit Delcroix
Model-Based Control Strategies to Enhance Energy Flexibility in Electrically Heated School
Buildings
Reprinted from: *Buildings* **2022**, *12*, 581, doi:10.3390/buildings12050581 . . . . **120**

Javier Arroyo, Fred Spiessens and Lieve Helsen
Comparison of Model Complexities in Optimal Control Tested in a Real Thermally Activated
Building System
Reprinted from: *Buildings* **2022**, *12*, 539, doi:10.3390/buildings12050539 . . . . **141**

Etienne Saloux and Kun Zhang
Data-Driven Model-Based Control Strategies to Improve the Cooling Performance of
Commercial and Institutional Buildings
Reprinted from: *Buildings* **2023**, *13*, 474, doi:10.3390/buildings13020474 . . . . **166**

Malek Almobarek, Kepa Mendibil and Abdalla Alrashdan
Predictive Maintenance 4.0 for Chilled Water System at Commercial Buildings: A Systematic
Literature Review
Reprinted from: *Buildings* **2022**, *12*, 1229, doi:10.3390/buildings12081229 . . . . **190**

Malek Almobarek, Kepa Mendibil, Abdalla Alrashdan and Sobhi Mejjaouli
Fault Types and Frequencies in Predictive Maintenance 4.0 for Chilled Water System at
Commercial Buildings: An Industry Survey
Reprinted from: *Buildings* **2022**, *12*, 1995, doi:10.3390/buildings12111995 . . . . **219**

Malek Almobarek, Kepa Mendibil and Abdalla Alrashdan
Predictive Maintenance 4.0 for Chilled Water System at Commercial Buildings: A
Methodological Framework
Reprinted from: *Buildings* **2023**, *13*, 497, doi:10.3390/buildings13020497 . . . . **234**

About the Editor

Etienne Saloux

Etienne Saloux is a Research Scientist at CanmetENERGY, Natural Resources Canada, Varennes (Québec, Canada). He holds a bachelor's degree in mechanical engineering from ISAE-Supméca, France (2010), and a Ph.D. from Polytechnique Montréal, Canada (2014). Prior to his position at CanmetENERGY, he worked as a postdoctoral fellow at Polytechnique Montréal (Canada) and Sherbrooke University (Canada). His main research activities encompass topics such as energy systems, energy efficiency, renewables, exergy assessment, and advanced controls for buildings and communities. His past and current research deals with the optimization of the design and the control of heating, ventilation, and air conditioning (HVAC) systems; thermal energy storage; solar collectors; heat pumps; photovoltaic panels; and organic Rankine cycles. In particular, he is currently actively involved in the areas of virtual energy meters, supervisory control strategies, and energy flexibility, as well as in projects dealing with operational data and targeting the use of data analytics and artificial intelligence techniques to improve building system operation.

Preface

Despite the development of increasingly efficient technologies and the increasing amount of available data from building automation systems and connected devices, buildings are still far from reaching their performance potential due to inadequate system controls and suboptimal operation sequences. To assist researchers and practitioners in the field of building controls, this reprint tackles the development of tools and algorithms that may facilitate the adoption of advanced control methods in the building control industry. It eventually aims to help bridge the gap between theory and practical implementations in actual buildings. More specifically, this Special Issue covers various topics such as virtual sensing of indoor air pollutants; prediction models for indoor air temperature as well as building heating and cooling loads; local and supervisory control strategies and predictive maintenance algorithms to improve setpoint tracking and optimize the operation of building heating and cooling systems. I would like to express my gratitude to MDPI and Buildings journal as well as the editorial staff for their excellent work, the reviewers for their dedication to peer review, and all the 25 authors for their contribution to this reprint.

Etienne Saloux
Editor

Editorial

Practical Applications of Model Predictive Control and Other Advanced Control Methods in the Built Environment: An Overview of the Special Issue

Etienne Saloux

CanmetENERGY, Natural Resources Canada, Varennes, QC J3X 1P7, Canada; etienne.saloux@nrcan-rncan.gc.ca

Abstract: This paper summarizes the results of a Special Issue focusing on the practical applications of model predictive control and other advanced control methods in the built environment. This Special Issue contains eleven publications and deals with various topics such as the virtual sensing of indoor air pollutants and prediction models for indoor air temperature and building heating and cooling loads, as well as local and supervisory control strategies. The last three publications tackle the predictive maintenance of chilled water systems. Most of these publications are field demonstrations of advanced control solutions or promising methodologies to facilitate the adoption of such control strategies, and they deal with existing buildings. The Special Issue also contains two review papers that provide a comprehensive overview of practical challenges, opportunities, and solutions to improve building operations. This article concludes with a discussion of the perspectives of advanced controls in the built environment and the increasing importance of data-driven solutions.

Citation: Saloux, E. Practical Applications of Model Predictive Control and Other Advanced Control Methods in the Built Environment: An Overview of the Special Issue. *Buildings* **2024**, *14*, 534. https://doi.org/10.3390/buildings14020534

Received: 18 January 2024
Accepted: 4 February 2024
Published: 17 February 2024

1. Foreword

Many buildings are adopting increasingly complex heating, ventilation, and air conditioning (HVAC) systems to improve their performance and contribute to the decarbonization of the building sector. However, inadequate system controls and suboptimal operation sequences prevent buildings from reaching their full performance potential. Advanced controls are thus required to maximize building performance at low or no capital cost. The field of building operation is undergoing a paradigm shift; buildings are evolving from being passive and reactive, simply responding after a change occurs, to being dynamic and proactive, detecting and correcting inefficiencies, anticipating changes, and adjusting the operation accordingly. The ever-increasing availability of operational data on buildings and the significant advancements in modelling capabilities represent an untapped opportunity to better manage and operate buildings [1].

Advanced control methods such as model-based controls or model predictive controls are widely acknowledged as effective solutions for improving building operation. Despite extensive investigation in the past, widespread adoption has yet to be realized. Existing buildings are inherently imperfect and may present obstacles to the implementation of advanced controls that might not be encountered in simulation studies; these include, among others flaws and inefficiencies in existing controls, a lack of data for critical variables and communication issues with the building automation system or external services [1]. These barriers may significantly affect the development of advanced control strategies and, thus, their deployment in actual buildings. This Special Issue has collected publications that deal with existing buildings and address some of these gaps. It ultimately aims to encourage research and foster the adoption of practical data-driven solutions to improve the operation of existing buildings.

2. The Papers

The Special Issue presents eleven publications tackling various topics related to advanced controls in buildings that are summarized in this section. They have been organized based on the scale of the application, from the sensor level to the supervisory control and predictive maintenance of heating and cooling systems. Figure 1 illustrates the various topics explored in the Special Issue with a word cloud plot based on titles and abstracts.

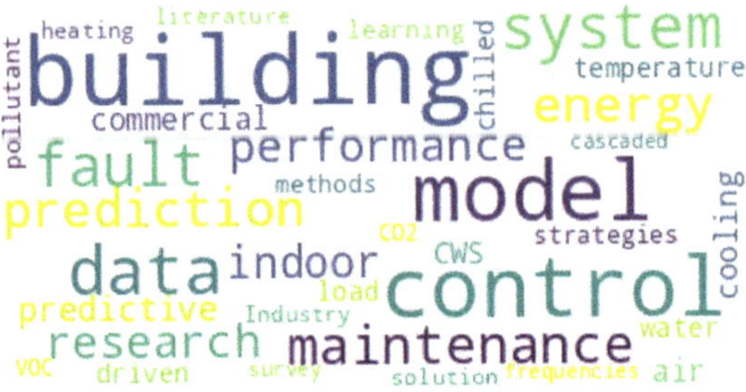

Figure 1. Word cloud plot based on titles and abstracts of research papers published in the Special Issue.

The opening article by Saloux et al. [1] is a perspective paper focusing on current research trends in the field of building operation. The authors explored pressing challenges and identified inefficient local controls, inadequate data availability and quality, communication issues with the building automation system, and the lack of guidelines and standards tailored for controls as critical obstacles to the widespread adoption of advanced controls. According to the authors, cost-effective solutions, successful case studies, and better training and engagement between the industry and research communities could also foster the uptake of advanced controls. They also discussed promising opportunities and highlighted the increasingly important role of modelling for various control applications, as well as the inevitable paradigm shift from building energy efficiency towards decarbonization. Data-driven and data analytics methods (e.g., automated fault detection and diagnosis, predictive maintenance, key performance indicators, virtual sensors and virtual energy meters, thermal and electric energy use breakdown, and automated data labelling), as well as high-performance local controls, occupant-centric controls, and advanced supervisory controls, including predictive control, were identified as promising avenues for improving building performance.

Gabriel and Auer [2] tackled the critical topic of human health and well-being and investigated the virtual sensing of indoor air pollutants using deep learning models. This virtual sensor is an alternative to physical sensors and relies on a long short-term memory (LSTM) neural network model to estimate indoor air pollutant concentrations such as particulate matter (PM), volatile organic compounds (VOCs), and CO_2 concentrations from building automation system data (e.g., temperature, humidity, illumination, noise, motion, and window state), as well as weather conditions and outdoor pollution data. The case study building is a high-rise office building located in Munich (Germany). The authors found that the proposed virtual sensing method is suitable for determining PM and VOCs and is less accurate but still reasonable for estimating CO_2 levels; they also demonstrated the potential for generalization and transfer learning.

Norouzi et al. [3] studied the application of deep learning algorithms to predict indoor air temperatures in educational buildings as part of a broader project toward the development of a digital twin for HVAC systems. They explored five algorithms (extra trees, random forest, multilayer perceptron, LSTM, and convolutional neural networks) along with different sliding windows (i.e., inputs at previous time steps; up to 120 min) and forecast horizons (up to 60 min). They tested their model on seven zones of a university building on the campus of the British Columbia Institute of Technology in Burnaby (British Columbia, Canada). The results showed that deep learning algorithms outperformed tree-based algorithms and can reach an average root mean square error of 0.16 °C.

Irshad et al. [4] developed a novel approach to predict residential building heating and cooling loads in hot arid climates. This approach relies on a shuffled shepherd red deer optimization linked self-systematized intelligent fuzzy reasoning-based neural network. The authors tested their model for the climate of Al-Dhahran (Saudi Arabia) and generated typical heating and cooling load profiles for residential buildings in such climates; for this purpose, they developed a database of 70 buildings with specific characteristics and systems. Simulations showed that the proposed model outperformed conventional methods in terms of various accuracy metrics.

Price et al. [5] presented the implementation of cascaded control architectures for air handling unit chilled water valves at three university campus buildings located at Texas A&M University in College Station (TX, USA). This cascaded control aims to overcome the often-overlooked issue of actuator hunting and targets better tracking and more consistent performance of control loops. This cascaded system only requires one additional line of code to existing control routines in the building automation system and was tested for more than a year in the case study buildings. Field implementation showed 2.2–4.4% energy savings, in addition to reduced operational costs for maintenance and controller retuning.

Morovat et al. [6] investigated model-based control strategies, aiming to unlock the energy flexibility of electrically heated school buildings. These strategies build on data-driven grey-box models to evaluate the optimal duration of building preheating. They were tested in a school building, located near Montreal (QC, Canada), equipped with geothermal heat pumps, hydronic radiant floors, and energy storage systems. The authors tested different resistance–capacitance thermal network models and used the dynamic building energy flexibility index (BEFI) to evaluate the performance. Simulations showed that the energy flexibility can be improved by 40% to 65% during peak demand periods.

Arroyo et al. [7] compared the performance of a model-based predictive control strategy using different types of control-oriented modelling paradigms (white-box, grey-box, and black-box models). The strategy was implemented during a 26-week period in a 6 m^2 test building zone equipped with a thermally active building system (TABS) and located in the Arenberg campus of the KU Leuven University in Heverlee (Belgium). The authors found that there was no significant correlation between prediction and control performance and that "a better prediction performance does not necessarily indicate an improved control performance"; the white-box model performed worse in prediction but led to better MPC performance. They eventually suggested using a modelling approach that combines both physics-based and data-driven methods.

Saloux and Zhang [8] evaluated the performance of three data-driven model-based control strategies to improve the cooling performance of commercial and institutional buildings: (a) chiller sequencing, (b) free cooling, and (c) supply air temperature reset. These strategies rely on simple yet accurate models, calibrated with operational data, and aim to be readily implementable in existing buildings using simple control rules. They were applied to an existing 36,000 m^2 commercial building in Montreal (QC, Canada) and simulations showed that the three measures together could reduce building cooling energy by 12% and cooling system electric energy by 33%.

Almobarek et al. [9] performed a systematic review of the literature on predictive maintenance applications of chilled water systems (CWS) and focused on two aspects: (1) the identification of operational faults and (2) the methods to better predict them. The authors covered chillers, cooling towers, circulating pumps, and terminal units and pinpointed the lack of studies tackling the entire CWS (rather than focusing on specific components); they also suggested that more attention should be given to cooling towers and pumps. The authors provided exhaustive lists of system faults and predictive tools for control, discussing operational parameters and data collection considerations. They finally identified research gaps such as the lack of information about the fault type or the data collection and the need for maintenance programs to go beyond fault detection and prediction and address the faults in the actual system.

Almobarek et al. [10] conducted an industry survey to complement the aforementioned systematic review outputs [9]. This survey targets the identification and frequencies of more operational faults and fault solutions for chilled water systems. The authors contacted 761 maintenance officers in different commercial buildings in Riyadh (Saudi Arabia) and compiled a total of 304 responses. The authors presented exhaustive tables of the faults and their solutions for each component (chillers, cooling towers, pumps, and terminal units). They also investigated the optimal data sampling time and history for the major faults, aiming to create better datasets to train machine learning models for predictive maintenance applications.

Finally, the closing article by Almobarek et al. [11] addressed the need for maintenance programs raised in the aforementioned system review [9] to implement predictive maintenance in existing buildings. They proposed a methodological framework to describe the requirements (drawings, measuring devices, and operational data), develop machine learning algorithms, and conduct quality control (ensure proper operation and correct faults). Within this framework, a decision tree model was developed to predict faults and was implemented in a university building in Riyadh (Saudi Arabia); the results showed that the developed model improved the fault prediction by more than 20% in all chilled water system components compared to the existing control system.

3. Perspectives and Conclusions

The eleven articles published in the Special Issue showcase just a few of the practical data-driven solutions that can be developed to encourage the widespread adoption of advanced controls in buildings. In the context of climate change, building performance targets are slightly reoriented from energy efficiency to decarbonization, where energy flexibility plays a pivotal role [1]. Virtual sensors [2] are emerging as a credible, cost-effective alternative to physical sensors for tracking critical variables, while data-driven modelling is becoming the cornerstone of many advanced control strategies [1]. Ongoing efforts to improve models for indoor air temperature [3] and building energy loads [4] are and will remain essential to strengthening confidence in data-driven solutions. These solutions could be applied at various scales [1], from setpoint tracking [5] to the adjustment of building indoor conditions [6] and the optimization of heating and cooling system operations [6–8,10,11], with different levels of complexity, such as model-based controls [6,8], predictive control [7], and predictive maintenance [10,11].

These publications illustrate the opportunities to optimize the operation of real-world buildings but also the efforts needed to take building controls to the next level. A tremendous amount of work is still required in various areas to improve the current scientific knowledge and enable the widespread adoption of practical advanced control solutions on a large scale.

Acknowledgments: The author would like to express his gratitude to the *Buildings* journal for enabling the exchange of recent findings related to advanced controls in the built environment and for inviting him to edit this Special Issue on a topic that is becoming increasingly important and will remain critical in the near future.

Conflicts of Interest: The author declares no conflicts of interest.

References

1. Saloux, E.; Zhang, K.; Candanedo, J.A. A Critical Perspective on Current Research Trends in Building Operation: Pressing Challenges and Promising Opportunities. *Buildings* **2023**, *13*, 2566. [CrossRef]
2. Gabriel, M.; Auer, T. LSTM Deep Learning Models for Virtual Sensing of Indoor Air Pollutants: A Feasible Alternative to Physical Sensors. *Buildings* **2023**, *13*, 1684. [CrossRef]
3. Norouzi, P.; Maalej, S.; Mora, R. Applicability of Deep Learning Algorithms for Predicting Indoor Temperatures: Towards the Development of Digital Twin HVAC Systems. *Buildings* **2023**, *13*, 1542. [CrossRef]
4. Irshad, K.; Zahir, M.H.; Shaik, M.S.; Ali, A. Buildings' Heating and Cooling Load Prediction for Hot Arid Climates: A Novel Intelligent Data-Driven Approach. *Buildings* **2022**, *12*, 1677. [CrossRef]
5. Price, C.; Park, D.; Rasmussen, B.P. Cascaded Control for Building HVAC Systems in Practice. *Buildings* **2022**, *12*, 1814. [CrossRef]

6.   Morovat, N.; Athienitis, A.K.; Candanedo, J.A.; Delcroix, B. Model-Based Control Strategies to Enhance Energy Flexibility in Electrically Heated School Buildings. *Buildings* **2022**, *12*, 581. [CrossRef]
7.   Arroyo, J.; Spiessens, F.; Helsen, L. Comparison of Model Complexities in Optimal Control Tested in a Real Thermally Activated Building System. *Buildings* **2022**, *12*, 539. [CrossRef]
8.   Saloux, E.; Zhang, K. Data-Driven Model-Based Control Strategies to Improve the Cooling Performance of Commercial and Institutional Buildings. *Buildings* **2023**, *13*, 474. [CrossRef]
9.   Almobarek, M.; Mendibil, K.; Alrashdan, A. Predictive Maintenance 4.0 for Chilled Water System at Commercial Buildings: A Systematic Literature Review. *Buildings* **2022**, *12*, 1229. [CrossRef]
10.   Almobarek, M.; Mendibil, K.; Alrashdan, A.; Mejjaouli, S. Fault Types and Frequencies in Predictive Maintenance 4.0 for Chilled Water System at Commercial Buildings: An Industry Survey. *Buildings* **2022**, *12*, 1995. [CrossRef]
11.   Almobarek, M.; Mendibil, K.; Alrashdan, A. Predictive Maintenance 4.0 for Chilled Water System at Commercial Buildings: A Methodological Framework. *Buildings* **2023**, *13*, 497. [CrossRef]

 *buildings*

Review

A Critical Perspective on Current Research Trends in Building Operation: Pressing Challenges and Promising Opportunities

Etienne Saloux [1,*], Kun Zhang [1,2] and José A. Candanedo [1,3]

[1] CanmetENERGY: Natural Resources Canada, Varennes, QC J3X 1P7, Canada; kun.zhang@etsmtl.ca (K.Z.); jose.candanedo@usherbrooke.ca (J.A.C.)
[2] Department of Mechanical Engineering, École de Technologie Supérieure, Montréal, QC H3C 1K3, Canada
[3] Department of Civil and Building Engineering, Université de Sherbrooke, Sherbrooke, QC J1K 5N4, Canada
* Correspondence: etienne.saloux@nrcan-rncan.gc.ca

Abstract: Despite the development of increasingly efficient technologies and the ever-growing amount of available data from Building Automation Systems (BAS) and connected devices, buildings are still far from reaching their performance potential due to inadequate controls and suboptimal operation sequences. Advanced control methods such as model-based controls or model-based predictive controls (MPC) are widely acknowledged as effective solutions for improving building operation. Although they have been well-investigated in the past, their widespread adoption has yet to be reached. Based on our experience in this field, this paper aims to provide a broader perspective on research trends on advanced controls in the built environment to researchers and practitioners, as well as to newcomers in the field. Pressing challenges are explored, such as inefficient local controls (which must be addressed in priority) and data availability and quality (not as good as expected, despite the advent of the digital era). Other major hurdles that slow down the large-scale adoption of advanced controls include communication issues with BAS and lack of guidelines and standards tailored for controls. To encourage their uptake, cost-effective solutions and successful case studies are required, which need to be further supported by better training and engagement between the industry and research communities. This paper also discusses promising opportunities: while building modelling is already playing a critical role, data-driven methods and data analytics are becoming a popular option to improve buildings controls. High-performance local and supervisory controls have emerged as promising solutions. Energy flexibility appears instrumental in achieving decarbonization targets in the built environment.

Keywords: advanced controls; data analytics; decarbonization; flexibility; model predictive control; building operation

Citation: Saloux, E.; Zhang, K.; Candanedo, J.A. A Critical Perspective on Current Research Trends in Building Operation: Pressing Challenges and Promising Opportunities. *Buildings* **2023**, *13*, 2566. https://doi.org/10.3390/buildings13102566

Academic Editors: Maxim A. Dulebenets and Lucio Soibelman

Received: 16 June 2023
Revised: 11 September 2023
Accepted: 9 October 2023
Published: 11 October 2023

1. Introduction

1.1. Energy Context in the Built Environment

Buildings are major energy end-users and are responsible for 40% of the world's total energy consumption, 60% of the world's electricity, and 30% of greenhouse gas (GHG) emissions [1]. Despite initiatives targeting energy efficiency, energy use is expected to further increase in the future as a result of the combined impact of economic development and the change in consumption patterns, as well as an increase in the world's population projected to rise from 7.6 billion in 2019 up to 9.7 billion by 2050 [2]. In Canada, the residential sector and the commercial and institutional buildings sector account for 13% and 12% of the country's end-use demand, respectively [3]. Ongoing initiatives to electrify buildings aim to reduce GHG emissions by 40–45% in 2030 compared to 2005 and become net-zero by 2050 [4]. As a result of these developments and other trends, Canada's total electricity demand is projected to increase by 47% in 2050 compared to 2021 levels [5], thus presenting daunting challenges to the current and projected electric grid infrastructure. Buildings could play a pivotal role in addressing these challenges.

The main energy usage in buildings corresponds to heating, ventilation, and air conditioning (HVAC); along with domestic hot water, these account for around 50% of building energy end-use, the rest being shared with cooking, lighting, and other equipment (appliances, other plug-in devices) [6]. Space cooling currently represents only around 6% of the total building energy demand; however, cooling energy is projected to increase considerably and may become the primary energy usage in buildings [2].

Furthermore, buildings existing today will represent 70% of the Canadian building stock in 2050 [7]. Therefore, apart from new high-performance buildings, it is also critical to tackle existing buildings' issues through major design and controls retrofits.

1.2. Current Controls in Buildings

Traditionally, buildings have been controlled considering the comfort of their occupants as the primary goal. Therefore, sensors have been installed mainly to ensure that indoor environment quality requirements (temperature, humidity, air quality, etc.) are satisfied, not necessarily considering energy efficiency. However, given the increasing importance of the energy context, efforts have been made to improve the operation of buildings to reduce energy use and related costs while satisfying indoor environmental quality (such as thermal comfort).

The operational data commonly available in commercial and institutional buildings are collected from metering devices installed in critical locations of the HVAC system. These data might include temperature, relative humidity, pressure, flow rate, valve positions, damper openings, equipment on/off and modulation, current readings, lighting, and occupancy levels. All this information can be used by the Building Automation System (BAS) for local controls to guarantee satisfactory indoor environment quality while reducing energy usage. However, local controls have been generally developed to optimize the operation of pieces of equipment, not necessarily to optimize performance at the system level. A new generation of tools has emerged to leverage available building data to further improve building performance and reduce energy usage, such as Building Energy Management Systems (BEMS) [8,9] or Energy Management and Information Systems (EMIS) with more advanced capabilities [10,11]:

- Energy Information Systems (EIS) targeting performance tracking;
- Fault Detection and Diagnostic (FDD);
- Automated System Optimization (ASO).

From an energy efficiency perspective, buildings commonly perform beneath their potential: unfortunately, as long as thermal comfort is met, inefficient operation often goes undetected. Improving building controls, such as introducing better control sequences or correcting inefficiencies, could help reduce a building's annual energy use by up to 30% [12], electric peak loads by up to 20% [12], and maintenance costs by 20% [13]. Recommissioning and ongoing commissioning initiatives have emerged to bridge the gap between design and in-operation performance. They could yield 5–15% annual energy savings with typical payback periods lower than 3 years [14]. Fault detection and diagnostic tools become critical in this context to detect and correct inefficiencies when they occur. Several commercial products that are available on the market address the optimization of building operations and controls [15]. The majority focus on EIS and FDD tools, and only a few of them target ASO [10]. These tools could provide energy savings of up to 9% on average while being cost-effective with a return on investment of less than 2 years [10].

Buildings are smoothly transitioning towards data-centric operations with an increased utilization of sub-hourly data and artificial intelligence (AI) techniques, as well as a focus on occupants and sustainability objectives. This paradigm shift is similar to that undergone by many industry sectors with the concepts of Industry 4.0 [16] and Industry 5.0 [17]. However, significant work remains to be achieved to accomplish such a shift.

1.3. Paper Objective

Within this context, the existing research has focused on different pathways to further increase building performance through better controls. However, significant practical hurdles (inefficient local controls, low data availability and quality, communication issues, etc.) are encountered when attempting to deploy the findings of advanced control research; this factor has slowed down their widespread adoption by building owners and operators. This paper aims to provide a critical perspective on current challenges of advanced controls in commercial and institutional buildings and to explore opportunities that promise groundbreaking solutions for controls in the built environment.

Previous initiatives have tackled similar topics, but the focus was rather on specific applications: metadata schemas [18], data analytics [19], FDD [20], model-based predictive controls (MPC) [9,21–23], reinforcement learning [24], occupant-centric controls (OCC) [25,26], predictive maintenance [27], peak load management [28], building energy flexibility [29], strategies for building energy management systems (including MPC, demand side management, optimization, and FDD) [8], and data-driven building operations (with a focus on metadata, FDD, OCC, key performance indicators, virtual energy meters, and load disaggregation) [30]. Some other work emphasized the perspectives, challenges, and opportunities but also concentrated on specific aspects: industry engagement [31], data requirements for MPC [32], reinforcement learning [33], and OCC [34]. In this paper, the objective is to provide a broader overview of research trends on advanced controls in the built environment with a focus on practical considerations such as challenges faced to deploy advanced controls and opportunities to facilitate their adoption.

It is not the intent of the authors to provide an exhaustive review of the vast field of building operation nor to dive into technical details in one specific application. Rather, the content and structure of this perspective paper builds upon the authors' experience and personal assessment to shed light on current challenges and future directions in building operation research at a higher level. Each topic is based on recent state-of-the-art publications, preferably review papers, which were found through conventional channels (e.g., search engines such as ScienceDirect, MDPI, Google Scholar, and ResearchGate) and appropriate keyword searching. A large variety of keywords were used in combination with one another using Boolean operators ("and", "or"); some examples are "buildings", "control-oriented models", "data-driven", "data labelling", "energy flexibility", "fault detection and diagnostics", "key performance indicators", "load disaggregation", "metadata", "modelling", "occupancy", "predictive control", "reinforcement learning", and "virtual energy meters". Some of the material presented was inspired by fruitful discussions in workshops and conferences with colleagues, collaborators, and peer researchers.

1.4. Scope

In this paper, we focus on advanced controls that are seen as a means towards better building performance (energy usage, costs, peak demand, etc.) through improved operation. Advanced controls in this paper refer to various techniques and tools that could help improve the way current controls are designed. Figure 1 shows a conceptual diagram of what we considered "advanced controls" in this study. It includes both supervisory and local controls, as well as data-driven and model-based controls. We do not intend to provide a new definition of advanced controls, but rather to delimit the scope of the paper.

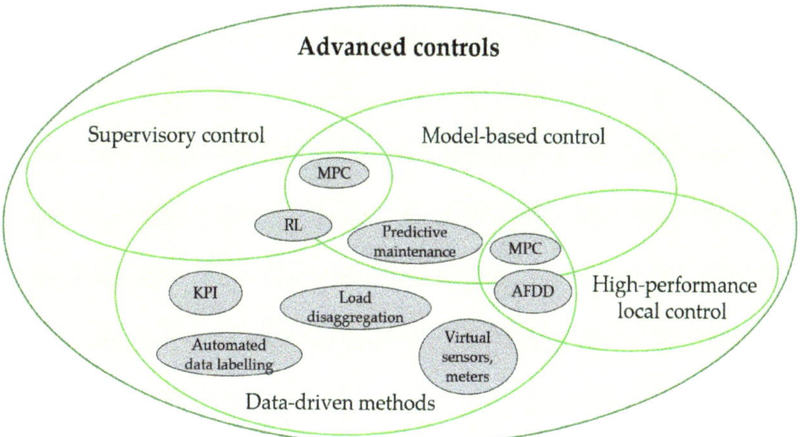

AFDD = Automated Fault Detection and Diagnosis; KPI = Key Performance Indicator; MPC = Model Predictive Control; RL = Reinforcement Learning

Figure 1. Conceptual diagram of what we consider as advanced controls in this study; applications in grey are discussed in this article.

Figure 2 complements Figure 1 by providing a word-cloud of research papers reviewed in this article; titles and abstracts were used for this purpose.

Figure 2. Word-cloud plots based on titles and abstracts of research papers reviewed in this article.

Finally, Abbreviations provides a list of the abbreviations that have been used throughout the paper for better readability.

2. Pressing Challenges

This section tackles challenges related to advanced controls in the built environment, specifically (in no particular order) local control inefficiencies, data availability and quality, communication issues, the need for guidelines and standards, the need for successful case studies and cost-effective solution sets, and insufficient engagement and training. These challenges are summarized in Figure 3 and discussed in detail in the corresponding subsections.

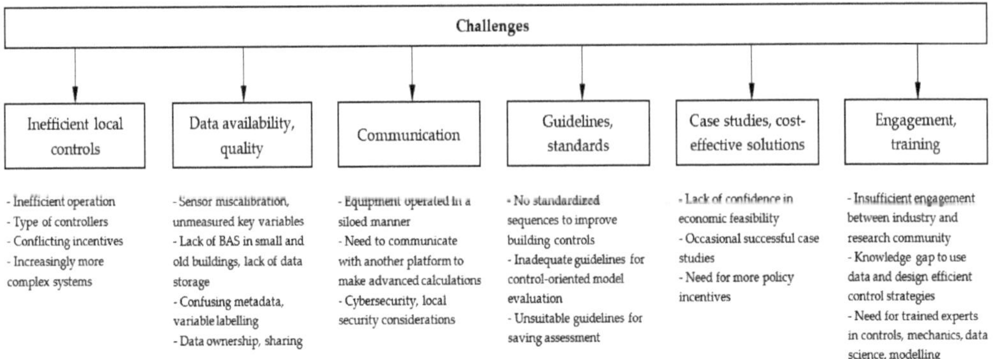

Figure 3. Challenges related to advanced controls in the built environment.

2.1. Inefficient Local Controls

2.1.1. Causes of Inefficient Local Controls

One of the main challenges for deploying advanced controls is the inefficient local controls currently seen in many buildings, which can be caused by various reasons.

- **Lack of simple energy-efficient control rules**: some buildings are still operated without simple energy-efficient control rules such as night setback of indoor air temperature in winter and air or water temperature reset strategies due to a lack of awareness, training, or time of building operators; a compelling example of such inefficiencies has been provided by Gunay et al. [35] for 14 office buildings.
- **Faulty operation and performance degradation**: inefficient operation could also be caused by performance degradation with time [27], faulty operation, or low-efficiency operation, although no fault has been detected.
- **Suboptimal electric power management**: better managing building electric power is also a key aspect for current and future operations and some buildings still show significant differences between average electric load and peak load, which can be costly when the utility rate depends on total energy use and peak power demand [36].
- **Inadequate occupant-centric controls**: occupancy is also a key aspect that affects building performance. Most buildings are generally operated based on full occupancy, for instance, to determine fresh air requirements; however, the actual occupancy of commercial buildings rarely exceeds 50% [30], unlocking energy efficiency opportunities (e.g., lower air-change per hour). Moreover, occupancy can show strong variability, especially with the rise of telework practices since the beginning of the COVID-19 pandemic.

Any new advanced control projects targeting the implementation of advanced control algorithms, such as model-based predictive controls, might need to tackle these issues first before adding another complexity layer with supervisory controls. Recommissioning, ongoing commissioning, and predictive maintenance generally address simple energy-efficient control rules, faulty operation, and performance degradation with time. FDD algorithms are typically used to detect and diagnose such inefficient behaviour to provide corrective actions. Hard faults consist of physical failure of mechanical equipment such as sensors and actuators (stuck, leaks, broken components, fouling) whereas soft faults are related to controller tuning errors, programming mistakes, poor installation, and non-optimal commissioning [37].

2.1.2. Type of Controllers

Another aspect that slows down the deployment of advanced controls is the type of controllers. While DDC controls are suitable for advanced controls, a significant number of buildings—generally relatively old—still have pneumatic controllers [38], which limits the amount of available operational data and allows less flexibility for control modifications. On a related topic, the current infrastructure of existing building control platforms might show significant flaws and may not necessarily enable the integration of complex control algorithms [39]. The deployment of recent research in advanced controls then becomes laborious.

2.1.3. Conflicting Incentives

Although a building could operate in fault-free modes, lessons learned from previous projects showed us that conflicting incentives could also affect the performance of the building. When a building participates in a utility's demand response program or is operated in the context of peak power demand charge, a dual-fuel hydronic heating system (i.e., electric and gas boilers) could favour the operation of gas boilers, although less efficient, at the expense of the electric boiler, leading to a significant increase in GHG emissions locally. Increasing ventilation in the context of COVID-19 is another compelling example. While essential to reduce the spread of SARS-CoV-2 in closed environments, ventilation might have been increased more than necessary [40]. Awad et al. [41] investigated the implications of COVID-19 for electricity use in 27 government buildings. They found that peak loads were reduced in almost all buildings; nonetheless, the average base load was increased in 43% of buildings as a possible consequence of increased ventilation. Overall, the change in annual energy use varied between -33% and $+25\%$.

2.1.4. Complex Integrated Energy Systems

Finally, buildings are being increasingly equipped with even more complex HVAC systems (e.g., heat pumps) and systems for on-site energy generation from renewable sources (e.g., photovoltaic) and for energy storage, either electrical (e.g., battery) or thermal (e.g., water tanks, ice banks). These buildings require sophisticated controls to optimally coordinate their operation and maximize operating efficiency and energy savings.

2.2. Data Availability and Quality Issues

2.2.1. Challenges Related to Data Availability and Quality

Sub-hourly operational data have become increasingly available in commercial and institutional buildings and represent an untapped opportunity to help better manage and operate buildings. However, this comes along with several challenges related to availability and quality:

- Sensors are not often recalibrated.
- Lack of submetering such as equipment electric current (fans, pumps), CO_2, and occupancy sensors, while some key variables might not be measured or directly available (e.g., critical temperatures, flow rates, etc.).
- Installing new sensors is relatively costly and involves practical issues such as installation, maintenance, and recalibration [42].

Since instrumentation has primarily been installed for monitoring and control purposes (not so much for energy efficiency), there is also a lack of submetering or tools to determine how the energy is used in the building [30]. Such information could help provide electric and thermal energy use breakdown in buildings to understand where the energy is used and where advanced controls and energy conservation measures could have more impact.

2.2.2. BAS for Small and Medium Commercial Buildings

Building automation systems generally record data in large commercial buildings where building operation is complex and needs to be automated appropriately to avoid excessive costs. Such an infrastructure allows for the gathering of fine-grained operational data at sub-hourly intervals. In contrast, these data are generally less likely to be available in small commercial buildings since they are under-served by energy conservation tools, given their dispersion and lower payback potential. Small and medium commercial buildings (<50,000 ft^2) account for 94% of commercial buildings in the United States, 50% of commercial floorspace, and 44% of energy consumption, but only 13% have a BAS compared to 71% in larger buildings [43]. Similarly, older buildings with pneumatic controls not only offer less flexibility for controls modifications but also provide fewer operational data, thus severely limiting advanced control opportunities.

2.2.3. Data Storage

In BAS, operational data are generally collected and saved in trend logs for a few hours or a few days, but not in the long term. In this case, data storage devices must be installed, which could be costly. Although this additional cost enables the usage of data analytics tools, building owners might be reluctant to install such storage devices if the benefits are not clear. The notion of *data warehouses* and, more recently, *data lakes* have emerged in the past decades [44,45] as long-term data storage and management solutions. Building data are generally stored in a data warehouse, which is a large repository where data are stored in a well-structured manner and used for decision-making through data analytics [45]. With the increasing amount of available data from various sources, data lakes have appeared as centralized storage repositories, enabling the storage of raw, unprocessed data, including unstructured, semi-structured, or structured data [45]. Although different in terms of structure, both data warehouses and data lakes aim to support decision-making through data analytics, visualization, and machine learning.

2.2.4. Metadata and Variable Labelling

Another important issue with operational data in buildings is metadata management and variable labelling, which slows down the large-scale deployment of data analytics tools and advanced controls. Hard-coded names, which are prone to inconsistencies and mistakes, are still mainstream in most existing buildings. The description of control points is generally of poor quality in terms of consistency, completeness, and usefulness, while it does not follow any standards and heavily depends on control solution vendors and technicians [30,46]. There have been several initiatives to tackle metadata schemas and ontologies [18], and a few have emerged as potential solutions, such as

- Haystack schema [47];
- Brick schema [48];
- Google Digital Building ontology [18];
- Real Estate Core and Smart applications reference ontology (SAREF) [18].

Work is still required to provide a unified approach for data semantic information. ASHRAE Standard 223P is a step in this direction [49].

2.2.5. Data Ownership and Data Sharing

Finally, the question of data ownership and data sharing is becoming more and more prevalent. Since several users could benefit from building operational data (the building owner, the building occupants, the service provider, or the utility), the question of "who owns the data?" remains. Ecobee, a Canadian home automation company, proposed a volunteer-based solution to this issue with the "Donate Your Data" program [50] and made available the smart thermostat data of over 10,000 anonymized residential buildings for research purposes [51,52]. However, this approach might be more difficult to deploy in commercial and institutional buildings. Building owners might not want to share the data due to potential security and privacy issues, but they might want to know what benefits they could get by sharing the data. Nonetheless, initiatives exist, such as the collection of sub-hourly measurement datasets of six real buildings for advanced control applications for energy use and indoor climate research purposes [53] or the Real Time Energy Management (RTEM) Incentive Program supported by New York State Energy Research and Development Authority (NYSERDA) [54], which allowed to make available operational data from over 200 commercial and institutional buildings in New York State [55]. Jin et al. [56] have also collected detailed information about 33 open datasets of city-level building energy use from eight countries. Data granularity varies from 1 year down to 15 min and information includes energy use intensity or electricity consumption, among others. Similarly, Li et al. [29] have identified 16 building datasets suitable for building-demand responses or building-to-grid services that are based on real operational data, hardware-in-the-loop setups, and numerical simulations.

2.3. Inadequate Communication with the Control System

2.3.1. BAS Design in Siloed Manner

One key issue slowing down the massive deployment of advanced controls is inadequate communication with the control system. Current BAS have been designed in a siloed manner, where each subsystem (HVAC, lighting, security, power, etc.) intends to improve its performance independently; however, they usually compete with each other and there is a need for optimizing all the systems at the same time [39]. Master controllers could be used for this purpose to provide the main operational direction (e.g., perform building preheating) while letting local controllers adjust their operations accordingly. Data from other sources (e.g., occupancy sensors, electric vehicle chargers) that are not directly integrated into the BAS or control algorithms add another layer of complexity to operating buildings.

2.3.2. BAS Control and Computational Capabilities

Current BAS have also been designed to deal with simple if-then rules, but not necessarily to support advanced calculations based on a large amount of operational data to train computationally intensive artificial intelligence models or to run optimization routines frequently, whereas software tools (e.g., Matlab R2023a, Python 3.8.8, R 4.2.1) could be quite powerful for such purposes. To enhance computational ability and resources, cloud and edge computing techniques could be leveraged. Additional communication infrastructure might thus be required to exploit the capabilities of current BAS for controlling buildings, along with the capabilities of more advanced data analytic tools for handling large amounts of data, building complex models, and optimization routines. This might complexify the communication between the BAS and the advanced calculations modules, which could physically be hosted in a workstation, in the same local network as a workstation, or in the cloud. Moreover, different communication protocols (e.g., BACnet, Modbus, LonWorks) may be used by different devices, which creates barriers to interoperability.

2.3.3. Cybersecurity

Finally, lessons learned from previous projects showed us that security considerations must be included during the design of advanced controls and should be addressed sooner rather than later, such as the existence of a firewall or VPN. Cybersecurity has become an increasing concern for smart connected buildings, especially for cloud-based tools. In contrast, some buildings could show particularly stringent security restrictions such as no access to the internet, which could make access to weather forecasts (for instance) and cloud computing more cumbersome.

2.4. Lack of Guidelines and Standards

2.4.1. Buildings Controls Improvement

In addition to previous challenges, more guidelines, standards, and codes tailored for control applications are needed. For instance, there is a lack of guidelines to improve building controls. The initiative of ASHRAE Guideline 36 (G36) [57], although primarily U.S. focused and not necessarily applicable to cold climates, aims to provide a recipe for high-performance sequences for local controls for air-side systems (mainly variable air volume systems and terminal units) but also for water-side systems.

2.4.2. Model Accuracy Assessment

Advanced control strategies such as predictive control might also take advantage of control-oriented models to make informed decisions and provide optimal operation recommendations or to directly operate the building. However, there are no model accuracy assessment guidelines specifically tailored for controls. ASHRAE Guideline 14 [58] provides calibration criteria for physics-based energy simulation models using hourly simulation time steps; a model can be considered calibrated if the coefficient of determination (R2) is higher than 0.75, the normalized mean bias error (NMBE) is lower than 10%, and the coefficient of variation of the root mean square error (CV-RMSE) is lower than 30% [59]. Such numbers generally apply to main energy or temperature variables (e.g., electricity use, gas consumption, indoor air temperature) and provide a reasonable gauge for the evaluation of control-oriented model accuracy. Nonetheless, other aspects specific to controls should also be addressed:

- The applicability to other types of variables: fan power, electric boiler power, electric baseload;
- The impact of sensor uncertainty and data quality [60];
- The robustness to extrapolation, especially for black-box models, which are purely data-driven models; this aspect is further discussed in Section 3.1.

2.4.3. Savings Assessment

There is also a lack of guidelines about the assessment of savings following the implementation of advanced control strategies. The International Performance Measurement and Verification Protocol (IPMVP) [61] prescribes four principles for energy savings evaluations, yet it remains a question of how these principles could be applied to advanced controls due to the lack of detailed guides. Signature models for energy, cost, and thermal comfort could be developed based on sub-hourly to hourly [62], daily [63–66], and monthly [67] behaviours, but there are also other options, such as alternating operations between reference and advanced controls, and then comparing the two scenarios [68]. Research initiatives investigating the level of complexity required for the benchmarking model (from linear regression to advanced machine learning models) can also be reported, such as in [69].

2.5. Lack of Successful Case Studies and Cost-Effective Solutions

2.5.1. Economic Feasibility of Advanced Controls

One of the key barriers to the adoption of advanced controls lies in the economic feasibility of such solutions and the market awareness and confidence in realizing cost benefits. It is worth mentioning that the economic feasibility of advanced controls significantly depends on the current building operation. As discussed in Section 2.1.1, if the original strategy shows several inefficiencies or flaws, the new strategy will most likely provide more savings and be more cost-effective; conversely, an already well-operated building will show lower savings with advanced controls. This observation makes the expected performance difficult to estimate, whereas savings generated with a similar control strategy could greatly vary from one building to another.

The Smart Buildings Analytics Campaign [10] was organized in this context to prove the business case for building analytics: 85 software tools from 40 different EMIS vendors have been installed in more than 6500 buildings from 9 different market sectors and 104 commercial organization across the U.S. After two years of EMIS installation, they estimated energy savings (whole building level, for all fuels) and found [10,70]:

- **EIS tools**: savings ranging from −15% to 22% with a median of 3% and a top quartile (best practice) of 11–22%.
- **FDD tools**: savings ranging from 1% to 28% with a median of 9% and a top quartile (best practice) of 15–28%.

For both EIS and FDD, the simple payback period is 2 years. Results were not reported for ASO since it was not prevalent in the study with only two participants [10]. However, estimated savings can still be found in the literature for ASO: energy savings of 0–11% and payback lower than 6.5 years [10,71]. Despite these initiatives, the increasing amount of solution providers and the difficulty in distinguishing the differences between them slow down their adoption; a primer built upon the Smart Buildings Analytics Campaign has been designed to help owners better plan and use EMIS tools [70].

At the research level, advanced controls such as predictive controls have been widely investigated at the simulation level, but practical implementation remains relatively rare. Some model-based predictive control field demonstrations were reviewed in [72]. Energy savings greatly vary from one study to another, with values ranging from 4–7% up to 70%, with an average of 26% in reported results. Similar numbers were obtained for cost savings, whereas demand cost could reach up to 10–30%. Thermal comfort was also improved when reported. It is worth mentioning that the cost-effectiveness of advanced control strategies is always related to the original control strategy.

2.5.2. Policy Drivers

There is also a need for more policy incentives to encourage the deployment of advanced controls and intelligent buildings, even though some initiatives are noticeable. Requiring advanced controls in building certification programs could accelerate the widespread adoption of such tools. A *Smart Readiness Index* (SRI) is a standard scheme developed in the European Union for rating a building's ability to adopt smart technologies and services [73]. Successful case studies could provide solid foundations for the energy and economic benefits of advanced controls, which could foster their inclusion in building codes. Targets and mandatory reporting for building energy use and GHG emissions could also be policy drivers, such as the local law 97 in New York City [74], which sets carbon emissions caps for most buildings over 25,000 ft^2, starting in 2024 and becoming increasingly stringent over time (40% emission reduction by 2030, 80% by 2050).

2.6. Lack of Engagement and Training

2.6.1. Engagement between Industry and Research

The deployment of advanced controls requires industry engagement. There is a broad consensus that academic researchers and industry practitioners need to be more engaged with each other [31]; this observation applies to the building industry but also to other fields. Samad et al. [31] conducted a survey to evaluate industry engagement and perceptions regarding control research; the authors emphasized the value of rudimentary control research to facilitate industry uptake, even though advanced control technologies such as data analytics, fault detection and diagnosis, and model predictive control are perceived by practitioners as potentially impactful soon. They also mentioned that it is not necessarily in the company's best interest to publish the implementation of control because of confidentiality issues or the lack of incentives for dissemination; this lack of transparency makes the adoption of advanced controls more difficult to track.

2.6.2. Operators' Training

Even though building operators are willing to optimize building operations and are aware of potential savings, they generally lack the time to spend on such an activity. They also lack the knowledge on how to deploy energy-efficient control strategies and how to efficiently leverage operational data and control-oriented models to improve building performance. Data analytics, modelling, and control optimization could provide insights into building operation, but such tools require adequate training to be correctly used.

The development of advanced controls in buildings combines system knowledge with advanced data analytic methods and advanced modelling techniques. Control company employees often have a background in electrical engineering in general but not necessarily in HVAC or mechanical systems. With the advent of the digital age, data-driven approaches based on analytics and modelling will also become predominant and a general understanding of these concepts will become a must-have. Therefore, there is a need for trained experts not only in the field of controls and mechanical systems but also in data science and modelling. Industry can play a critical role along with academia in making newly graduated building technicians, operators, engineers, and researchers more employable and valuable in a fast-evolving, technology-driven world [31].

3. Promising Opportunities

This section discusses promising opportunities related to advanced controls in the built environment, namely, building models, data-driven and data analytics methods, high-performance local controls and advanced supervisory controls, and decarbonization through flexibility. They are shown in Figure 4 and are discussed in the corresponding subsections.

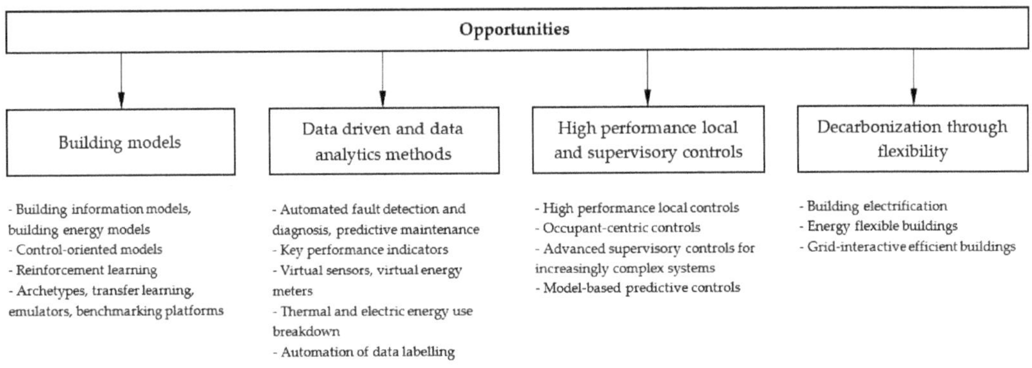

Figure 4. Opportunities related to advanced controls in the built environment.

3.1. Building Models to Support Operational Decision Making

Operational data analysis is critical to better understand how buildings behave and to improve controls. While the analysis of raw data already provides invaluable insights, it also builds the foundations for more advanced approaches where modelling plays a key role. Building models can respond to various control needs and the type of model depends on the application. For instance, models could be tailored for performance tracking, testing of hypothetical scenarios, forecasting of building performance for optimization purposes, and generation of measured and unmeasured synthetic data.

3.1.1. Building Information Modelling and Building Energy Modelling

Various types of building models have been developed in the past and could be useful for control applications. *Building information modelling* (BIM) provides an integrated approach to the management of information for built assets over their lifecycles and aims to facilitate collaboration between disciplines such as architecture, engineering, construction, operations, and maintenance [75]. BIM generally builds upon a rigorously detailed 3D model of the building and gathers relevant information from the design to the operation phase. *Building energy modelling* (BEM), although not necessarily seamlessly integrated into BIM [76], has been extensively used for decades for design purposes and has evolved into dynamic and highly detailed models [77]. Based on physical principles (i.e., mass and energy balance, heat transfer equations), BEM allows for the simulation of the behaviour of a building as a function of numerous variables and parameters such as weather conditions, building geometry and characteristics, internal loads, system schedules and occupancy patterns, generally estimated from design information, educated assumptions, and modeller's experience. These models are mainly developed in a laborious manual process. They have been used for designing new buildings as well as in energy audits, certification programs, and recommissioning studies for existing buildings. In this latter case, numerous (even up to thousands of) parameters can generally be calibrated to make simulation results match monthly energy bills and, in some cases, operational data. Therefore, the calibration of such models is a challenging but essential step to building confidence in model utilization since there have been significant discrepancies between simulated and measured energy use [78]. An important drawback of most BEM tools is that these modelling platforms were mainly developed for design and are not necessarily suitable for controls, making the testing of control strategies even more challenging.

3.1.2. Control-Oriented Models

Control-oriented models (COM), i.e., models accurate enough for decision-making for a specific goal but simple enough to be easily incorporated in further calculations [63], have then appeared and can be distinguished among others by their simpler structure (more suitable for optimization), an appropriate selection of input variables (including controllable variables and disturbances such as weather and occupancy), and a shorter time-scale (usually minutes to days, rather than one year) [79]. These simplified yet accurate enough models are generally calibrated using sub-hourly operational data, but their structure makes them more easily generalized from one building to another, compared with BEM. These models are applicable for different scales, from a specific zone to the whole building, and target various outputs such as thermal behaviour (e.g., indoor air temperature, thermal comfort), energy usage, and system performance [32]. Control-oriented models are generally divided into three categories [21,32,80]:

- White-box models;
- Black-box models;
- Grey-box models.

Whereas white-box models are physics-based, for which new generation building performance simulation software such as Modelica-based tools are clear examples [72,81], black-box models, such as machine learning models, solely rely on operational data. They aim to find the relationships between a set of inputs and outputs. Grey-box models appear as a compromise between white-box and black-box models since they are based on physical principles but are calibrated with operational data. Table 1 summarizes the differences between white-box, grey-box, and black-box approaches for control-oriented models. However, there is no clear distinction between these categories (white, grey, black) but rather a continuum with different "shades of grey" [80]: a resistance-capacitance (RC) thermal network, whose parameters have been estimated from domain knowledge could be categorized as a white-box model, whereas state-space models having only parameters with physical meaning could be considered as grey-box models [32].

Table 1. Distinction between white-box, grey-box, and black-box approaches.

Modelling Type	White-Box	Grey-Box	Black-Box
Techniques	Physics-based models	Resistance-capacitance thermal networks	Artificial Intelligence, time-series, state-space models
Typical software	EnergyPlus 23.1.0, TRNSYS 18, Modelica * v4.0.0	Modelica * v4.0.0, Matlab R2023a, Python * 3.8.8, R * 4.2.1	Matlab R2023a, Python * 3.8.8, R * 4.2.1
Principles	Based on physical principles generally coupled with design data	Based on physical principles generally coupled with operational data	Solely based on operational data, without any physics insights

Table 1. *Cont.*

Modelling Type	White-Box	Grey-Box	Black-Box
Benefits	+ Based on design information + Detailed physics-based simulation + Suitable for testing detailed scenarios (e.g., equipment, occupation) + Controls at a more granular level	+ Compromise between white-box and black-box models + Requires smaller datasets + Flexible and robust to extrapolation	+ Ease of development + Suitable for AI techniques, entirely based on data + Knowledge of building systems and operation not required
Drawbacks	- Usually calibrated with energy bills, rarely with operational data - Low flexibility, adaptability - Development requires significant efforts	- Model structure determined on a case-by-case basis - Diverse estimations required (parameter, system state, etc.) - Less detailed than white-box models, less accurate than black-box models	- Requires larger datasets - High overfitting likelihood - Less reliable for extrapolation (i.e., operating conditions not seen during model development)
Examples of applications	Hypothetical scenario testing, virtual sensors and meters, FDD, MPC	Virtual sensors and meters, FDD, predictive maintenance, MPC, load disaggregation, load forecasting	FDD, predictive maintenance, MPC, load disaggregation, load forecasting, automated data labelling, measurement and verification

* These programming languages are generally used with compatible packages or libraries for modelling.

3.1.3. Reinforcement Learning

Another control approach has recently emerged, *Reinforcement Learning* (RL), as a substitute for conventional control techniques [24,33,82]. RL is an agent-based AI algorithm where agents are trained based on given environmental conditions to take actions that optimize an objective function. Depending on the performance resulting from a given action, a reward function is used to either penalize or encourage it and aims to help agents improve their decision-making process. These agents generally rely on machine learning algorithms and do not require knowledge of building systems and operation; however, they are time- and data-demanding (years of data are required) [33] and face several practical challenges such as model dimensionality, latency, and result interpretability [82].

3.1.4. Software Tools, Control-Oriented Archetypes and Transfer Learning

Different initiatives have been conducted to exploit modelling capabilities for control purposes. An extensive review of software tools for building modelling, simulation, and control for MPC applications has been conducted by Drgoňa et al. [21] and tackles building energy simulation tools, control-oriented modelling tools, MPC design tools, and solvers. Kazmi et al. [83] have focussed on model requirements as well as popular modelling techniques and software packages for load forecasting. Grey-box models were also identified as a pathway for the systematic application of advanced controls in buildings through the notion of control-oriented archetypes [80]. These archetypes are *"reduced-order models that can provide a generic representation for a common zone or building geometry and mechanical system configuration, and which can thus provide enough information to evaluate the effect of control sequences in the short-term and thus inform decision-making"*. This approach is seen as a potential breakthrough in generating versatile control-oriented models to develop general control policies and strategies for typical buildings and to facilitate the dissemination of model-based solutions. On a relatively similar topic, transfer learning [84] has gained in popularity and aims to answer the question: "how can one building benefit from another building's modelling and controls project?" Simply said, transfer learning intends to transfer models trained for highly instrumented buildings to buildings with limited available data. Four main applications were identified: building load prediction, occupancy detection and activity recognition, building dynamics modelling, and energy systems control [84,85].

3.1.5. Large-Scale Comparison of Control Strategies and Models

Furthermore, building models and advanced control strategies have recently elicited great research interest and new models and strategies have been investigated for buildings. However, they have mainly been applied to specific case study buildings and datasets, which makes a fair comparison between them difficult to perform. In this context, the tool BOPTEST (building optimization testing framework) has been developed and aims to enable rapid, repeatable deployment of common building emulators representing different system types [86]. These emulators enable testing and comparing different control strategies in standardized conditions. Similarly, the ASHRAE Great Energy Predictor III competition [87,88] was a machine learning contest for long-term prediction with an application to measurement and verification. This competition allowed for the testing ofthe performance of diverse machine learning techniques on the same datasets, composed of over 20 million points of training data from 2380 energy meters collected for 1448 buildings from 16 sources. Likewise, *CityLearn*, an OpenAI Gym environment, has been developed to provide a benchmark platform that helps facilitate and standardize the evaluation of RL agents [82]. It was the foundation of the CityLearn Challenge, where five teams competed in training the best RL agent targeting the management of a microgrid of nine buildings [89]. On a related topic, the M4 competition [90] has evaluated 61 forecasting methods on 100,000 time-series data from a wide range of domains such as industry, finance, and demographic. The project ADRENALIN is another recent initiative of data-driven smart building competitions [91].

3.1.6. Ties between Modelling Approaches

As shown in Figure 5, the abovementioned modelling techniques could be linked together to improve building controls. Among others, BIM could be used to partially automate the development of BEMs; BIMs could support the management of architectural information and operational data, which will facilitate the development of COMs; and BEMs could work as virtual testbeds to test control strategies developed with COMs. In addition, COMs could also rely on a combination of white-box, grey-box, and black-box models.

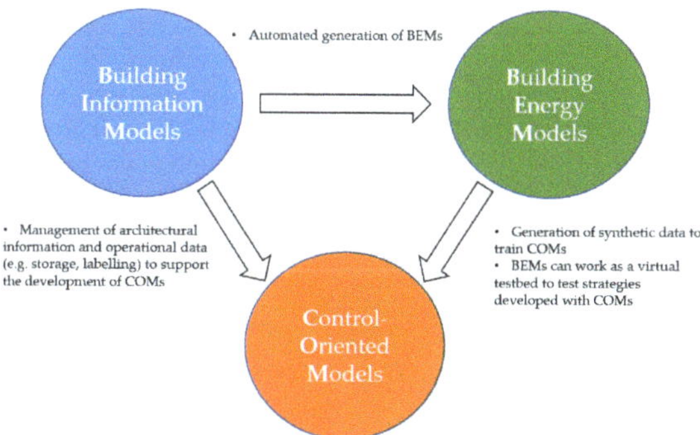

Figure 5. Schematic of some potential links between building information models, building energy models, and control-oriented models.

3.2. Data-Driven and Data Analytics Methods to Improve Building Operation

Operational data represents essential information to investigate in-depth the actual operation and performance of building heating and cooling systems, and it often reveals that systems should be operated differently. However, it is still unclear for practitioners how to leverage operational data to optimize building performance. While FDD commercial products have already shown good promise [10], other methods based on data-driven and data analytics methods could complement these tools to improve building operation and performance.

3.2.1. Fault Detection and Diagnostics, and Predictive Maintenance

FDD commercial tools generally use a rule-based approach to detect operational faults. For instance, current temperature readings and setpoints are compared and an alarm is triggered when the difference becomes too large and exceeds a pre-defined threshold. Moreover, FDD tools usually focus on the first D in the acronym, which is detection. The second D, i.e., the diagnosis, or at least the *correct* diagnosis rather than several potential diagnoses [11], deserves more attention. Research has focussed on automating this process (Automated FDD or AFDD). Models can be developed to supplement the rule-based approach and make the detection more robust [19]. These models, either knowledge-based or data-driven [20], could then help flag performance degradation while providing insights to correctly diagnose faults. Other initiatives, such as automated fault correction, can be found in the literature to reduce the reliance on human intervention while making the diagnosis more proactive [92]. The access to fault datasets for existing buildings during normal operation, rather than faults artificially generated (simulation, experiments), remains a challenge for FDD applications [19,20]. Similarly to FDD, predictive maintenance also plays an important role; unlike preventive measures planned heuristically, predictive maintenance relies on models to predict when a component requires maintenance or replacement [19]. These models, which learn from operational data, identify conditions that may lead to failure events, thus enabling corrective measures that can increase equipment lifetime and save costs. Typical faults requiring maintenance are refrigerant leakage, heat exchanger fouling, pump clogging, damper jam, and coil blockage [27]. FDD and predictive maintenance tools could be incorporated into a broader predictive maintenance strategy, where a BIM model could be used along with operational data to facilitate day-to-day building operation [93].

3.2.2. Key Performance Indicators

Key Performance Indicators (KPI) could help quantify building energy performance, provide more comprehensive operational insights, and track performance degradation over time at the component, system, and whole building level [30,94]. These KPIs have been commonly used at the *whole building* level for performance rating (e.g., energy use intensity, carbon footprint) or at the *component level* for code compliance purposes (e.g., coefficient of performance, heating seasonal performance factor) [94]. KPIs at the *system level* are far less common; they quantify the performance of an entire system providing building services, considering the contributions and impacts of all its components. System-level KPIs can point directly to the system or group of components that should be prioritized in a building. They could thus highlight performance issues, which would remain undetected otherwise (e.g., longer HVAC operation time than required, excessive perimeter heating) [30]. Li et al. [94] developed a comprehensive set of system-level KPIs and calculated them for 16 U.S. DOE commercial prototype buildings using EnergyPlus building models. However, the adoption of these system-level KPIs for existing buildings is still in its early stages [94]; this can be partially explained by the following:

- The lack of sensors and submeters;
- The cost-effectiveness of implementing such KPIs;
- The unknown typical values that can be expected;
- The lack of knowledge of the appropriate KPIs to apply for a given situation.

The development of KPIs becomes even more crucial to assess advanced performance, such as building electric peak load management [28] and energy flexibility [95]. Another benefit of data-driven KPI is the ability to capture the actual system performance. For instance, two theoretically identical systems, such as heat pumps [65] or chillers [96], could show different performance curves in real-life performance; this difference could be further exploited in operation to maximize building performance. KPI could also be used for FDD applications by comparing measurements with the expected performance [20].

3.2.3. Virtual Energy Meters and Load Disaggregation

Another application is the development of *virtual sensors* and *virtual energy meters* to infer unmeasured variables, as discussed in Section 2.2.1. Virtual sensors are mathematical models that predict a variable using measurements available from other installed sensors; similarly, virtual energy meters predict the energy consumption of a building zone or one of its systems [60]. Although more operational data are available, some key variables are still not measured and virtual sensing or metering is a cost-effective solution as a substitution for installing new physical devices, which can be costly and face practical issues [60]. Li et al. [97] reviewed virtual sensing technologies used in buildings. They investigated various virtual sensors, including air and refrigerant mass flow rates, air temperature (outdoor, mixed, supply, return), refrigerant pressure, air humidity ratio, as well as compressor volumetric efficiency, chiller efficiency, valve leakage, and coil capacity.

In addition to the calculation of unmeasured variables, virtual energy meters allow to determine energy flows within a building—for instance, per floor, per air handling unit (AHU) or per variable air volume (VAV) box—and contribute to a better understanding of thermal energy-use breakdown in buildings [98]. Virtual energy meters also unlock many opportunities, such as [60]

- Performance tracking (daily and weekly patterns and schedules, nature of the load, energy distribution, performance degradation),
- Energy-use mapping (aiming at better matching building energy usage with actual occupant's needs),
- Inefficiency detection (by analyzing the load distribution),
- Operation optimization (using gained information for improving building modelling and optimization through advanced controls).

Virtual energy meters are not the only method to break down building energy usage. Load disaggregation methods have emerged as promising techniques to estimate energy end-uses for specific systems or equipment from total electric and thermal energy usage measured by a single meter [30,60]. By means of models, electric and/or thermal load can be disaggregated into different categories such as space heating and cooling, ventilation (fresh air), distribution (AHUs), and/or occupant-controlled (lighting, plug loads), among others [99–101]. Physical submetering of energy usage is still an option to encourage occupants to change their behaviour to save energy [102], although it is relatively costly.

3.2.4. Automated Data Labelling

A key barrier to understanding and analyzing building datasets at a large scale is the lack of standardized variable nomenclature. To reduce the effort required for data interpretation, inference techniques could be used to label data and generate descriptive names without the need for intensive human labour. Diverse automated methods have arisen to infer contextual information from operational data. For instance, Waterworth et al. [103] developed a novel neural language processing method that aims to automatically segment sensor metadata into tokens (i.e., words and sub-words), which is further used for tagging; data from over 182 buildings were used in this study. Chen et al. [104] proposed a method to classify variables according to their type (e.g., indoor air temperature and setpoint, air flow and setpoint, and damper position) only based on numerical values; the approach was applied to zone-level metadata in two office buildings. Mishra et al. [105] presented a unified architecture, which uses time-series data and raw variable names and builds upon rule-based logic and machine learning techniques to apply Haystack tags to variables; this architecture was applied to three commercial retail buildings and one office building. These methods showcase the various possibilities to address the challenge of automating data labelling in buildings. In the future, large language models and other AI techniques may facilitate data management and interpretation.

3.3. High-Performance Local Controls and Advanced Supervisory Controls

3.3.1. High-Performance Control Sequences

Inefficient local controls are a significant barrier to the adoption of advanced controls. The recently released *ASHRAE Guideline 36* [57] inventories various high-performance control sequences to enhance HVAC air-side and water-side system operation. These standardized sequences aim to reduce the time dedicated to developing and implementing local controls and could already help achieve significant energy savings while improving indoor air quality. Zhang et al. [106] simulated new control sequences based on G36 for multi-zone VAV systems under different climates, hours of operation, and internal loads. They obtained energy savings between 2% and 75%, with an average of 31% for a medium-sized office building and found that supply air temperature reset, duct static pressure reset, and zone airflow control contributed the most to these savings. Saloux and Zhang [96] evaluated the impact of a supply air temperature reset strategy based on outdoor air enthalpy for an existing large commercial building; they found that the building cooling load could be decreased by 13%, resulting in a 9% reduction in cooling system electric energy. Nassif et al. [107] built upon G36 to develop new strategies to reset supply air temperature and zone minimum airflow rate setpoints for typical VAV systems; the authors showed a potential of 2–6% fan energy decrease and 8–34% heating load reduction. Lu et al. [108] compared the performance of G36 for supply air temperature and static difference pressure setpoints with intelligent controllers (optimization-based controller and deep reinforcement-learning-based controller) for a simulated medium-sized office building with a multi-zone variable air volume cooling system. G36 sequences demonstrated for this case study similar performance in terms of energy efficiency and thermal comfort compared to the intelligent controllers.

3.3.2. Occupant-Centric Controls

In addition to these energy efficiency measures, the integration of occupant behaviours into building controls has elicited an increasing interest and has led to the development of occupant-centric controls [34,109]. A large range of applications can be found: temperature and humidity setbacks, demand controlled ventilation, temperature setpoint adaptation to occupants, and illuminance adaptation to occupants (preferences, vacancy off) [25,26]. The data required for these controls—such as occupant presence (binary patterns), count, and activity (thermostat use behaviour, comfort feedback)—show high temporal and spatial variability [25]. As a result, gathering occupants' feedback data through web, mobile, and wearable applications has become crucial. In this respect, mobile applications [110] and "smartwatches" [111] capable of retrieving occupant's thermal comfort feedback and platforms to improve workspace allocation based on the occupants' activity [112] provide new data sources. A critical review of field implementations of occupant-centric building controls has been conducted by Park et al. [26]. The impact of the global COVID-19 pandemic has disrupted the way buildings are occupied. Adapting building controls to individual preferences while maintaining an energy-efficient building operation may emerge as a key feature of the next generation of controls.

3.3.3. Supervisory Controls for Complex Integrated Energy Systems

With the advent of increasingly more efficient technologies, building heating and cooling systems have become more and more complex and the optimal design and control of these systems are far from being straightforward [113]. Advanced control strategies are thus required to exploit their full potential and advanced supervisory controls are promising solutions. Unlike local controls that ensure the operation of individual devices, advanced supervisory controls target system-level operations and aim to fulfill the role of the master controller discussed in Section 2.3.1. While individual components could be competing to maximize their own performance, advanced supervisory controls use a bird's-eye-view to optimize the system's performance as a whole, which incorporates heating and cooling systems, on-site energy generation from renewable sources, and energy storage devices. Although most buildings still use conventional methods, new technologies have emerged and could play a more significant role in commercial and institutional buildings in the near future:

- Heating and cooling systems generally consist of rooftop units, electric and natural gas boilers, electrical baseboards, connection to district heating, mechanical chillers, etc. Technologies such as evaporative cooling [114], air-source and ground-source heat pumps [95], and high-temperature heat pumps (currently mainly for industrial applications [115]) will become increasingly present.

- On-site energy generation systems from renewable sources may include photovoltaic and photovoltaic/thermal systems [116], wind turbines [117], solar collectors [118], as well as organic Rankine cycles [114,119] (mainly used for waste-heat recovery at the moment [120]).

- Regarding energy storage devices, batteries represent the main electrical storage device option; electric vehicles and associated charging stations could provide some opportunities for electric load management as well [121]. Radiant floors [95], water tanks, and geothermal fields are common options for thermal energy storage; electrically heated thermal storage [95] is becoming more and more popular. Phase change materials, although mainly used in the residential sector [122], could emerge as a potential solution for commercial buildings [123]. For cold storage, ice banks have shown encouraging promise [124,125].

These numerous systems add complexity to building controls on top of the already existing variety in HVAC system configurations (e.g., air handling units, fan coil units, terminal units, variable refrigerant flow), which make each building practically unique. Figure 6 shows some of the possible heating, cooling, and distribution systems, along with electricity generation and energy storage devices that could be found in commercial and institutional buildings. This is where building modelling (see Section 3.1) could be a game changer in testing different control strategies and determining the optimal operation of integrated systems. It could be achieved through model-based controls (MBC) [96], targeting the use of models to estimate building performance and optimize the operation, as well as model predictive control (see Section 3.3.4), also based on models but taking advantage of available forecasts to anticipate changes and adjust accordingly. Reinforcement learning [33] is another type of control technique based on models to address supervisory controls. Furthermore, such controls could be applied at different levels, from the equipment level (e.g., zone supplied by dedicated heat pump and radiant floor [95]) to the central plant (e.g., chilled water plant [96]) and the building level [64]. The use of control-oriented archetypes [80] and building emulators to test control strategies [86] could facilitate the development of general models and control solutions to eventually foster supervisory controls deployment.

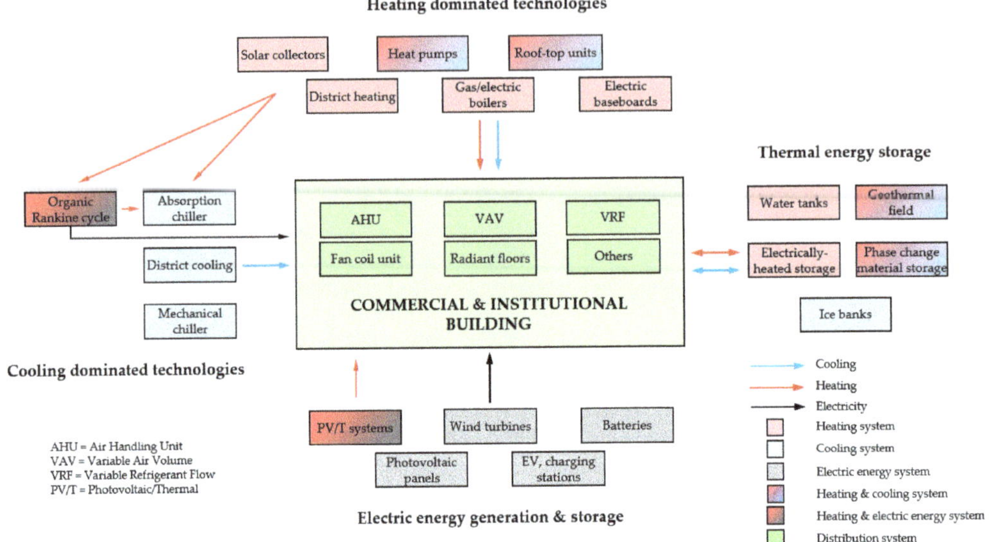

Figure 6. Schematic of some technologies which could be used for heating, cooling, and electricity generation in commercial and institutional buildings.

3.3.4. Model Predictive Control

In addition to inefficient local controls, buildings are currently mainly operated in a reactive manner and simply respond after a change occurs. Advanced controls such as model-based predictive control can enable buildings to be proactive by anticipating changes and adjusting accordingly to satisfy a given objective. Such strategies could build on various forecasts such as weather conditions, occupancy patterns, dynamic energy costs, and grid carbon intensity to satisfy objectives such as the reduction in energy use, energy costs, peak demand, GHG emissions, or the improvement of indoor environment quality and thermal comfort. For example, by anticipating solar gains in a zone, the heating power could be reduced beforehand to avoid overheating and thermal discomfort. Similarly, a building could be preheated at night in anticipation of a cold morning to avoid excessive electric peak load in the morning. This topic of model-based predictive control has raised much interest for the past decades and has shown promising results. Several review papers can be found in the literature on this topic:

- Drgoňa et al. [57] performed a thorough review of MPC for the built environment, from the structure and formulation to the deployment and performance assessment.
- MPC for HVAC systems has been reviewed by Afram and Janabi-Sharifi [22], Serale et al. [126], and more recently by Yao and Shekhar [23] and Taheri et al. [127].
- MPC for energy storage has been investigated by Thieblemont et al. [128] and Yu et al. [129].
- Péan et al. [130] have focused on MPC for heat pumps.
- Mirakhorli and Dong [131] and Park et al. [26] concentrated on occupancy behaviour-based MPC and occupant-centric building controls.

Control-oriented modelling has been acknowledged as the cornerstone of MPC, and its development takes a significant share of the effort required for developing and implementing MPC strategies: up to 70% of project costs are attributable to model creation and calibration [132]. Whereas white-box models are favoured for simulation-based MPC studies, black-box and grey-box models are used more in actual experiments [32]. A recent publication suggested shifting from model-centric MPC to data-centric MPC to address the heterogeneity among buildings and the need for model customization for each building [133]. MPC faces other practical challenges that hold up massive deployment, such as data availability and processing, MPC scheme, qualification of control engineers, and risk mitigation of MPC implementations [9,134]. Nonetheless, a few initiatives can be reported. Vallianos et al. [135] investigated the aggregated effect on the grid of model predictive control strategies applied to 7500 houses in Canada using the Ecobee dataset [50], whereas Deng et al. [55] evaluated the potential of data-driven predictive controls for 78 commercial buildings using the data available in the RTEM database [54].

3.4. Decarbonization and the Need for Energy Flexibility

3.4.1. From Energy Efficiency to Decarbonization

Current control strategies have mainly focused on minimizing building energy use, energy costs (especially under a real-time electricity pricing rate), and electric peak load [72]. In the context of climate change and GHG emission reduction targets, these objectives have been slightly reoriented towards building decarbonization, which requires a new paradigm shift. The electrification of buildings is seen as an essential means to achieve it, especially in countries where electricity is already generated from low-carbon sources (e.g., hydroelectricity) or where renewable energy shows a possibly reasonable penetration rate. To reduce the current and future burden on the electric grid's infrastructure, buildings should not be optimized in a siloed manner anymore but must be part of the Smart Grid and must become good grid citizens.

3.4.2. Energy Flexible Buildings

The concept of energy flexibility [136] is becoming critical, and buildings can play a significant role. IEA EBC Annex 67 Energy Flexible Buildings defined the energy flexibility of a building as the *"ability to manage its demand and generation according to local climate conditions, user needs and energy network requirements"* [137]. Built on this concept, the U.S. DOE introduced the notion of *grid-interactive efficient buildings* (GEB), which are *"energy efficient buildings with smart technologies characterized by the active use of distributed energy resources to optimize energy use for grid services, occupant needs and preferences, climate mitigation, and cost reductions in a continuous and integrated way"* [138]. Simply said, buildings should be able to manipulate their electric load to improve grid reliability: they should reduce their energy usage during the grid's peak hours (downward flexibility), whereas they should help store the grid's energy surplus when available (upward flexibility, e.g., from solar plants during sunny days, from wind farms during windy days). Buildings can provide different grid services [138]:

- Efficiency (consistent load reduction; for instance, more efficient HVAC systems);
- Load shedding (temporary load reduction such as lighting dimming);
- Load shifting (load shifted to off-peak periods such as building preheating);
- Power modulation (e.g., grid frequency and control system voltage);
- Power generation (such as on-site photovoltaic).

To provide these services, multicarrier systems combining different energy vectors (e.g., electricity, natural gas, oil, biomass) are particularly of interest and show high flexibility potential with their ability to switch from one energy source to another (e.g., dual fuel boiler, electrically driven heat pump with natural gas fired boiler) [136]. Many indicators have been proposed in the past to quantify building energy flexibility. Li et al. [29] have reviewed 48 of them that are related to peak power and energy load shedding, peak power and energy rebound, valley filling, load shifting, energy storage capability, demand response costs savings, and environmental impact, among others.

3.4.3. Role of Advanced Controls to Enhance Flexibility

In the meantime, at the building level, advanced controls are essential to ensure the continued operation of critical building services, and to improve building resilience and climate-readiness. Therefore, advanced control strategies offer considerable opportunities to unlock the flexibility potential of buildings [139] and might take advantage of dynamic price signals and the electric grid's real-time carbon intensity to minimize building GHG emissions while maximizing profitability and satisfying building objectives such as thermal comfort and GHG emission reduction targets. MPC is an appropriate measure to achieve higher flexibility in buildings. As a complementary measure, policy incentives should tackle the carbon intensity of buildings and the electric grid as a whole, for example, by applying dynamic (e.g., hourly) GHG emission factors. The flexibility potential for building communities harnessed by aggregators and for district energy systems could even be more attractive with additional thermal energy storage capabilities and aggregated benefits [136].

4. Hierarchy of Advanced Controls

Sections 2 and 3 dealt with pressing challenges and promising opportunities in building operation where various topics were tackled and targeted a large number of applications at different scales. This section aims to organize these topics together and develop a hierarchy in advanced controls opportunities and requirements. Inspired by the energy management hierarchy of needs developed by Nexus Labs for smart building platforms [140], Figure 7 shows a hierarchy of advanced controls opportunities and requirements for buildings.

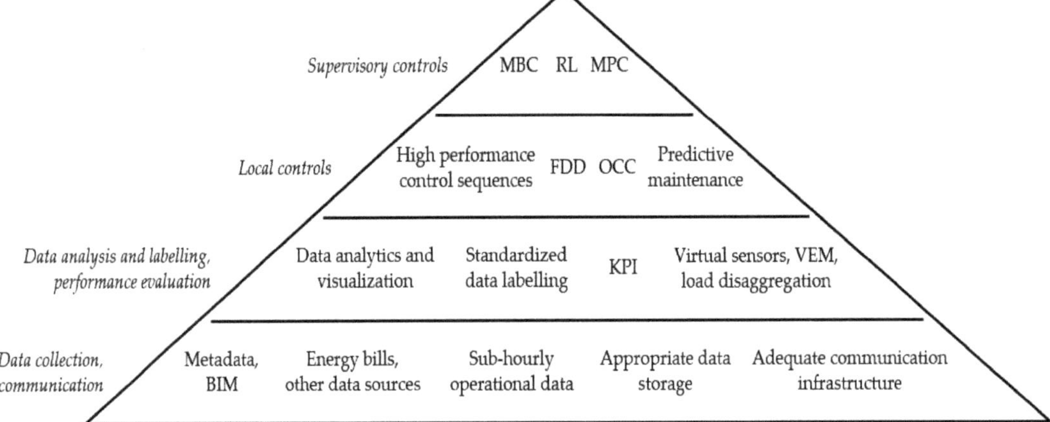

Figure 7. Hierarchy in advanced controls opportunities and requirements.

The lower level represents enablers for advanced controls in buildings, which are mainly related to available data and communication considerations. It includes the access to general building information such as metadata, BIM (Section 3.1.1), energy bills, and other data sources; the availability and storage of sub-hourly operational data (Section 2.2); and an adequate communication infrastructure (Section 2.3). The second level corresponds to data analysis and performance evaluation. It refers to data analytics and visualization tools, but it also covers proper data labelling (Section 3.2.4) as well as virtual sensors, energy meters, and load disaggregation to monitor key variables that are not measured (Section 3.2.3), and finally, key performance indicators to track building and system performance (Section 3.2.2). The third level tackles the improvement of local controls and includes various applications, from fault detection and diagnostics and predictive maintenance (Section 3.2.1) to high performance control sequences (Section 3.3.1) and occupant-centric controls (Section 3.3.2). Finally, the higher level builds upon local controls improvements and optimizes the operation of complex integrated energy systems to maximize the performance of the building heating and cooling systems as a whole. It could rely on MBC (Section 3.3.3) and RL (Section 3.1.3) but also MPC (Section 3.3.4) by leveraging available forecasts such as weather, occupancy, energy prices, and the grid's carbon intensity (Section 3.3). Flexibility (Section 3.4) could be incorporated in the objective function to transition from building energy efficiency to building decarbonization.

Such a categorization is certainly no easy task and relies on the authors' experience and personal assessment. The goal of Figure 7 is to emphasize the main trends, although there might be no clear distinctions between categories and levels (Figure 1), similarly to the different "shades of grey" in control-oriented models (Section 3.1.2). For instance, RL and MPC could be used for local controls (e.g., setpoint tracking) but they show higher potential for supervisory controls to optimally operate complex integrated building systems (setpoint optimization, central plant control). MBC could be used to improve local controls as well, but we see it more in facilitating the development and adoption of supervisory control solutions (Section 3.3.3). FDD, OCC, and predictive maintenance algorithms might be already based on models and could be seen as MBC to some extend; similarly, some MPC strategies can be occupant-centric [62]. KPI are shown in Figure 7 in the second level for performance evaluation, but it could be used as well for FDD (Section 3.2.2). Data labelling, virtual sensors, and virtual energy meters could be seen as enablers (lower level); nonetheless, advanced controls could still be developed even if variable nomenclature is not standardized, or virtual energy meters are not yet available.

5. Conclusions

Commercial and institutional buildings are currently not operated at their full potential, resulting in suboptimal performance that could be addressed by more appropriate control strategies. Although the potential of advanced controls such as MPC is proven, massive deployment in existing buildings is still not a common practice. This article has focussed on pressing challenges and promising opportunities in building controls. It aims to provide researchers, practitioners, and newcomers in the field with a state-of-the-art overview of the situation. The numerous, significant current challenges are grouped into six categories:

1. Existing local control flaws that need to be addressed in order of priority. Old control infrastructure is a crucial hurdle for adopting increasingly more complex mechanical systems.
2. With the advent of the digital era, operational data have become increasingly available for buildings and are of paramount importance for advanced controls; however, data are not necessarily available or appropriately stored in every building, key variables are not necessarily well measured (or measured at all), variable labelling is often confusing, and data could simply not be shared.
3. Current building automation systems have been designed in a siloed manner and communication with the control system is no easy task, nor can it support advanced calculations; cybersecurity and local security considerations add another layer of complexity.
4. There is, in general, a lack of guidelines tailored for control applications that address standardized high-performance building control sequences, the evaluation of control-oriented model performance, and the assessment of energy and economic savings.
5. Economic feasibility and expected savings of advanced controls still need to be widely acknowledged; successful case studies are rare, and more policy incentives are required to encourage adoption.
6. Engagement between industry and the research community still needs to be improved; however, practitioners may face a knowledge gap and require appropriate training to keep up with a fast-evolving, technology-driven world.

On the other hand, four main streams of opportunities were identified:

1. Building modelling is the cornerstone of advanced control strategies; BIM and BEM, control-oriented models, and reinforcement learning can serve several objectives, whereas different initiatives, such as archetypes, emulators, benchmarking platforms, and transfer learning, aim to scale and generalize existing models and control strategies.
2. Data-driven and data analytics methods allow us to provide invaluable insights into building operations and facilitate the development of advanced controls; this includes the development of KPIs, virtual sensors and energy meters, load disaggregation methods, and automated data labelling.

3. Attempts at high-performance local controls and advanced supervisory controls have shown encouraging prospects; work should be continued towards this direction. MPC promises to make buildings more proactive by anticipating changes rather than being reactive and simply responding after a change occurs.

4. In the context of climate change, decarbonization of buildings is vital. It requires a new paradigm shift: advanced control strategies could help make buildings become energy flexible, more resilient, and good citizens of the smart grid.

Author Contributions: Conceptualization, E.S., K.Z. and J.A.C.; methodology, E.S.; investigation, E.S.; writing—original draft preparation, E.S.; writing—review and editing, E.S., K.Z. and J.A.C.; visualization, E.S., K.Z. and J.A.C. All authors have read and agreed to the published version of the manuscript.

Funding: The authors want to gratefully acknowledge the financial support of Natural Resources Canada through the Office of Energy Research and Development.

Data Availability Statement: No new data were created or analyzed in this study. Data sharing is not applicable to this article.

Acknowledgments: The authors would like to thank our past and current colleagues and collaborators for the fruitful discussions, which have certainly contributed to this article. Internal and external reviewers are also acknowledged for their valuable feedback.

Conflicts of Interest: The authors declare no conflict of interest.

Abbreviations

AFDD	Automatic fault detection and diagnosis
AI	Artificial intelligence
ASO	Automated system optimization
BAS	Building automation system
BEMS	Building energy modelling
BEMS	Building energy management system
BIM	Building information modelling
COM	Control-oriented modelling
EIS	Energy information system
EMIS	Energy management and information system
FDD	Fault detection and diagnosis
HVAC	Heating, ventilation, and air conditioning
IPMVP	International performance measurement and verification protocol
G36	ASHRAE Guideline 36
GEB	Grid-interactive efficient buildings
GHG	Greenhouse gases
KPI	Key performance indicator
MBC	Model-based control
MPC	Model predictive control
OCC	Occupant-centric control
RL	Reinforcement learning
RTEM	Real time energy management
VAV	Variable air volume

References

1. United Nations Environment Programme. Energy Efficiency for Buildings. Available online: https://www.renewableinstitute.org/images/unep%20info%20sheet%20-%20ee%20buildings.pdf (accessed on 1 December 2022).

2. Santamouris, M.; Vasilakopoulou, K. Present and Future Energy Consumption of Buildings: Challenges and Opportunities towards Decarbonisation. *E-Prime—Adv. Electr. Eng. Electron. Energy* **2021**, *1*, 100002. [CrossRef]

3. Government of Canada, C.E.R. CER—Provincial and Territorial Energy Profiles—Canada. Available online: https://www.cer-rec.gc.ca/en/data-analysis/energy-markets/provincial-territorial-energy-profiles/provincial-territorial-energy-profiles-canada.html (accessed on 1 December 2022).

4. Natural Resources Canada, N.R. Green Buildings. Available online: https://natural-resources.canada.ca/energy-efficiency/green-buildings/24572 (accessed on 27 April 2023).

5. Government of Canada, C.E.R. CER—Welcome to Canada's Energy Future. 2021. Available online: https://www.cer-rec.gc.ca/en/data-analysis/canada-energy-future/2021/ (accessed on 1 December 2022).

6. González-Torres, M.; Pérez-Lombard, L.; Coronel, J.F.; Maestre, I.R.; Yan, D. A Review on Buildings Energy Information: Trends, End-Uses, Fuels and Drivers. *Energy Rep.* **2022**, *8*, 626–637. [CrossRef]

7. *Catharine The Future Passes through Old Buildings—Canadian Climate Institute—Blog*; Canadian Climate Institute: Ottawa, ON, Canada. Available online: https://climateinstitute.ca/the-future-passes-through-old-buildings/ (accessed on 1 December 2020).

8. Mariano-Hernández, D.; Hernández-Callejo, L.; Zorita-Lamadrid, A.; Duque-Pérez, O.; Santos García, F. A Review of Strategies for Building Energy Management System: Model Predictive Control, Demand Side Management, Optimization, and Fault Detect & Diagnosis. *J. Build. Eng.* **2021**, *33*, 101692. [CrossRef]

9. Killian, M.; Kozek, M. Ten Questions Concerning Model Predictive Control for Energy Efficient Buildings. *Build. Environ.* **2016**, *105*, 403–412. [CrossRef]

10. Kramer, H.; Lin, G.; Curtin, C.; Crowe, E.; Granderson, J. *Proving the Business Case for Building Analytics*; Lawrence Berkeley National Laboratory: Berkeley, CA, USA, 2020. [CrossRef]

11. Lin, G.; Kramer, H.; Granderson, J. Building Fault Detection and Diagnostics: Achieved Savings, and Methods to Evaluate Algorithm Performance. *Build. Environ.* **2020**, *168*, 106505. [CrossRef]

12. Fernandez, N.; Katipamula, S.; Wang, W.; Xie, Y.; Zhao, M.; Corbin, C. *Impacts of Commercial Building Controls on Energy Savings and Peak Load Reduction*; Pacific Northwest National Laboratory, U.S. Department of Energy: Washington, DC, USA, 2017.

13. U.S. Green Building Council. Benefits of Green Building. Available online: https://www.usgbc.org/articles/benefits-green-building (accessed on 16 September 2022).

14. Natural Resources Canada. Recommissioning for Existing Buildings. Available online: https://www.nrcan.gc.ca/energy-efficiency/buildings/existing-buildings/recommissioning/20705 (accessed on 1 December 2022).

15. Continental Automated Buildings Association (CABA). Intelligent Building Energy Management Systems. 2020. Available online: https://www.caba.org/wp-content/uploads/2020/12/CABA-IBEMS-Report-2020-FULL-WEB.pdf (accessed on 1 December 2022).

16. Jan, Z.; Ahamed, F.; Mayer, W.; Patel, N.; Grossmann, G.; Stumptner, M.; Kuusk, A. Artificial Intelligence for Industry 4.0: Systematic Review of Applications, Challenges, and Opportunities. *Expert Syst. Appl.* **2023**, *216*, 119456. [CrossRef]

17. Barata, J.; Kayser, I. Industry 5.0—Past, Present, and Near Future. *Procedia Comput. Sci.* **2023**, *219*, 778–788. [CrossRef]

18. Pritoni, M.; Paine, D.; Fierro, G.; Mosiman, C.; Poplawski, M.; Saha, A.; Bender, J.; Granderson, J. Metadata Schemas and Ontologies for Building Energy Applications: A Critical Review and Use Case Analysis. *Energies* **2021**, *14*, 2024. [CrossRef]

19. Gunay, B.H.; Shen, W.; Newsham, G. Data Analytics to Improve Building Performance: A Critical Review. *Autom. Constr.* **2019**, *97*, 96–109. [CrossRef]

20. Chen, J.; Zhang, L.; Li, Y.; Shi, Y.; Gao, X.; Hu, Y. A Review of Computing-Based Automated Fault Detection and Diagnosis of Heating, Ventilation and Air Conditioning Systems. *Renew. Sustain. Energy Rev.* **2022**, *161*, 112395. [CrossRef]

21. Drgoňa, J.; Arroyo, J.; Cupeiro Figueroa, I.; Blum, D.; Arendt, K.; Kim, D.; Ollé, E.P.; Oravec, J.; Wetter, M.; Vrabie, D.L.; et al. All You Need to Know about Model Predictive Control for Buildings. *Annu. Rev. Control* **2020**, *50*, 190–232. [CrossRef]

22. Afram, A.; Janabi-Sharifi, F. Theory and Applications of HVAC Control Systems—A Review of Model Predictive Control (MPC). *Build. Environ.* **2014**, *72*, 343–355. [CrossRef]

23. Yao, Y.; Shekhar, D.K. State of the Art Review on Model Predictive Control (MPC) in Heating Ventilation and Air-Conditioning (HVAC) Field. *Build. Environ.* **2021**, *200*, 107952. [CrossRef]

24. Fu, Q.; Han, Z.; Chen, J.; Lu, Y.; Wu, H.; Wang, Y. Applications of Reinforcement Learning for Building Energy Efficiency Control: A Review. *J. Build. Eng.* **2022**, *50*, 104165. [CrossRef]

25. Nagy, Z.; Gunay, B.; Miller, C.; Hahn, J.; Ouf, M.; Lee, S.; Hobson, B.W.; Abuimara, T.; Bandurski, K.; André, M.; et al. Ten Questions Concerning Occupant-Centric Control and Operations. *Build. Environ.* **2023**, *242*, 110518. [CrossRef]

26. Park, J.Y.; Ouf, M.M.; Gunay, B.; Peng, Y.; O'Brien, W.; Kjærgaard, M.B.; Nagy, Z. A Critical Review of Field Implementations of Occupant-Centric Building Controls. *Build. Environ.* **2019**, *165*, 106351. [CrossRef]

27. Almobarek, M.; Mendibil, K.; Alrashdan, A. Predictive Maintenance 4.0 for Chilled Water System at Commercial Buildings: A Systematic Literature Review. *Buildings* **2022**, *12*, 1229. [CrossRef]

28. Darwazeh, D.; Duquette, J.; Gunay, B.; Wilton, I.; Shillinglaw, S. Review of Peak Load Management Strategies in Commercial Buildings. *Sustain. Cities Soc.* **2021**, *77*, 103493. [CrossRef]

29. Li, H.; Johra, H.; de Andrade Pereira, F.; Hong, T.; Le Dréau, J.; Maturo, A.; Wei, M.; Liu, Y.; Saberi-Derakhtenjani, A.; Nagy, Z.; et al. Data-Driven Key Performance Indicators and Datasets for Building Energy Flexibility: A Review and Perspectives. *Appl. Energy* **2023**, *343*, 121217. [CrossRef]

30. Abuimara, T.; Hobson, B.W.; Gunay, B.; O'Brien, W. A Data-Driven Workflow to Improve Energy Efficient Operation of Commercial Buildings: A Review with Real-World Examples. *Build. Serv. Eng. Res. Technol.* **2022**, *43*, 517–534. [CrossRef]

31. Samad, T.; Bauer, M.; Bortoff, S.; Di Cairano, S.; Fagiano, L.; Odgaard, P.F.; Rhinehart, R.R.; Sánchez-Peña, R.; Serbezov, A.; Ankersen, F.; et al. Industry Engagement with Control Research: Perspective and Messages. *Annu. Rev. Control* **2020**, *49*, 1–14. [CrossRef]

32. Zhan, S.; Chong, A. Data Requirements and Performance Evaluation of Model Predictive Control in Buildings: A Modeling Perspective. *Renew. Sustain. Energy Rev.* **2021**, *142*, 110835. [CrossRef]

33. Wang, Z.; Hong, T. Reinforcement Learning for Building Controls: The Opportunities and Challenges. *Appl. Energy* **2020**, *269*, 115036. [CrossRef]

34. O'Brien, W.; Wagner, A.; Schweiker, M.; Mahdavi, A.; Day, J.; Kjærgaard, M.B.; Carlucci, S.; Dong, B.; Tahmasebi, F.; Yan, D.; et al. Introducing IEA EBC Annex 79: Key Challenges and Opportunities in the Field of Occupant-Centric Building Design and Operation. *Build. Environ.* **2020**, *178*, 106738. [CrossRef]

35. Gunay, H.B.; Ouf, M.; Newsham, G.; O'Brien, W. Sensitivity Analysis and Optimization of Building Operations. *Energy Build.* **2019**, *199*, 164–175. [CrossRef]

36. Date, J.; Candanedo, J.A.; Athienitis, A. Predictive Setpoint Optimization of a Commercial Building Subject to a Winter Demand Penalty Affecting 12 Months of Utility Bills. In Proceedings of the 15th IBPSA Conference, San Francisco, CA, USA, 7–9 August 2017.

37. Torabi, N.; Gunay, H.B.; O'Brien, W.; Barton, T. Common Human Errors in Design, Installation, and Operation of VAV AHU Control Systems—A Review and a Practitioner Interview. *Build. Environ.* **2022**, *221*, 109333. [CrossRef]

38. Brambley, M.R.; Haves, P.; McDonald, S.C.; Torcellini, P.; Hansen, D.G.; Holmberg, D.; Roth, K. *Advanced Sensors and Controls for Building Applications: Market Assessment and Potential R&D Pathways*; Pacific Northwest National Lab. (PNNL): Richland, WA, USA, 2005.

39. Nexus Labs. Nexus Lore: The Smart Building Blocks. 2022. Available online: https://www.nexuslabs.online/nexus-lore/ (accessed on 1 December 2022).

40. Zheng, W.; Hu, J.; Wang, Z.; Li, J.; Fu, Z.; Li, H.; Jurasz, J.; Chou, S.K.; Yan, J. COVID-19 Impact on Operation and Energy Consumption of Heating, Ventilation and Air-Conditioning (HVAC) Systems. *Adv. Appl. Energy* **2021**, *3*, 100040. [CrossRef]

41. Awad, H.; Ashouri, A.; Bahiraei, F. Implications of COVID-19 for Electricity Use in Commercial Smart Buildings in Canada—A Case Study. In Proceedings of the 12th eSim Building Simulation Conference, Ottawa, ON, Canada, 21–24 June 2022.

42. Saloux, E.; Zhang, K. Virtual Energy Metering of Whole Building Cooling Load from Both Airside and Waterside Measurements. In Proceedings of the 12th eSim Building Simulation Conference, Ottawa, ON, Canada, 21–24 June 2022.

43. Nexus Labs. The Untapped 87%: Simplifying Controls Technology for Small Buildings. 2021. Available online: https://www.nexuslabs.online/untapped-87/ (accessed on 1 December 2022).

44. Giebler, C.; Gröger, C.; Hoos, E.; Schwarz, H.; Mitschang, B. Leveraging the Data Lake: Current State and Challenges. In *Big Data Analytics and Knowledge Discovery*; Ordonez, C., Song, I.-Y., Anderst-Kotsis, G., Tjoa, A.M., Khalil, I., Eds.; Springer International Publishing: Cham, Switzerland, 2019; pp. 179–188. [CrossRef]

45. Nambiar, A.; Mundra, D. An Overview of Data Warehouse and Data Lake in Modern Enterprise Data Management. *Big Data Cogn. Comput.* **2022**, *6*, 132. [CrossRef]

46. Fierro, G.; Pritoni, M.; Abdelbaky, M.; Lengyel, D.; Leyden, J.; Prakash, A.; Gupta, P.; Raftery, P.; Peffer, T.; Thomson, G.; et al. Mortar: An Open Testbed for Portable Building Analytics. *ACM Trans. Sen. Netw.* **2019**, *16*, 1–31. [CrossRef]

47. Project Haystack. Available online: https://project-haystack.org/ (accessed on 13 December 2022).

48. BrickSchema. Available online: https://brickschema.org/ (accessed on 13 December 2022).

49. ASHRAE's BACnet Committee. Project Haystack and Brick Schema Collaborating to Provide Unified Data Semantic Modeling Solution | Ashrae.Org. Available online: https://www.ashrae.org/about/news/2018/ashrae-s-bacnet-committee-project-haystack-and-brick-schema-collaborating-to-provide-unified-data-semantic-modeling-solution (accessed on 13 December 2022).

50. Ecobee. Donate Your Data Smart Wi-Fi Thermostats. Available online: https://www.ecobee.com/donate-your-data/ (accessed on 5 December 2022).

51. John, C.; Vallianos, C.; Candanedo, J.; Athienitis, A. Estimating Time Constants for over 10,000 Residential Buildings in North America: Towards a Statistical Characterization of Thermal Dynamics. In Proceedings of the 7th International Building Physics Conference 2018, Syracuse, NY, USA, 23–26 September 2018.

52. Huchuk, B.; O'Brien, W.; Sanner, S. A Longitudinal Study of Thermostat Behaviors Based on Climate, Seasonal, and Energy Price Considerations Using Connected Thermostat Data. *Build. Environ.* **2018**, *139*, 199–210. [CrossRef]

53. Sartori, I.; Walnum, H.T.; Skeie, K.S.; Georges, L.; Knudsen, M.D.; Bacher, P.; Candanedo, J.; Sigounis, A.-M.; Prakash, A.K.; Pritoni, M.; et al. Sub-Hourly Measurement Datasets from 6 Real Buildings: Energy Use and Indoor Climate. *Data Brief* **2023**, *48*, 109149. [CrossRef] [PubMed]

54. New York State Energy Research and Development Authority (NYSERDA). Real Time Energy Management (RTEM) Program—Project Dashboard. Available online: https://www.nyserda.ny.gov/All-Programs/Real-Time-Energy-Management/Project-Dashboard (accessed on 27 April 2023).

55. Deng, Z.; Wang, X.; Jiang, Z.; Zhou, N.; Ge, H.; Dong, B. Evaluation of Deploying Data-Driven Predictive Controls in Buildings on a Large Scale for Greenhouse Gas Emission Reduction. *Energy* **2023**, *270*, 126934. [CrossRef]

56. Jin, X.; Zhang, C.; Xiao, F.; Li, A.; Miller, C. A Review and Reflection on Open Datasets of City-Level Building Energy Use and Their Applications. *Energy Build.* **2023**, *285*, 112911. [CrossRef]

57. *Guideline 36-2021*; High-Performance Sequences of Operation for HVAC Systems. American Society of Heating, Refrigeration and Air Conditioning Engineers ASHRAE: Atlanta, GA, USA, 2021.

58. *Guideline 14-2014*; Measurement Of Energy, Demand, and Water Savings. American Society of Heating, Refrigeration and Air Conditioning Engineers ASHRAE: Atlanta, GA, USA, 2014.

59. Ruiz, G.R.; Bandera, C.F. Validation of Calibrated Energy Models: Common Errors. *Energies* **2017**, *10*, 1587. [CrossRef]
60. Saloux, E.; Zhang, K. Towards Integration of Virtual Meters into Building Energy Management Systems: Development and Assessment of Thermal Meters for Cooling. *J. Build. Eng.* **2022**, *65*, 105785. [CrossRef]
61. International Performance Measurement and Verification Protocol Committee. *International Performance Measurement and Verification Protocol: Concepts and Options for Determining Energy and Water Savings*; U.S. Department of Energy: Oak Ridge, TN, USA, 2002; Volume I.
62. Hobson, B.W.; Gunay, H.B.; Ashouri, A.; Newsham, G.R. Occupancy-Based Predictive Control of an Outdoor Air Intake Damper: A Case Study. In Proceedings of the 11th eSim Building Simulation Conference, Vancouver, BC, Canada, 14–16 June 2020; p. 8.
63. Cotrufo, N.; Saloux, E.; Hardy, J.M.; Candanedo, J.A.; Platon, R. A Practical Artificial Intelligence-Based Approach for Predictive Control in Commercial and Institutional Buildings. *Energy Build.* **2020**, *206*, 109563. [CrossRef]
64. Saloux, E.; Cotrufo, N.; Candanedo, J. A Practical Data-Driven Multi-Model Approach to Model Predictive Control: Results from Implementation in an Institutional Building. In Proceedings of the 6th International High Performance Buildings Conference 2021, West Lafayette, IN, USA, 24–28 May 2021.
65. De Coninck, R.; Helsen, L. Practical Implementation and Evaluation of Model Predictive Control for an Office Building in Brussels. *Energy Build.* **2016**, *111*, 290–298. [CrossRef]
66. Arroyo, J.; Spiessens, F.; Helsen, L. Comparison of Model Complexities in Optimal Control Tested in a Real Thermally Activated Building System. *Buildings* **2022**, *12*, 539. [CrossRef]
67. Freund, S.; Schmitz, G. Implementation of Model Predictive Control in a Large-Sized, Low-Energy Office Building. *Build. Environ.* **2021**, *197*, 107830. [CrossRef]
68. Jain, A.; Smarra, F.; Reticcioli, E.; D'Innocenzo, A.; Morari, M. NeurOpt: Neural Network Based Optimization for Building Energy Management and Climate Control. *Proc. Mach. Learn. Res.* **2020**, *120*, 445–454.
69. Berkeley Lab. Building Technology & Urban Systems Division Data Analytics for Commercial Buildings: Advanced Measurement & Verification (M&V) Research. Available online: https://buildings.lbl.gov/emis/assessment-automated-mv-methods (accessed on 3 May 2023).
70. U.S. Department of Energy. *A Primer on Organizational Use of Energy Management and Information Systems (EMIS)*, 2nd ed.; Lawrence Berkeley National Lab. (LBNL): Berkeley, CA, USA, 2021.
71. Pachuta, S.; Dean, J.; Kandt, A.; Nguyen Cu, K. *Field Validation of a Building Operating System Platform*; U.S. General Services Administration: Washington, DC, USA, 2022. Available online: https://www.gsa.gov/governmentwide-initiatives/climate-action-and-sustainability/center-for-emerging-building-technologies/published-findings/energy-management/energy-management-information-system-with-automated-system-optimization (accessed on 15 December 2022).
72. Zhang, K.; Prakash, A.; Paul, L.; Blum, D.; Alstone, P.; Zoellick, J.; Brown, R.; Pritoni, M. Model Predictive Control for Demand Flexibility: Real-World Operation of a Commercial Building with Photovoltaic and Battery Systems. *Adv. Appl. Energy* **2022**, *7*, 100099. [CrossRef]
73. European Commission. Smart Readiness Indicator. Available online: https://energy.ec.europa.eu/topics/energy-efficiency/energy-efficient-buildings/smart-readiness-indicator_en (accessed on 13 December 2022).
74. Urban Green Council. All about Local Law 97. Available online: https://www.urbangreencouncil.org/content/projects/all-about-local-law-97 (accessed on 15 December 2022).
75. Teng, Y.; Xu, J.; Pan, W.; Zhang, Y. A Systematic Review of the Integration of Building Information Modeling into Life Cycle Assessment. *Build. Environ.* **2022**, *221*, 109260. [CrossRef]
76. Fernald, H.; Hong, S.; O'Brien, L.; Bucking, S. BIM to BEM Translation Workflows and Their Challenges: A Case Study Using a Detailed BIM Model. In Proceedings of the 10th eSim Building Simulation Conference, Montréal, QC, Canada, 9–10 May 2018; Volume 10, pp. 482–491.
77. Reeves, T.; Olbina, S.; Issa, R.R.A. Guidelines for Using Building Information Modeling for Energy Analysis of Buildings. *Buildings* **2015**, *5*, 1361–1388. [CrossRef]
78. Chong, A.; Gu, Y.; Jia, H. Calibrating Building Energy Simulation Models: A Review of the Basics to Guide Future Work. *Energy Build.* **2021**, *253*, 111533. [CrossRef]
79. Candanedo, J.A.; Dehkordi, V.R.; Lopez, P. A Control-Oriented Simplified Building Modelling Strategy. In Proceedings of the 13th Conference of International Building Performance Simulation Association, Chambéry, France, 26–28 August 2013.
80. Candanedo, J.A.; Vallianos, C.; Delcroix, B.; Date, J.; Saberi Derakhtenjani, A.; Morovat, N.; John, C.; Athienitis, A.K. Control-Oriented Archetypes: A Pathway for the Systematic Application of Advanced Controls in Buildings. *J. Build. Perform. Simul.* **2022**, *15*, 433–444. [CrossRef]
81. Blum, D.; Wetter, M. MPCPy: An Open-Source Software Platform for Model Predictive Control in Buildings. In Proceedings of the 15th Conference of International Building Performance Simulation, San Francisco, CA, USA, 7–9 August 2017.
82. Nweye, K.; Liu, B.; Stone, P.; Nagy, Z. Real-World Challenges for Multi-Agent Reinforcement Learning in Grid-Interactive Buildings. *Energy AI* **2022**, *10*, 100202. [CrossRef]
83. Kazmi, H.; Fu, C.; Miller, C. Ten Questions Concerning Data-Driven Modelling and Forecasting of Operational Energy Demand at Building and Urban Scale. *Build. Environ.* **2023**, *239*, 110407. [CrossRef]
84. Pinto, G.; Wang, Z.; Roy, A.; Hong, T.; Capozzoli, A. Transfer Learning for Smart Buildings: A Critical Review of Algorithms, Applications, and Future Perspectives. *Adv. Appl. Energy* **2022**, *5*, 100084. [CrossRef]

85. Pinto, G.; Messina, R.; Li, H.; Hong, T.; Piscitelli, M.S.; Capozzoli, A. Sharing Is Caring: An Extensive Analysis of Parameter-Based Transfer Learning for the Prediction of Building Thermal Dynamics. *Energy Build.* **2022**, *276*, 112530. [CrossRef]

86. Blum, D.; Arroyo, J.; Huang, S.; Drgoňa, J.; Jorissen, F.; Walnum, H.T.; Chen, Y.; Benne, K.; Vrabie, D.; Wetter, M.; et al. Building Optimization Testing Framework (BOPTEST) for Simulation-Based Benchmarking of Control Strategies in Buildings. *J. Build. Perform. Simul.* **2021**, *14*, 586–610. [CrossRef]

87. Miller, C.; Kathirgamanathan, A.; Picchetti, B.; Arjunan, P.; Park, J.Y.; Nagy, Z.; Raftery, P.; Hobson, B.W.; Shi, Z.; Meggers, F. The Building Data Genome Project 2, Energy Meter Data from the ASHRAE Great Energy Predictor III Competition. *Sci. Data* **2020**, *7*, 368. [CrossRef] [PubMed]

88. Miller, C.; Arjunan, P.; Kathirgamanathan, A.; Fu, C.; Roth, J.; Park, J.Y.; Balbach, C.; Gowri, K.; Nagy, Z.; Fontanini, A.D.; et al. The ASHRAE Great Energy Predictor III Competition: Overview and Results. *Sci. Technol. Built Environ.* **2020**, *26*, 1427–1447. [CrossRef]

89. Nagy, Z.; Vázquez-Canteli, J.R.; Dey, S.; Henze, G. The Citylearn Challenge 2021. In Proceedings of the 8th ACM International Conference on Systems for Energy-Efficient Buildings, Cities, and Transportation, Coimbra, Portugal, 17–18 November 2021; pp. 218–219.

90. Makridakis, S.; Spiliotis, E.; Assimakopoulos, V. The M4 Competition: 100,000 Time Series and 61 Forecasting Methods. *International J. Forecast.* **2020**, *36*, 54–74. [CrossRef]

91. SDU Center for Energy Informatics. ADRENALIN, Data-Driven Smart Buildings: Data Sandbox and Competition. Available online: https://www.sdu.dk/en/forskning/centreforenergyinformatics/research-projects/adrenalin (accessed on 14 June 2023).

92. Pritoni, M.; Lin, G.; Chen, Y.; Vitti, R.; Weyandt, C.; Granderson, J. From Fault-Detection to Automated Fault Correction: A Field Study. *Build. Environ.* **2022**, *214*, 108900. [CrossRef]

93. Hosamo, H.H.; Svennevig, P.R.; Svidt, K.; Han, D.; Nielsen, H.K. A Digital Twin Predictive Maintenance Framework of Air Handling Units Based on Automatic Fault Detection and Diagnostics. *Energy Build.* **2022**, *261*, 111988. [CrossRef]

94. Li, H.; Hong, T.; Lee, S.H.; Sofos, M. System-Level Key Performance Indicators for Building Performance Evaluation. *Energy Build.* **2020**, *209*, 109703. [CrossRef]

95. Morovat, N.; Athienitis, A.K.; Candanedo, J.A.; Delcroix, B. Model-Based Control Strategies to Enhance Energy Flexibility in Electrically Heated School Buildings. *Buildings* **2022**, *12*, 581. [CrossRef]

96. Saloux, E.; Zhang, K. Data-Driven Model-Based Control Strategies to Improve the Cooling Performance of Commercial and Institutional Buildings. *Buildings* **2023**, *13*, 474. [CrossRef]

97. Li, H.; Yu, D.; Braun, J.E. A Review of Virtual Sensing Technology and Application in Building Systems. *HVACR Res.* **2011**, *17*, 619–645.

98. Darwazeh, D.; Gunay, B.; Duquette, J.; O'Brien, W. A Virtual Meter-Based Visualization Tool to Present Energy Flows in Multiple Zone Variable Air Volume Air Handling Unit Systems. *Build. Environ.* **2022**, *221*, 109275. [CrossRef]

99. Gunay, B.; Shi, Z.; Wilton, I.; Bursill, J. Disaggregation of Commercial Building End-Uses with Automation System Data. *Energy Build.* **2020**, *223*, 110222. [CrossRef]

100. Xiao, Z.; Gang, W.; Yuan, J.; Zhang, Y.; Fan, C. Cooling Load Disaggregation Using a NILM Method Based on Random Forest for Smart Buildings. *Sustain. Cities Soc.* **2021**, *74*, 103202. [CrossRef]

101. Zaeri, N.; Ashouri, A.; Gunay, H.B.; Abuimara, T. Disaggregation of Electricity and Heating Consumption in Commercial Buildings with Building Automation System Data. *Energy Build.* **2022**, *258*, 111791. [CrossRef]

102. Lee, J.; Kim, T.W.; Koo, C. A Novel Process Model for Developing a Scalable Room-Level Energy Benchmark Using Real-Time Bigdata: Focused on Identifying Representative Energy Usage Patterns. *Renew. Sustain. Energy Rev.* **2022**, *169*, 112944. [CrossRef]

103. Waterworth, D.; Sethuvenkatraman, S.; Sheng, Q.Z. Advancing Smart Building Readiness: Automated Metadata Extraction Using Neural Language Processing Methods. *Adv. Appl. Energy* **2021**, *3*, 100041. [CrossRef]

104. Chen, L.; Gunay, H.B.; Shi, Z.; Shen, W.; Li, X. A Metadata Inference Method for Building Automation Systems with Limited Semantic Information. *IEEE Trans. Autom. Sci. Eng.* **2020**, *17*, 2107–2119. [CrossRef]

105. Mishra, S.; Glaws, A.; Cutler, D.; Frank, S.; Azam, M.; Mohammadi, F.; Venne, J.-S. Unified Architecture for Data-Driven Metadata Tagging of Building Automation Systems. *Autom. Constr.* **2020**, *120*, 103411. [CrossRef]

106. Zhang, K.; Blum, D.; Cheng, H.; Paliaga, G.; Wetter, M.; Granderson, J. Estimating ASHRAE Guideline 36 Energy Savings for Multi-Zone Variable Air Volume Systems Using Spawn of EnergyPlus. *J. Build. Perform. Simul.* **2022**, *15*, 215–236. [CrossRef]

107. Nassif, N.; Tahmasebi, M.; Ridwana, I.; Ebrahimi, P. New Optimal Supply Air Temperature and Minimum Zone Air Flow Resetting Strategies for VAV Systems. *Buildings* **2022**, *12*, 348. [CrossRef]

108. Lu, X.; Fu, Y.; O'Neill, Z. Benchmarking High Performance HVAC Rule-Based Controls with Advanced Intelligent Controllers: A Case Study in a Multi-Zone System in Modelica. *Energy Build.* **2023**, *284*, 112854. [CrossRef]

109. Hong, T.; Yan, D.; D'Oca, S.; Chen, C. Ten Questions Concerning Occupant Behavior in Buildings: The Big Picture. *Build. Environ.* **2017**, *114*, 518–530. [CrossRef]

110. Gupta, S.K.; Atkinson, S.; O'Boyle, I.; Drogo, J.; Kar, K.; Mishra, S.; Wen, J.T. BEES: Real-Time Occupant Feedback and Environmental Learning Framework for Collaborative Thermal Management in Multi-Zone, Multi-Occupant Buildings. *Energy Build.* **2016**, *125*, 142–152. [CrossRef]

111. Jayathissa, P.; Quintana, M.; Sood, T.; Nazarian, N.; Miller, C. Is Your Clock-Face Cozie? A Smartwatch Methodology for the in-Situ Collection of Occupant Comfort Data. *J. Phys. Conf. Ser.* **2019**, *1343*, 012145. [CrossRef]

112. Sood, T.; Janssen, P.; Miller, C. Spacematch: Using Environmental Preferences to Match Occupants to Suitable Activity-Based Workspaces. *Front. Built Environ.* **2020**, *6*, 113. [CrossRef]
113. Saloux, E.; Teyssedou, A.; Sorin, M. Development of an Exergy-Electrical Analogy for Visualizing and Modeling Building Integrated Energy Systems. *Energy Convers. Manag.* **2015**, *89*, 907–918. [CrossRef]
114. Dong, H.-W.; Kim, B.-J.; Yoon, S.-Y.; Jeong, J.-W. Energy Benefit of Organic Rankine Cycle in High-Rise Apartment Building Served by Centralized Liquid Desiccant and Evaporative Cooling-Assisted Ventilation System. *Sustain. Cities Soc.* **2020**, *60*, 102280. [CrossRef]
115. Bamigbetan, O.; Eikevik, T.M.; Nekså, P.; Bantle, M. Review of Vapour Compression Heat Pumps for High Temperature Heating Using Natural Working Fluids. *Int. J. Refrig.* **2017**, *80*, 197–211. [CrossRef]
116. Hasan, J.; Fung, A.S.; Horvat, M. A Comparative Evaluation on the Case for the Implementation of Building Integrated Photovoltaic/Thermal (BIPV/T) Air Based Systems on a Typical Mid-Rise Commercial Building in Canadian Cities. *J. Build. Eng.* **2021**, *44*, 103325. [CrossRef]
117. Chong, W.T.; Yip, S.Y.; Fazlizan, A.; Poh, S.C.; Hew, W.P.; Tan, E.P.; Lim, T.S. Design of an Exhaust Air Energy Recovery Wind Turbine Generator for Energy Conservation in Commercial Buildings. *Renew. Energy* **2014**, *67*, 252–256. [CrossRef]
118. Peris Pérez, B.; Ávila Gutiérrez, M.; Expósito Carrillo, J.A.; Salmerón Lissén, J.M. Performance of Solar-Driven Ejector Refrigeration System (SERS) as Pre-Cooling System for Air Handling Units in Warm Climates. *Energy* **2022**, *238*, 121647. [CrossRef]
119. Dumont, O.; Quoilin, S.; Lemort, V. Experimental Investigation of a Reversible Heat Pump/Organic Rankine Cycle Unit Designed to Be Coupled with a Passive House to Get a Net Zero Energy Building. *Int. J. Refrig.* **2015**, *54*, 190–203. [CrossRef]
120. Saloux, E.; Sorin, M.; Nesreddine, H.; Teyssedou, A. Thermodynamic Modeling and Optimal Operating Conditions of Organic Rankine Cycles (ORC) Independently of the Working Fluid. *Int. J. Green Technol.* **2019**, *5*, 9–22. [CrossRef]
121. Cedillo, M.H.; Sun, H.; Jiang, J.; Cao, Y. Dynamic Pricing and Control for EV Charging Stations with Solar Generation. *Appl. Energy* **2022**, *326*, 119920. [CrossRef]
122. Nair, A.M.; Wilson, C.; Huang, M.J.; Griffiths, P.; Hewitt, N. Phase Change Materials in Building Integrated Space Heating and Domestic Hot Water Applications: A Review. *J. Energy Storage* **2022**, *54*, 105227. [CrossRef]
123. Sevault, A.; Bøhmer, F.; Næss, E.; Wang, L. Latent Heat Storage for Centralized Heating System in a ZEB Living Laboratory: Integration and Design. *IOP Conf. Ser. Earth Environ. Sci.* **2019**, *352*, 012042. [CrossRef]
124. Candanedo, J.A.; Dehkordi, V.R.; Stylianou, M. Model-Based Predictive Control of an Ice Storage Device in a Building Cooling System. *Appl. Energy* **2013**, *111*, 1032–1045. [CrossRef]
125. Kang, X.; Wang, X.; An, J.; Yan, D. A Novel Approach of Day-Ahead Cooling Load Prediction and Optimal Control for Ice-Based Thermal Energy Storage (TES) System in Commercial Buildings. *Energy Build.* **2022**, *275*, 112478. [CrossRef]
126. Serale, G.; Fiorentini, M.; Capozzoli, A.; Bernardini, D.; Bemporad, A. Model Predictive Control (MPC) for Enhancing Building and HVAC System Energy Efficiency: Problem Formulation, Applications and Opportunities. *Energies* **2018**, *11*, 631. [CrossRef]
127. Taheri, S.; Hosseini, P.; Razban, A. Model Predictive Control of Heating, Ventilation, and Air Conditioning (HVAC) Systems: A State-of-the-Art Review. *J. Build. Eng.* **2022**, *60*, 105067. [CrossRef]
128. Thieblemont, H.; Haghighat, F.; Ooka, R.; Moreau, A. Predictive Control Strategies Based on Weather Forecast in Buildings with Energy Storage System: A Review of the State-of-the Art. *Energy Build.* **2017**, *153*, 485–500. [CrossRef]
129. Yu, Z.; Huang, G.; Haghighat, F.; Li, H.; Zhang, G. Control Strategies for Integration of Thermal Energy Storage into Buildings: State-of-the-Art Review. *Energy Build.* **2015**, *106*, 203–215. [CrossRef]
130. Péan, T.Q.; Salom, J.; Costa-Castelló, R. Review of Control Strategies for Improving the Energy Flexibility Provided by Heat Pump Systems in Buildings. *J. Process Control* **2019**, *74*, 35–49. [CrossRef]
131. Mirakhorli, A.; Dong, B. Occupancy Behavior Based Model Predictive Control for Building Indoor Climate—A Critical Review. *Energy Build.* **2016**, *129*, 499–513. [CrossRef]
132. Henze, G.P. Model Predictive Control for Buildings: A Quantum Leap? *J. Build. Perform. Simul.* **2013**, *6*, 157–158. [CrossRef]
133. Zhan, S.; Quintana, M.; Miller, C.; Chong, A. From Model-Centric to Data-Centric: A Practical MPC Implementation Framework for Buildings. In Proceedings of the 9th ACM International Conference on Systems for Energy-Efficient Buildings, Cities, and Transportation, Boston, MA, USA, 9–10 November 2022; pp. 270–273.
134. Cigler, J.; Gyalistras, D.; Siroky, J.; Tiet, V.-N.; Ferkl, L. Beyond Theory: The Challenge of Implementing Model Predictive Control in Buildings. In Proceedings of the 11th Rehva World Congress, Prague, Czech Republic, 16–19 June 2013.
135. Vallianos, C.; Candanedo, J.; Athienitis, A. Application of a Large Smart Thermostat Dataset for Model Calibration and Model Predictive Control Implementation in the Residential Sector. *Energy* **2023**, *278*, 127839. [CrossRef]
136. Li, R.; Satchwell, A.J.; Finn, D.; Christensen, T.H.; Kummert, M.; Le Dréau, J.; Lopes, R.A.; Madsen, H.; Salom, J.; Henze, G.; et al. Ten Questions Concerning Energy Flexibility in Buildings. *Build. Environ.* **2022**, *223*, 109461. [CrossRef]
137. Jensen, S.Ø.; Marszal-Pomianowska, A.; Lollini, R.; Pasut, W.; Knotzer, A.; Engelmann, P.; Stafford, A.; Reynders, G. IEA EBC Annex 67 Energy Flexible Buildings. *Energy Build.* **2017**, *155*, 25–34. [CrossRef]
138. U.S. Department of Energy. *A National Roadmap for Grid-Interactive Efficient Buildings*; Office of Energy Efficiency and Renewable Energy, Building Technologies Office: Washington, DC, USA, 2021.

139. Kathirgamanathan, A.; De Rosa, M.; Mangina, E.; Finn, D.P. Data-Driven Predictive Control for Unlocking Building Energy Flexibility: A Review. *Renew. Sustain. Energy Rev.* **2021**, *135*, 110120. [CrossRef]
140. Nexus Labs. The Energy Management Hierarchy of Needs. Available online: https://www.nexuslabs.online/content/the-energy-management-hierarchy-of-needs (accessed on 28 June 2023).

buildings

Article

LSTM Deep Learning Models for Virtual Sensing of Indoor Air Pollutants: A Feasible Alternative to Physical Sensors

Martin Gabriel * and Thomas Auer *

Building Technology and Climate Responsive Design, TUM School of Engineering and Design, Technical University of Munich, Arcisstraße 21, 80333 Munich, Germany
* Correspondence: martin.gabriel@tum.de (M.G.); thomas.auer@tum.de (T.A.)

Abstract: Monitoring individual exposure to indoor air pollutants is crucial for human health and well-being. Due to the high spatiotemporal variations of indoor air pollutants, ubiquitous sensing is essential. However, the cost and maintenance associated with physical sensors make this currently infeasible. Consequently, this study investigates the feasibility of virtually sensing indoor air pollutants, such as particulate matter, volatile organic compounds (VOCs), and CO_2, using a long short-term memory (LSTM) deep learning model. Several years of accumulated measurement data were employed to train the model, which predicts indoor air pollutant concentrations based on Building Management System (BMS) data (e.g., temperature, humidity, illumination, noise, motion, and window state) as well as meteorological and outdoor pollution data. A cross-validation scheme and hyperparameter optimization were utilized to determine the best model parameters and evaluate its performance using common evaluation metrics (R^2, mean absolute error (MAE), root mean square error (RMSE)). The results demonstrate that the LSTM model can effectively replace physical indoor air pollutant sensors in the examined room, with evaluation metrics indicating a strong correlation in the testing set (MAE; CO_2: 15.4 ppm, $PM_{2.5}$: 0.3 µg/m^3, VOC: 20.1 IAQI; R^2; CO_2: 0.47, $PM_{2.5}$: 0.88, VOC:0.87). Additionally, the transferability of the model to other rooms was tested, with good results for CO_2 and mixed results for VOC and particulate matter (MAE; CO_2: 21.9 ppm, $PM_{2.5}$: 0.3 µg/m^3, VOC: 52.7 IAQI; R^2; CO_2: 0.45, $PM_{2.5}$: 0.09, VOC:0.13). Despite these mixed results, they hint at the potential for a more broadly applicable approach to virtual sensing of indoor air pollutants, given the incorporation of more diverse datasets, thereby offering the potential for real-time occupant exposure monitoring and enhanced building operations.

Keywords: machine learning; deep learning; virtual sensing; LSTM; IAQ; monitoring

Citation: Gabriel, M.; Auer, T. LSTM Deep Learning Models for Virtual Sensing of Indoor Air Pollutants: A Feasible Alternative to Physical Sensors. *Buildings* **2023**, *13*, 1684. https://doi.org/10.3390/buildings13071684

Academic Editor: Etienne Saloux

Received: 27 May 2023
Revised: 28 June 2023
Accepted: 29 June 2023
Published: 30 June 2023

1. Introduction

Indoor air pollutants are of different sizes and types, are harmful at different concentrations, and have different intake pathways and effects. The major groups of pollutants are inorganic gases, organic gases, particulates, microbial pollutants, and viral and bacterial infections [1]. Controlling indoor air pollutant exposure is especially relevant since we spend up to 87% of our time in buildings [2], rendering them the most important environments. In efforts to make buildings more energy efficient, they have become better sealed and indoor spaces more dependent on HVAC systems [2]. Studies have shown that bad indoor air quality leads to a multitude of different symptoms and health impacts. The gravity of these impacts depends on the pollutants, their concentration, the exposure time, and individual factors such as age, constitution, and health [3]. Most frequently, occupants experience tiredness, burning eyes, headaches, and concentration problems [3]. Prolonged exposure may also lead to respiratory syndromes and immune system reactions such as asthma, especially in vulnerable groups such as children or elderly persons [3]. According to a study from the WHO, air pollution is a significant health threat and a primary environmental factor in causing premature deaths in Europe [4]. Exposure to fine particulate matter has been linked to over 400,000 premature deaths in European countries [5]. Therefore, many countries are taking steps to reduce indoor air pollutant concentration by enforcing exposure limits. An in-detail summary of exposure limits in different countries is given in Abdul et al. [6]. The European Union largely adopted the exposure limits suggested by the WHO in [7]. Effective strategies for reducing indoor air pollutant exposure include source control, which involves identifying and minimizing sources of pollution, mitigation measures such as removing pollutants and introducing clean air, and monitoring indoor air pollutant concentrations [1]. These measures are situated at different points of the building life cycle: source control is relevant in the planning and construction phase of buildings, and mitigation and monitoring are required during the use of the building. Building codes progressively ensure source control in new buildings by restricting harmful, pollutant-emitting materials. However, the majority of existing building stock was built without awareness of indoor air quality concerns. Therefore, improving indoor air quality in existing buildings is key. Monitoring and mitigation measures are especially relevant in the non-residential sector since occupants have little influence on the indoor environment, as opposed to residential buildings. Therefore, the following study will focus on non-residential typologies in the existing building stock; specifically, typology—with high occupant density and prolonged exposure will be part of the study.

2. Indoor Air Pollutants in the Non-Residential Building Stock

Several studies have examined the spatiotemporal distribution of indoor air pollutants within rooms in non-residential building stock. Szigeti et al. [8] examined the spatiotemporal distribution of particulate matter in European office buildings and found a significant variation between buildings. Within buildings, temporal variation is more pronounced than spatial distribution variations [8]. According to Szigeti et al. [8], occupants may be exposed to significantly different pollutant concentrations in different rooms within a building. Li et al. [9] examined the spatiotemporal distribution of particulate matter within one room (workshop) with localized sources and found high spatial and temporal variations within a single room. Sahu et al. [10] examined the distribution of indoor air pollutants—CO_2, particulate matter, VOC—in a multi-story library with shared air volume. They found high spatial and temporal variations within the library, with temporal variations mainly driven by the number of occupants and spatial variation more pronounced between the different stories. Studies Szigeti et al. [8], Li et al. [9], and Sahu et al. [10] show that indoor air pollutants show significant spatiotemporal variations. Therefore, a continuous and spatiotemporally high-resolution monitoring is required in order to evaluate occupant exposure and control ventilation units. The state of research in continuous and spatiotemporal high-resolution indoor air pollutant monitoring was analyzed, looking at fifteen studies, monitoring indoor air pollutants in non-residential buildings: [9,11–24]. Over all studies, the most measured pollutant is particulate matter, examined by 14 out of 15 publications. In 8 of 15 studies measured one or multiple volatile organic compounds, and 7 out of 15 studies measured CO_2. Carbon monoxide and nitrogen dioxide were assessed in 5 out of 15 studies, and ozone and sulfur dioxide in 2 studies. Several other pollutants as nitrogen oxide and fungi, are only considered in 1 publication. The studies evaluated the importance of different pollutants in non-residential buildings and found that carbon monoxide and radon only accumulate in particular spaces, such as kitchens with gas ovens and basements, and are insignificant in typical non-residential buildings [11,25]. Irga et al. [14] found the levels of nitrogen oxide, volatile organic compounds, fungi, and sulfur dioxide to be harmless in 11 office buildings. According to Challoner et al. [22], the most problematic pollutant in non-residential buildings is fine particulate matter, exceeding health thresholds in 10% of the measured time. Additionally, carbon dioxide regularly exceeds the threshold of 1000 ppm in naturally ventilated buildings [15]. Likewise, volatile organic compounds are effectively controlled by mechanical ventilation systems but can reach problematic concentrations if the ventilation system is switched off or buildings are naturally ventilated [15]. The reviewed articles exclusively use on-site measurement technology since laboratory analysis is not able to gather continuous, high-resolution measurements. Most studies used MOS technology VOC sensors, NDIR technology CO_2 sensors, and OPC technology particulate matter sensors.

3. Virtual Sensing for Indoor Air Pollutants

Ubiquitous monitoring of indoor air pollutants requires durable, low-cost, low-maintenance, low-energy sensor equipment. NDIR and OPC technologies are optical measurement principles requiring fragile components and suffer from measurement drift and longevity issues due to the build-up of foreign particles in the measurement chambers [26]. Furthermore, optical measurement methods have higher energy consumption and are less suited for battery operation. Even though MOS-based VOC sensors are low cost and low energy, MOS-based VOC measurements are prone to drift and suffer from reproducibility issues [27]. These drawbacks necessitate careful maintenance, diligent monitoring for failures, and thorough post-processing of data, which is often overlooked in building operations. A mitigation of these shortcomings is presented in previous studies [28,29] that developed calibration models to improve accuracy and reduce the drift of IAP sensors. However, those systems have not found their way into practice yet. Therefore, alternatives to the measurement of particulate matter (PM), volatile organic compounds (VOC), and carbon dioxide (CO_2) have to be found.

Virtual sensing of PM, VOC, and CO_2 is an alternative to ubiquitous sensor deployment. Virtual sensing "aims to approximate unmeasured physical quantities in a dynamic system using existing sensor information. This is especially beneficial when important locations of the system are difficult to instrument, or the cost of sensors is very high" [30]. "A virtual sensor uses low-cost measurements and mathematical models to estimate a difficult to measure or expensive quantity" [31]. The models are based on related physical measurements, control signals, operation information, and design information [32]. Virtual sensing finds widespread application in the domains of process control, automotive, avionics, and robotics [31]. However, with the exponential rise of available data points through developments in IoT and cost reduction of sensors, virtual sensing has been increasingly adopted in the building industry [32]. The application of virtual sensing in the building industry is manifold. Buildings gather many data points, and nearly every physical sensor can provide additional information for virtual sensing. Li et al. [31] gives an example of the potential of virtual sensing in buildings: "A 'smart' lighting fixture could provide power, lighting, and heat gain outputs based on the input control signal. A 'smart' window could provide estimates of heat gain and even solar radiation based on low-cost measurements and a model". Application scenarios in buildings include HVAC operation monitoring [33,34], indoor infiltration rate [35], zone temperature distribution [36], zone occupancy estimation [37] and indoor air pollutant monitoring [38–40]. Generally, virtual sensors can be differentiated into three application scenarios: replacement and backup, observation, and assistance [32]. Replacement and backup virtual sensors are deployed in parallel to their physical counterparts and, by computing the residuals between physical and virtual measurement, are able to detect sensor faults or calibration drifts [41], and can replace their physical counterparts if needed [32]. Observation virtual sensors estimate a data point without their physical counterpart using other measurements and mathematical models [32]. Assistance virtual sensors do not estimate a physical quantity but are integrated into other virtual sensors to improve accuracy. The output of assistance virtual sensors is often normalized [32]. Virtual sensors can further be differentiated regarding their modeling method and the underlying measurement characteristics [31]. Modeling methods are white-box, grey-box, and black-box models [31]. Measurement characteristics can be differentiated in transient (e.g., power usage, indoor temperature) or steady state (e.g., system failure state) data [31]. White and grey-box models require in-depth knowledge of the building. These approaches are infeasible for older buildings due to unavailable planning documents, undocumented changes, and performance deterioration. One alternative would be a black-box model, which requires extensive measurement data.

To summarize, virtual sensing for indoor air pollutant prediction in non-residential typologies requires transient-state observational virtual sensors created using a black-box modeling method.

Other studies have already examined the applicability of virtual sensing for indoor air pollutant prediction. A study by Gabriel and Auer [42] used multi-layer perceptron (MLP) artificial neural networks and support vector machines (SVM) to create an observational virtual particulate matter sensor based on available Building Management System (BMS) data (temperature, pressure, humidity, sound, illumination, window opening state, and printer power consumption). Six months of measurement data were used to train and test the two machine-learning models. Gabriel and Auer [42] found that MLPs performed best, and the results indicated that physical particulate matter sensors could be replaced by virtual sensors based on BMS data. Kusiak et al. [41] created virtual replacement sensors for temperature, humidity, and CO_2 with four modeling approaches for the calibration and monitoring of physical sensors using HVAC, climate, and other indoor air pollutant data. MLPs were found to perform best in modeling the physical sensors. Kusiak et al. [41] conclude that the virtual sensors are able to detect failures of their corresponding physical sensors and replace them if necessary. Skoen et al. [38] used MLPs to create an observational virtual sensor using temperature and humidity as input for modeling CO_2. Skoen et al. [38] concludes that estimating CO_2 only based on temperature and humidity is difficult and requires additional measurements to support the black-box model. Leidinger et al. [43] created a virtual replacement sensor for selective VOC sampling of formaldehyde, benzene, and naphthalene using an array of low-cost MOS sensors as input. Leidinger et al. [43] used linear discriminant analysis to estimate the target variables. Under laboratory conditions, the study achieved a classification ratio of over 99%. However, in field tests, the classification ratio significantly dropped (83%) due to VOC emissions of the hardware [43]. A summary of Literature on Virtual Sensor Creation is given in Table 1. Research in other domains showed that long short-term memory (LSTM) recurrent neural networks are suited for time-series data in virtual sensor creation due to their ability to incorporate measurements from a lookback window into their model. LSTMs are recurrent neural networks specialized in time series data by incorporating memory cells in their network architecture, which enable them to identify and remember patterns in time series data [30]. In the building industry, LSTMs have already been applied to building load management for forecasting energy consumption [44,45] and predicting occupancy [46]. LSTMs have not yet been applied to modeling virtual indoor air pollutant sensors.

Table 1. Summary of Literature on Virtual Sensor Creation.

Study	Virtual Sensor Type	Methods Used	Main Findings
Gabriel and Auer [42]	Particulate Matter	MLP, SVM	MLPs performed better than SVM; results show the potential of virtual sensors to replace physical ones
Kusiak et al. [41]	Temperature, Humidity, CO_2	MLP, SVM, Pacereg, RBF	MLP outperformed other models; Virtual sensors can detect and replace failing physical sensors
Skoen et al. [38]	CO_2	MLP	Estimating CO_2 based only on temperature and humidity is challenging
Leidinger et al. [43]	VOC Sampling	Linear Discriminant Analysis	99% lab accuracy, 83% field accuracy due to hardware VOC emissions
Karijadi et al. [44] and Jang et al. [45]	Energy Consumption	LSTM	LSTMs have been successfully applied in energy consumption forecasting
Qolomany et al. [46]	Occupancy	LSTM	LSTM can be used for predicting occupancy

4. Study Definition

The literature indicates that monitoring occupant pollutant exposure and mitigating indoor air pollutant concentrations is important to ensure health and well-being. Since occupants in non-residential typologies have no or low possibility of intervention regarding indoor air quality, particular care has to be taken in providing adequate indoor conditions in these typologies. Due to the high spatiotemporal variation of indoor air pollutants in non-residential buildings, high-resolution monitoring is required. However, ubiquitous sensing of indoor air pollutants is infeasible due to the high cost and time required for installation, operation, and maintenance. Therefore, virtual sensing is suggested as an alternative to the ubiquitous deployment of physical sensors. Previous studies have already been conducted, applying virtual sensing to indoor air pollutant estimation. However, these studies mostly assessed only one air pollutant, even though exposure monitoring requires the estimation of multiple pollutants. Furthermore, all currently available studies build virtual sensors based on data from one zone in a single typology and do not consider the transferability of the models to other zones and/or typologies. Additionally, all studies reviewed here used less than a year of measurement data to build the models, thus introducing significant bias into the models. Despite the demonstrated effectiveness of LSTM in handling time-series problems across various domains, it is observed that none of the known virtual sensing approaches to indoor air pollutants have adopted LSTM as their modeling approach [47,48].

Therefore, our study examines the feasibility of observational virtual sensors for PM, VOC, and CO_2 based on an LSTM modeling approach. The study uses multiple years of accumulated measurement data from multiple zones and typologies to build the virtual-sensor model and check its transferability to other zones and typologies. The capability of the virtual sensor was evaluated independently for the room where the model was trained and on unknown rooms.

Figure 1 gives a visual overview of the study definition.

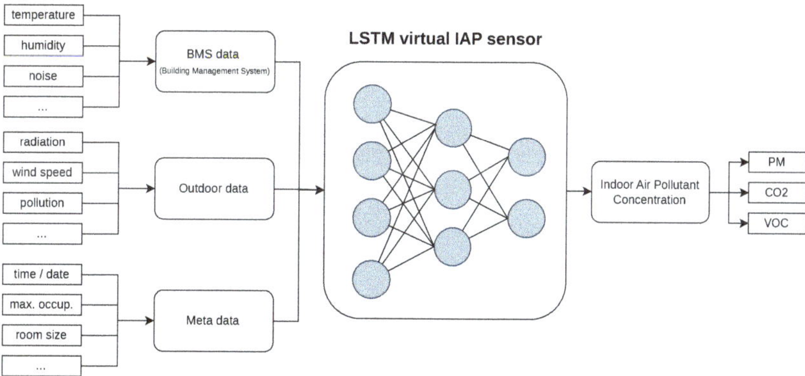

Figure 1. Flowchart of the study definition with a black box LSTM model and input/output components (Own representation).

5. Methods

In this study, we employed an LSTM model trained on collected measurement data to predict indoor air pollutant concentrations using BMS, outdoor meteorological, and outdoor pollution data as model inputs. In the following sections, we detail the methods used. Section 5.1 describes the steps performed in order to build the dataset, including the measurement equipment, the measurement setup, and the measurement location. Section 5.2 presents the steps to preprocess the data for machine learning model training. Section 5.3 encompasses the training of the models, while Section 5.4 focuses on model evaluation. Finally, the transferability tests are detailed in Section 5.5.

A graphical representation of the method is illustrated in Figure 2.

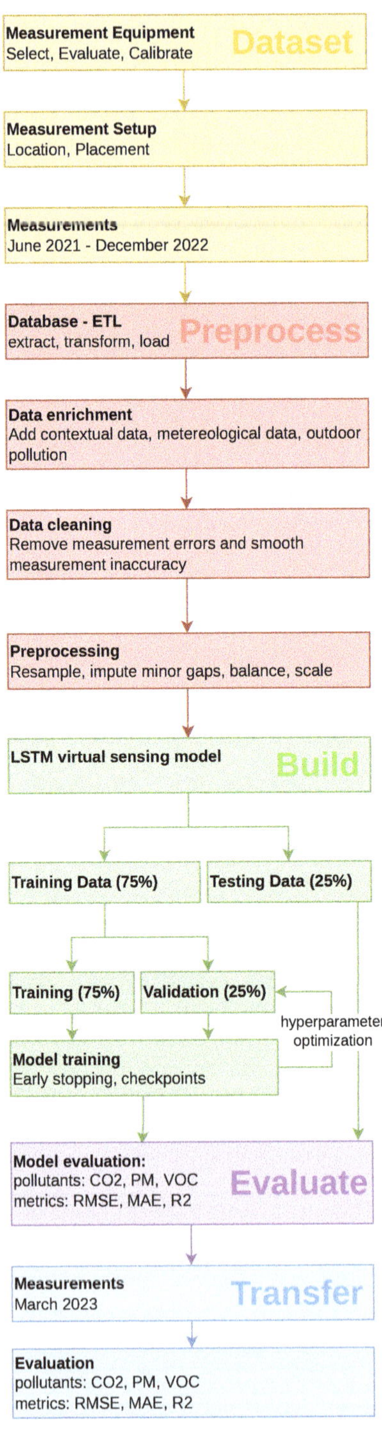

Figure 2. Flowchart of the implemented data processing and machine learning pipeline (Own representation).

5.1. Measurements

Measurement equipment was developed for indoor air pollutant (IAP) measurements. The measurement infrastructure for the BMS, meteorological, and outdoor pollution data was already in place. The BMS is implemented based on the LoRaWAN standard. LoRaWAN is a wireless IoT standard that achieves long-range communication with low power consumption, thus enabling battery-powered nodes. Due to the minimal installation effort and battery-powered nodes, it is applicable as a retrofit solution. The BMS data recorded measurements at a 1 min interval.

The IAP nodes are required to measure CO_2 concentrations in parts per million (ppm), particulate matter concentrations in micrograms per cubic meter ($\mu g/m^3$), and total volatile organic compound concentrations as Indoor Air Quality Index (IAQI). Furthermore, the IAP nodes must achieve continuous, automated measurements with a high sampling rate (10 s) over a prolonged period of time and should account for measurement drift by frequent recalibration. While the initial data sampling rate is 10 s, these measurements will be resampled to a one-minute interval later. This higher sampling rate allows for smoother and more reliable data, as it enables using more data points for each resampled data point. Due to the high volume of data collected, data must be stored centrally rather than locally on the measurement nodes. Therefore, a communication infrastructure supporting high data rates and low latencies was required. Since tuning periods will be needed in subsequent deployments, sensor costs must be low to achieve the goal of a ubiquitous deployment.

No currently available commercial system fulfilled these requirements. Therefore, custom indoor air pollutant nodes (see Figure 3) were developed in order to meet the requirements. Sensors were selected based on their evaluation in the literature. For particulate matter measurements, we selected the Sensirion SPS30 sensor (Sensirion AG, Switzerland, Stäfa) based on its evaluation in previous studies [49]. Ref. [49] ascertained a very strong correlation with the reference instrument for fine particulate matter [49]. The Sensirion SPS30 utilizes the optical particles counter measurement principle, which has been shown to have good accuracy in measuring particulate matter of varying diameters [27].

For VOC measurements, we chose the Sensirion SGP30 (Sensirion AG, Switzerland, Stäfa) sensor. The sensor employs a metal oxide sensing (MOS) element, which is able to detect a wide range of volatile organic compounds through changes in the material's resistance due to chemical reactions with the pollutants. However, due to their broad sensitivity, it is not possible to identify the pollutant concentrations of individual VOCs, which means the output value of these sensors is qualitative. However, ref. [49] evaluated a range of VOC MOS sensors under different pollution events and, in the case of the Sensirion SGP30, performed well compared to reference instruments, thus making it viable for a qualitative evaluation of VOC pollution.

For CO_2 measurements, we selected the Sensirion SCD30 (Sensirion AG, Switzerland, Stäfa) due to its proven accuracy [27]. This sensor uses the optical NDIR measurement principle, which is the common standard in accurately measuring CO_2 concentration [50]. The sensors were connected to a microcontroller, which performs continuous measurements in the defined interval, automatic recalibration, and upload the data to a central database via WiFi connectivity.

Figure 3. Custom-built IAP sensor node (Own representation).

Since multiple indoor air pollutant nodes would be deployed, it was important to reduce sensor bias. Therefore, a cross-calibration scheme was introduced in this study. Cross-calibration is a method used to reduce sensor bias and improve accuracy by comparing the readings of individual sensors to a chosen reference sensor. In our case, one sensor was selected as the reference, and all other sensors were calibrated to perform like the reference sensor. This approach ensures consistency among the sensor readings.

The cross-calibration procedure was conducted over a 24 h period, during which a wide range of environmental conditions were introduced to test sensor response over the entire measurement range. Based on the gathered data, calibration curves for each individual sensor are generated using regression analysis. By applying these calibration curves to the input data of the latter measurements, we could reduce the influence of sensor biases.

Table 2 gives an overview of the used measurement equipment.

Table 2. Overview of the measurement equipment and sensors used.

Property	Sensirion SPS30	Sensirion SGP30	Sensirion SCD30
Parameter	Particulate Matter	Volatile Organic Compounds (VOC)	Carbon Dioxide (CO_2)
Measurement Principle	Optical Particle Counter	Metal Oxide Sensing (MOS)	Optical NDIR
Evaluation Source	[27,49]	[49]	[27,50]
Measurement Interval	10 s	10 s	10 s
Use in Literature	[51,52]	[53,54]	[55,56]

Measurements were taken in a high-rise office building in the center of Munich with 23 stories and 130,000 m^2 floor area that accommodates about 2500 employees. The building is supplied with heating and cooling through thermally activated ceilings (concrete core activation) supplied by groundwater heat pumps. A central mechanical ventilation system supplies the building with fresh air introduced into the room through induction units and extracted through exhaust outlets in the center of the zones. The ventilation system is not designed to supply heating or cooling energy. The ventilation operates at a constant schedule of 1.6 air changes per hour between 5.15 am and 8 pm. In addition to the mechanical ventilation systems, rooms in the lower stories also have operable windows. All rooms have radiation-controlled shading systems that can be overridden by the occupants. The building is in close proximity to much-frequented roads and railway tracks.

The examined office (Office 1) is located on the third floor of the building. It has two external façades, which are orientated toward the northwest and southeast. The room provides workplaces for about thirty-five employees and features operable windows. Measurements were taken in Office 1 from June 2021 to December 2022 with three independent indoor air pollutant nodes. The placement of the IAP nodes is in accordance with the guidelines for monitoring indoor air pollutants of the United States Environmental Protection Agency (EPA):

- Installation of the nodes in the breathing zone (1.10 m height)
- More than 0.5 m away from walls, corners, and windows
- More than 1 m away from local pollutant sources and occupants
- Not in front or below air supply units
- Not exposed to direct sunlight

The floorpolan of the office as well as the sensor node setup is shown in Figure 4.

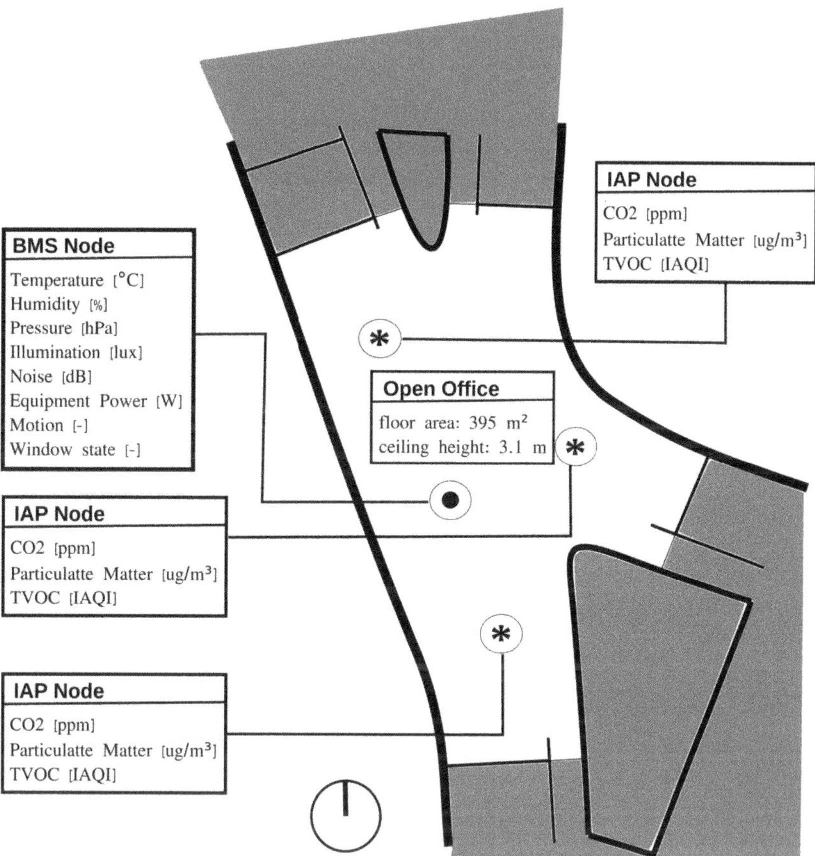

Figure 4. Floorplan illustrating sensor placement and room layout of Office 1 (Own representation).

5.2. Preprocessing

The following section outlines the steps that were taken to bring the raw datasets into a form that can be used as input for a machine-learning model. These include filtering or selecting relevant data, handling missing or corrupted values, normalizing the data, and splitting the data into training, validation, and test sets.

The initial data preparation involved extracting measurement data from IAP and the BMS node from the database and transforming the data from a long format to a wide format. This data was then loaded into a Pandas data frame for further processing. Pandas is a widely used library in Python for data analysis. It provides a structure for storing and manipulating the data in preparation for machine learning tasks.

The available measurement data was enriched by adding contextual and outdoor environmental data. For contextual data, date and time tags were added, with hours and days encoded as continuous sinus. Workdays, weekends, holidays, and seasons were added as boolean tags. Furthermore, information on the HVAC operation schedule, room size, and number of occupants were integrated into the dataset.

Additionally, outdoor environmental data from a local meteorological station was added to the dataset. The outdoor environmental data encompasses air temperature, ground temperature, dew point temperature, global and diffuse radiation, humidity, illumination, air pressure, precipitation, sunlight hours, wind- direction and -speed, as well as outdoor particulate matter concentration. The meteorological station is at a distance of 5 km.

We noted that measurements after power cycling the nodes, e.g., after a power outage, showed elevated values for temperature and humidity for a short timespan after. In order to avoid model bias, measurements up to 15 min after a power cycle were excluded from the data set. Furthermore, random measurement fluctuations due to sensor inaccuracies were removed programmatically from the data set by smoothing measurements.

The final pre-processing steps involved optimizing the dataset for machine learning. We resampled the data to a one-minute frequency, a balance between attaining high accuracy, capturing brief temporal fluctuations, and ensuring smooth, even data. Missing data up to 15 min was input due to transient sensor response post power cycling, which we found to normalize after this interval. Overall, the input data amounted to less than 16 h for the whole measurement period. Finally, the datasets were balanced and normalized using a min-max scaler for each feature.

5.3. LSTM Setup and Training Protocol

We opted for a deep learning approach utilizing a recurrent neural network architecture, specifically an LSTM with two hidden layers. Data were fed into the LSTM as a three-dimensional input tensor, with the first dimension representing the length of the input variables (temperature, humidity, etc.), the second dimension being the lookback period (number of past timesteps), and the third dimension representing the batch size, which indicates the number of input sequences processed concurrently during training and inference. The learning rate, batch size, lookback period, and the number of neurons in the two hidden layers were determined through hyperparameter optimization.

The model's hyperparameters were optimized using Bayesian optimization, a method that uses a Gaussian process objective function and utilizes probabilistic reasoning to optimize the model's hyperparameters with the goal of minimizing the model error. An early stopping function was implemented to prevent model overfitting by monitoring the validation loss and terminating model training if the validation loss did not improve for five consecutive runs. The overall training of the LSTM took 28 min on a GPU. An overview of the model input is provided in the Appendix A in Table A1, and the model output is shown in Table A2.

Data collected from Office 1 were used to train the machine learning model. To ensure that the model could generalize and predict indoor air pollutant concentrations, a cross-validation scheme was employed. This process involved reserving 25% of the data for testing purposes and using the remaining data to train the models. This training data were further divided into a training set (75%) and a validation set (25%), with the latter being employed to trigger the early stopping algorithm to prevent overfitting, determine the best epoch, and perform hyperparameter optimization.

5.4. Evaluation

Model predictions were evaluated using the set-aside testing dataset, employing the R^2, mean absolute error (MAE) and root mean squared error (RMSE) metrics for quantification. Metrics were calculated individually for each model, pollutant, and room. We selected the MAE, RMSE, and coefficient of determination (R^2) metrics to assess model performance, as they are widely used in model performance evaluation [57]. MAE and RMSE are not dimensionless and are expressed in the units of the evaluated target. The MAE metric output represents the mean absolute difference between predicted and true values for all tested timesteps. Due to its quadratic component in the RMSE calculation, larger errors are weighted more heavily than smaller ones [57]. Consequently, MAE provides a good indication of the overall error in target units, while RMSE indicates the number of high deviations. R^2 is a dimensionless metric that measures the proportion of the total variance in the dependent variable that is predictable from the independent variables. Smaller values for both RMSE and MAE signify a better fit, while a higher value for R^2 indicates a more accurate fit.

$$RMSE(y, \hat{y}) = \sqrt{\frac{\sum_{i=0}^{N-1}(y_i - \hat{y}_i)^2}{N}} \tag{1}$$

$$MAE(y, \hat{y}) = \frac{\sum_{i=0}^{N-1}|y_i - \hat{y}_i|}{N} \tag{2}$$

$$R^2(y, \hat{y}) = 1 - \frac{\sum_{i=0}^{N-1}(y_i - \hat{y}_i)^2}{\sum_{i=0}^{N-1}(y_i - \bar{y})^2}. \tag{3}$$

5.5. Transferability Testing

To evaluate the model's ability to predict indoor air pollutant concentrations in other rooms and environments, the trained and assessed model was transferred to an unseen office room (Office 2) in the same building with a different layout, occupancy patterns, density, and orientation.

Office 2 is located on the third floor of the same building as the previous room. It has one external façade, which is oriented towards the east. The room accommodates ten employees and features operable windows. Measurements were taken in Office 2 in March 2023 using one IAP node. The same BMS data points used in the training of the LSTM model in Office 1 are available in Office 2. The outdoor meteorological and pollution data were retrieved from the same source, as both rooms are located in the same building.

The locations of the nodes and the room layout are depicted in Figure 5.

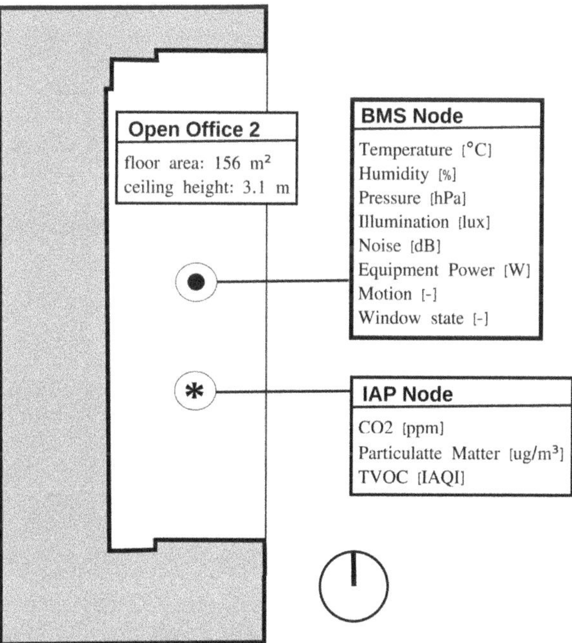

Figure 5. Floorplan illustrating sensor placement and room layout of Office 2 (Own representation).

The measurements of the IAP nodes were solely used for evaluation in Office 2. The trained model was used as is.

The model inputs, as specified in Table A1, were provided by the BMS-node as well as outdoor and metadata. The model then predicts the indoor air pollutant concentrations for each timestep (1 min). In the evaluation, the predicted values were then compared to the actual measurements of the IAP nodes for March 2023. As previously, three prediction metrics were calculated: MAE, RMSE, and R^2.

6. Results

In this section, we present and discuss the results of our machine-learning model. Section 6.1 presents the results of the model training and its evaluation metrics. Section 6.2 reports the results of transferring the model to Office 2.

6.1. Model Evaluation

Figure 6 displays the predictions of the trained LSTM model for the testing set in Office 1 (yellow) and the measured truth (blue) for each indoor air pollutant. The evaluation metrics are calculated individually for each pollutant and shown in the top-left corner of each plot. The testing was conducted for three months, from March 2022 to May 2022. A visual assessment of the time series plots reveals a high correlation between the truth and prediction. The most significant deviations between truth and prediction are identified for CO_2 predictions. In the case of CO_2, the model tends to slightly overestimate the CO_2 concentration during low concentration periods, whereas high concentration events show a closer fit. However, the model occasionally predicts pollutant peaks incorrectly during low concentration periods and vice versa. In the case of CO_2 predictions, they appear to be more accurate during the second half of the testing period. For particulate matter, the visual assessment shows an excellent fit between prediction and truth. All peaks are identified correctly. The time series plot demonstrates a slight underestimation of peaks and high pollution events by the prediction compared to the truth. In the case of VOC, the visual assessment of the time series plots reveals an excellent fit between prediction and truth. The model can detect all concentration peaks, even though VOC concentration is highly dynamic. However, a slight underestimation of pollutant peaks can be observed in the time series, especially during the first half of the testing period.

Table 3 summarizes the evaluation metrics (R^2, RMSE, MAE) for each pollutant. Overall, the model exhibits a low error for all pollutants, as demonstrated by the MAE and RMSE performance metrics. In the case of CO_2, the mean absolute error amounts to 15.4 ppm for the testing period, while a slightly increased RMSE value of 20.2 ppm indicates that no outliers significantly impact the model's predictions. The CO_2 measurements ranged from 380 to 560 ppm during the measurement period. For particulate matter, the errors amount to 0.3 and 0.5 $\mu g/m^3$ for MAE and RMSE, respectively, indicating consistently low error rates without outliers. The measurements ranged from 0 to 13 $\mu g/m^3$ during the measurement period. In the case of volatile organic compounds, MAE and RMSE errors amounted to 20.1 IAQI and 31.4 IAQI, respectively, demonstrating low error rates without major deviations. The measurements for VOC ranged from 0 to 450 IAQI. The R^2 performance metric is a statistical measure representing the goodness of fit of the LSTM model and indicates the percentage of variance in the truth data that can be explained by the LSTM model. In the case of CO_2, an R^2 value of 0.47 indicates that the model explains a substantial part of the variance in CO_2 concentration and has a reasonably good fit, providing meaningful predictions. For PM, a high R^2 of 0.88 was identified, indicating that the model explains a significant percentage of the variability in particulate matter measurements. Furthermore, it shows a very good fit and indicates that the model is highly predictive of PM. In the case of VOC, a high R^2 of 0.87 was identified, which shows that a significant percentage of VOC volatility is explained by the LSTM virtual sensing model. Furthermore, the R^2 indicates a very good fit for VOC and that the model is highly predictive.

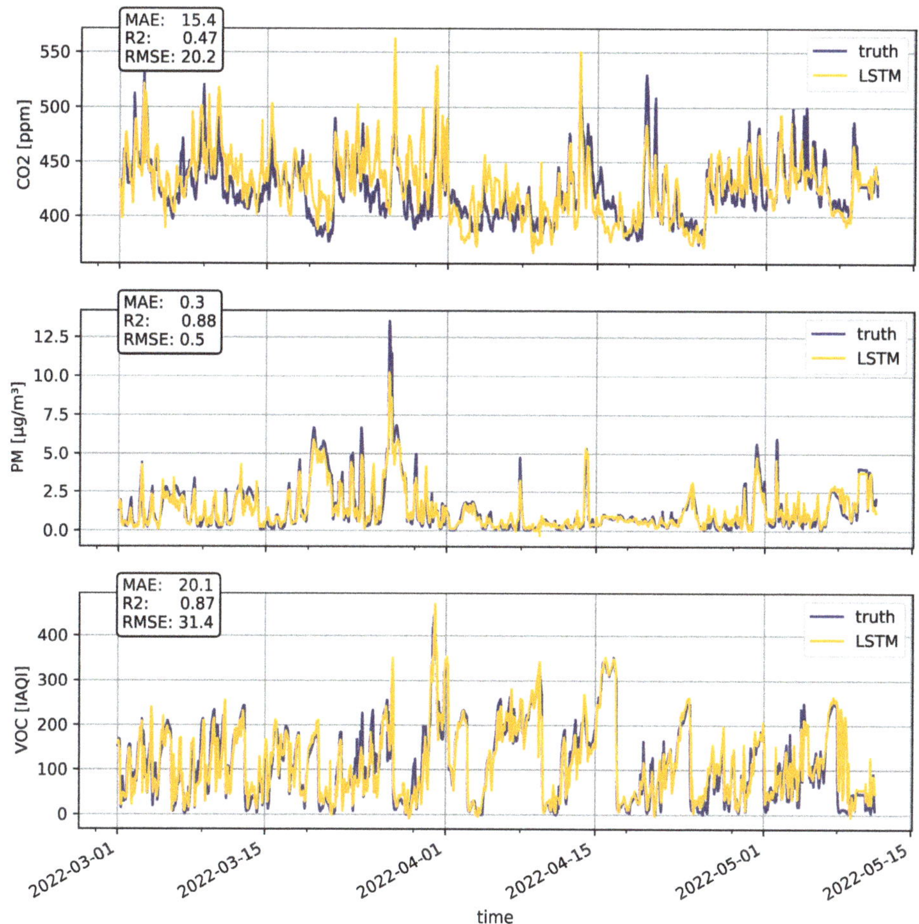

Figure 6. Comparison of virtual indoor air pollutant sensors (yellow) and physical indoor air pollutant sensors (blue) with overlayed evaluation metrics for Indoor Air Pollutants: VOC (**bottom**), PM (**middle**), and CO_2 (**top**) for Office 1 (Own representation).

Table 3. Evaluation metrics LSTM virtual sensing model in Office 1.

Pollutant	MAE	R^2	RMSE
CO_2	15.4	0.47	20.2
$PM_{2.5}$	0.3	0.88	0.5
VOC	20.1	0.87	31.4

The model successfully identified all pollutant peaks during the testing period, with the only error being a slight underestimation of peak concentrations. For CO_2, a less ideal but still satisfactory prediction result was achieved. This led to minor errors and a less accurate representation of the variability in actual concentrations, resulting in some erroneous predictions, such as misidentified pollutant peaks during the testing period. Nevertheless, the predictions yielded a mean absolute error within the range of measurement inaccuracies for most sensors.

The performance metrics of MAE = 15.4 ppm, RMSE = 20.2 ppm, and R^2 = 0.47 for CO_2 showed very low errors with insignificant outliers. The R^2 value indicated a reasonably good fit and predictive capability. For PM, the metrics MAE = 0.3 µg/m^3, RMSE = 0.5 µg/m^3, and R^2 = 0.88 signified a strong prediction capability with minimal errors and an excellent fit. Similar results were observed for VOC, with MAE = 20.1 IAQI, RMSE = 31.4 IAQI, and R^2 = 0.87. Overall, the LSTM model demonstrated strong performance in predicting indoor air pollutant concentrations, with some room for improvement in CO$_2$ predictions. Based on the findings from this study, the LSTM model shows promise to potentially replace physical sensors, contributing to more cost-effective and efficient air quality monitoring solutions.

6.2. Transferability Evaluation

Figure 7 displays the predictions of the trained LSTM model (yellow) for Office 2, as well as the measured actual values (blue) for each indoor air pollutant. The test took place in March 2023.

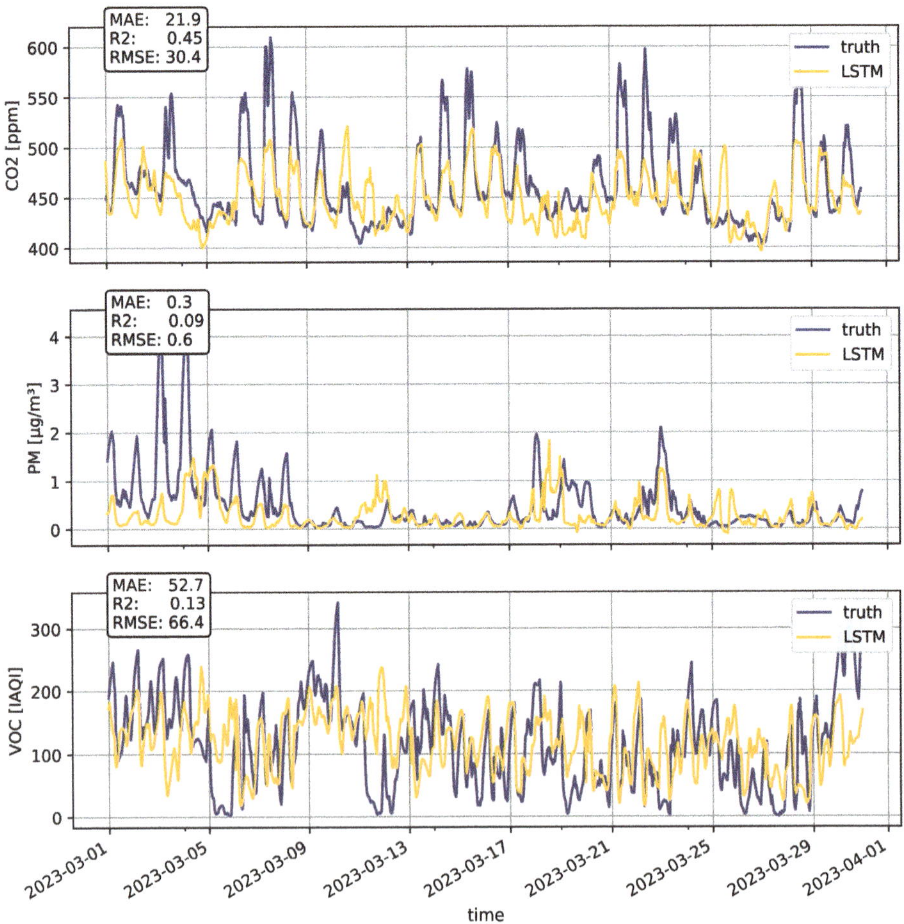

Figure 7. Comparison of virtual indoor air pollutant sensors (yellow) and physical indoor air pollutant sensors (blue) with overlayed evaluation metrics for Indoor Air Pollutants: VOC (**bottom**), PM (**middle**), and CO$_2$ (**top**) for Office 2 (Own representation).

Visual assessment of the time series plots revealed a correlation between actual values and predictions for all pollutants, albeit with varying degrees of fit. The highest correlation between actual and predicted values was observed for CO_2 predictions. The prediction model successfully identified pollutant peaks, albeit with underestimation. During low pollutant events, such as weekends or nights, the model results were less smooth and tended to overestimate variability in pollutant concentrations. Occasionally, the model predicted pollutant peaks under unpolluted conditions.

For particulate matter, the visual assessment showed a general fit between the magnitudes of predicted and actual concentrations. However, the prediction failed to detect some peaks and underestimated all others. In some cases, the prediction exhibited a phase shift, resulting in delayed identification of rising concentrations.

For VOC, a visual assessment of the time series plots revealed that the model could identify some concentration peaks. However, the model frequently and erroneously detected pollutant peaks when none were present.

Table 3 summarizes the evaluation metrics (R^2, RMSE, MAE) for each pollutant. Overall, the model exhibited very low errors for all pollutants, as evidenced by the MAE and RMSE performance metrics. For CO_2, the mean absolute error was 21.9 ppm during the testing period, while a slightly increased RMSE value of 30.4 ppm indicated that no outliers affected the model's predictions. CO_2 measurements ranged from 420 ppm to 610 ppm during the measurement period.

For particulate matter, errors amounted to 0.3 and 0.6 $\mu g/m^3$ for MAE and RMSE, respectively, indicating consistently low error rates without outliers. Measurements ranged from 0 to 4 $\mu g/m^3$ during the measurement period. For volatile organic compounds, MAE and RMSE errors were 52.7 IAQI and 66.4 IAQI, respectively, demonstrating very low error rates without significant deviations. Measurements for VOC ranged from 0 to 330 IAQI.

For CO_2, an R^2 value of 0.45 indicated that the model accounted for a substantial portion of the variability in CO_2 concentrations, exhibited a reasonably good fit, and provided meaningful predictions. For PM, a low R^2 of 0.09 suggested that the model explained a smaller percentage of the variability in particulate matter measurements, demonstrated a less accurate fit, and was less predictive. For VOC, a low R^2 of 0.13 indicated that the model explained a smaller percentage of VOC variability and was less accurate and less predictive. The evaluation metrics are summarized in Table 4.

Table 4. Evaluation metrics LSTM virtual sensing model transfer in Office 2.

Pollutant	MAE	R^2	RMSE
CO_2	21.9	0.45	30.4
$PM_{2.5}$	0.3	0.09	0.6
VOC	52.7	0.13	66.4

The LSTM-based virtual indoor air pollutant sensor was tested for Office 2 using the testing dataset for March 2023. The evaluation results indicated varying degrees of correlation between the actual and predicted pollutant concentrations. For CO_2, the model successfully identified pollutant peaks, albeit underestimated, and exhibited an MAE of 21.9 ppm, RMSE of 30.4 ppm, and R^2 of 0.45, indicating a reasonably good fit and predictive capabilities. For particulate matter and volatile organic compounds (VOC), the model showed less accurate predictions in terms of R^2 values; however, the MAE and RMSE errors remained low. For PM, despite the model's failure to detect some pollutant peaks and a low R^2 of 0.09, the MAE and RMSE were consistently low at 0.3 $\mu g/m^3$ and 0.6 $\mu g/m^3$, respectively, indicating a relatively low error rate without significant outliers. Similarly, for VOC, the model erroneously detected pollutant peaks in some cases and showed a low R^2 of 0.13. Yet, the MAE and RMSE remained low at 52.7 IAQI and 66.4 IAQI, respectively, demonstrating low error rates without major deviations. In conclusion, the LSTM model exhibits varying performance in predicting indoor air pollutant concentrations for Office 2, with better results for CO_2 predictions and low error rates in terms of MAE and RMSE for PM and VOC predictions. However, there is room for improvement in capturing the variability of PM and VOC concentrations, as indicated by the low R^2 values.

7. Discussion

The findings of this study indicate that machine learning models, particularly LSTM networks, are effective in predicting indoor air pollutants, especially particulate matter, and VOC, as demonstrated by the low error rates achieved in the testing set of Office 1. The testing results from Office 1 indicate certain limitations of the virtual sensing model in capturing the full range of variability in CO_2 concentrations. This limitation may be attributed to the model's reduced precision in predicting occupancy and occupant count. Skoen et al. [38] previously noted similar findings when applying Multi-Layer Perceptron (MLP) models for virtual sensing of CO_2, notably, even though the R^2 values from Skoen et al. (0.39) closely match the 0.47 achieved in this study, and a significantly lower Root Mean Square Error (RMSE) of 31.4 was obtained in this study, compared to Skoen et al.'s 122.85 [38]. This suggests that, despite the model's inability to capture full variability with Long Short-Term Memory (LSTM), the error margins remained relatively low, particularly when compared to other models. When the pre-trained models were applied to other rooms of identical typologies, they still exhibited predictive capacity. However, these models demonstrated a decreased ability to explain the variability of pollutant concentrations as well as increased errors. This suggests a limitation in model transferability to different rooms, with a significant decline in the model's predictive capability noticed, particularly in terms of capturing the ground truth variability. It is postulated that this decrease in performance is attributable to the limitations of the training dataset, which was exclusively trained in Office 1. Given that occupancy and numerous other dynamic factors influence indoor air pollutants, indoor environments can significantly differ from each other. They may also display vastly different pollutant dynamics, as previously demonstrated by Szigeti et al. [8]. It is anticipated that the model's performance will be reduced when applied to rooms in other buildings or those belonging to different typologies, as these environments may present conditions not encountered during model training. Consequently, it is crucial to enhance the transferability and performance of the virtual sensing LSTM model by generating larger and more diverse datasets.

While current results do not yet allow for a complete replacement of physical sensors with LSTM models, the promising predictions of IAP concentrations in the training room, along with the successful prediction of CO_2 levels in a separate office, demonstrate potential. The general application of this model is not yet feasible, but, given more diverse data, the outlook for the full replacement of physical sensors with such models becomes more attainable.

The use of machine learning techniques to create virtual sensors for monitoring indoor air pollutants has the potential to provide real-time data and improve building operations. Further research and development may lead to the use of virtual sensors for wider application in building environments, potentially allowing for the optimization of mechanical ventilation systems and operable window usage. It is important to continue exploring the potential of virtual indoor air pollutant sensors as a tool for improving indoor air quality and the overall comfort and health of building occupants.

Further research is needed in expanding the training data for LSTM models for virtual sensing of indoor air pollutants and testing their generalizability across various typologies and buildings in different climate zones. Additionally, future studies could investigate the integration of these models into Heating, Ventilation, and Air Conditioning (HVAC) systems and evaluate their performance when only a fraction of the given input data is available. This would help advance the practical implementation of virtual sensing in real-world scenarios and contribute to the field of indoor air quality monitoring.

8. Conclusions

This study demonstrates the potential of machine learning models, specifically LSTM networks, to accurately predict indoor air pollutant concentrations in a range of environments. By using a large dataset with several years of accumulated data, we were able to build a virtual indoor air pollutant sensor that exhibited strong performance in predicting indoor air pollutant concentrations for the room in which it was trained. The evaluation results indicated a very high correlation between the actual and predicted pollutant concentrations for particulate matter and VOC, with performance metrics such as MAE = 0.3 $\mu g/m^3$, RMSE = 0.5 $\mu g/m^3$, and R^2 = 0.88 for PM; and MAE = 20.1 IAQI, RMSE = 31.4 IAQI, and R^2 = 0.87 for VOC. These results show that the model was able to identify most pollutant peaks during the testing period with only a slight underestimation of peak concentrations. For CO_2, the model achieved less ideal but reasonable prediction results. The performance metrics of MAE = 15.4 ppm, RMSE = 20.2 ppm, and R^2 = 0.47 for CO_2 indicated very low errors with insignificant outliers, and the R^2 value suggested a reasonably good fit and predictive capabilities. However, the model was not able to explain the variability of the actual concentrations and showed some erroneous predictions, such as misidentified pollutant peaks during the testing period. Nevertheless, the predictions resulted in a mean absolute error within the range of the measurement inaccuracy of most sensors. When transferring the model to another room, the LSTM model demonstrated varying performance, with better results for CO_2 predictions and low error rates in terms of MAE and RMSE for PM and VOC predictions. Specifically, the CO_2 predictions exhibited a mean absolute error of 21.9 ppm, RMSE of 30.4 ppm, and R^2 of 0.45, indicating a reasonably good fit and predictive capabilities. However, there is room for improvement in capturing the variability of PM and VOC concentrations, as indicated by the low R^2 values of 0.09 for PM and 0.13 for VOC. Despite these challenges, the LSTM model shows its potential in generalizing its ability to predict indoor air pollutant concentrations in different rooms. To enhance the model's performance when transferring to other rooms, further research and optimization could focus on refining the LSTM architecture, incorporating additional features such as building materials, type of air distribution, and the distance of the nodes from vents and windows, or exploring other machine learning techniques to improve the model's ability to capture the variability of different pollutants. In summary, the LSTM-based virtual indoor air pollutant sensor presents a promising approach to monitoring air quality in indoor environments. With further refinement and optimization, this model could potentially replace physical sensors, contributing to more cost-effective and efficient air quality monitoring solutions. Ultimately, the development and deployment of accurate virtual sensing models can play a crucial role in addressing indoor air pollution, leading to improved public health and well-being.

Author Contributions: Conceptualization, M.G.; methodology, M.G.; software, M.G.; validation, M.G.; formal analysis, M.G.; investigation, M.G.; resources, M.G.; data curation, M.G.; writing—original draft preparation, M.G.; writing—review and editing, M.G.; visualization, M.G.; supervision, T.A.; project administration, M.G. All authors have read and agreed to the published version of the manuscript.

Funding: This research received no external funding.

Data Availability Statement: The data from the model evaluation is available upon request. The data used for creating the LSTM model, including the training, testing, and validation datasets, contain sensitive information and cannot be shared to ensure data privacy.

Conflicts of Interest: The authors declare no conflict of interest.

Abbreviations

The following abbreviations are used in this manuscript:

BMS	Building management system
IAP	Indoor air pollutants
IAQ	Indoor air quality
IoT	Internet of things
MAE	Mean absolute error
RMSE	Root mean squared error
LSTM	long short-term memory network
VOC	Volatile organic compounds
PM	Particulate matter
IAQI	Indoor air quality index
ppm	Parts per million
GPU	Graphics processing unit
HVAC	Heating, Ventilation, and Air Conditioning
MOS	Metal oxide sensing
OPC	Optial particle counter
NDIR	Non-dispersive infrared
SVM	Support vector machine
MLP	Multi-layer perceptron
ETL	Extract transfer load

Appendix A

Table A1. Input Features for LSTM Model.

Feature	Description	Dimension (Normalized)	Group
month_sin	Continuous sinusoidal encoding of month	0–1	Meta
hr_sin	Continuous sinusoidal encoding of hour	0–1	Meta
day_sin	Continuous sinusoidal encoding of day	0–1	Meta
workday	Boolean tag for workdays	0, 1	Meta
weekend	Boolean tag for weekends	0, 1	Meta
holiday	Boolean tag for holidays	0, 1	Meta
season	Boolean tags for each season	0, 1	Meta
hvac	Boolean tag for HVAC operation	0, 1	Indoor
room_size	Size of the room	0–1	Meta
occupants	Occupant density	0–1	Meta
temp	Outdoor air temperature	0–1	Outdoor
ground_temp	Outdoor ground temperature	0–1	Outdoor
dew_point_temp	Outdoor dew point temperature	0–1	Outdoor
global_rad	Outdoor global radiation	0–1	Outdoor
diffuse_rad	Outdoor diffuse radiation	0–1	Outdoor
humidity	Outdoor humidity	0–1	Outdoor
illumination	Outdoor illumination	0–1	Outdoor
air_pressure	Outdoor air pressure	0–1	Outdoor

Table A1. *Cont.*

Feature	Description	Dimension (Normalized)	Group
precipitation	Outdoor precipitation	0–1	Outdoor
wind_dir	Outdoor wind direction	0–1	Outdoor
wind_speed	Outdoor wind speed	0–1	Outdoor
particulate_matter	Outdoor particulate matter concentration	0–1	Outdoor
indoor_temp	Indoor air temperature	0–1	Indoor
indoor_humidity	Indoor humidity	0–1	Indoor
indoor_air_pressure	Indoor air pressure	0–1	Indoor
indoor_illum	Indoor illumination	0–1	Indoor
noise_level	Indoor noise level	0–1	Indoor
window_state	State of window (open/closed)	0, 1	Indoor
power_consumption	Power consumption of equipment	0–1	Indoor

Table A2. LSTM model output.

Output	Description	Dimension (Normalized)
pm	Particulate matter concentration	0–1
co2	CO_2 concentration	0–1
voc	Volatile organic compound concentration	0–1

References

1. Tham, K.W. Indoor air quality and its effects on humans—A review of challenges and developments in the last 30 years. *Energy Build.* **2016**, *130*, 637–650. [CrossRef]
2. Hasager, F.; Bjerregaard, J.D.; Bonomaully, J.; Knap, H.; Afshari, A.; Johnson, M.S. Indoor Air Quality: Status and Standards. *Air Pollut. Sources Stat. Health Eff.* **2021**, 135–162.
3. Berglund, B.; Brunekreef, B.; Knöppe, H.; Lindvall, T.; Maroni, M.; Mølhave, L.; Skov, P. Effects of indoor air pollution on human health. *Indoor Air* **1992**, *2*, 2–25. [CrossRef]
4. Henschel, S.; Chan, G.; World Health Organization. *Health Risks of Air Pollution in Europe-HRAPIE Project: New Emerging Risks to Health from Air Pollution-Results from the Survey of Experts*; WHO: Geneva, Switzerland, 2013.
5. Soares, A.G.O.G.J. *Air Quality in Europe—2020 Report*; Technical Report; European Environment Agency: Copenhagen, Denmark, 2020.
6. Abdul-Wahab, S.A.; En, S.C.F.; Elkamel, A.; Ahmadi, L.; Yetilmezsoy, K. A review of standards and guidelines set by international bodies for the parameters of indoor air quality. *Atmos. Pollut. Res.* **2015**, *6*, 751–767. [CrossRef]
7. World Health Organization. *WHO Guidelines for Indoor Air Quality: Selected Pollutants*; World Health Organization. Regional Office for Europe: Geneva, Switzerland, 2010.
8. Szigeti, T.; Dunster, C.; Cattaneo, A.; Spinazzè, A.; Mandin, C.; Le Ponner, E.; de Oliveira Fernandes, E.; Ventura, G.; Saraga, D.E.; Sakellaris, I.A.; et al. Spatial and temporal variation of particulate matter characteristics within office buildings—The OFFICAIR study. *Sci. Total Environ.* **2017**, *587*, 59–67. [CrossRef]
9. Li, J.; Li, H.; Ma, Y.; Wang, Y.; Abokifa, A.A.; Lu, C.; Biswas, P. Spatiotemporal distribution of indoor particulate matter concentration with a low-cost sensor network. *Build. Environ.* **2018**, *127*, 138–147. [CrossRef]
10. Sahu, V.; Gurjar, B.R. Spatio-temporal variations of indoor air quality in a university library. *Int. J. Environ. Health Res.* **2021**, *31*, 475–490. [CrossRef]
11. Zhang, H.; Srinivasan, R.; Ganesan, V. Low Cost, Multi-Pollutant Sensing System Using Raspberry Pi for Indoor Air Quality Monitoring. *Sustainability* **2021**, *13*, 370. [CrossRef]
12. Kim, J.; Kong, M.; Hong, T.; Jeong, K.; Lee, M. The effects of filters for an intelligent air pollutant control system considering natural ventilation and the occupants. *Sci. Total Environ.* **2019**, *657*, 410–419. [CrossRef]
13. Saraga, D.; Maggos, T.; Sadoun, E.; Fthenou, E.; Hassan, H.; Tsiouri, V.; Karavoltsos, S.; Sakellari, A.; Vasilakos, C.; Kakosimos, K. Chemical characterization of indoor and outdoor particulate matter (PM2. 5, PM10) in Doha, Qatar. *Aerosol Air Qual. Res.* **2017**, *17*, 1156–1168.

14. Irga, P.; Torpy, F. Indoor air pollutants in occupational buildings in a sub-tropical climate: Comparison among ventilation types. *Build. Environ.* **2016**, *98*, 190–199. [CrossRef]
15. Montgomery, J.F.; Storey, S.; Bartlett, K. Comparison of the indoor air quality in an office operating with natural or mechanical ventilation using short-term intensive pollutant monitoring. *Indoor Built Environ.* **2015**, *24*, 777–787. [CrossRef]
16. Ha, Q.P.; Metia, S.; Phung, M.D. Sensing data fusion for enhanced indoor air quality monitoring. *IEEE Sens. J.* **2020**, *20*, 4430–4441. [CrossRef]
17. Kang, J.; Hwang, K.I. A comprehensive real-time indoor air-quality level indicator. *Sustainability* **2016**, *8*, 881. [CrossRef]
18. Mendoza, D.; Benney, T.M.; Boll, S. Long-term analysis of the relationships between indoor and outdoor fine particulate pollution: A case study using research grade sensors. *Sci. Total. Environ.* **2021**, *776*, 145778. [CrossRef]
19. Tiele, A.; Esfahani, S.; Covington, J. Design and development of a low-cost, portable monitoring device for indoor environment quality. *J. Sens.* **2018**, *2018*. [CrossRef]
20. Spinazzè, A.; Campagnolo, D.; Cattaneo, A.; Urso, P.; Sakellaris, I.A.; Saraga, D.E.; Mandin, C.; Canha, N.; Mabilia, R.; Perreca, E.; et al. Indoor gaseous air pollutants determinants in office buildings—The OFFICAIR project. *Indoor Air* **2020**, *30*, 76–87. [CrossRef]
21. Saini, J.; Dutta, M.; Marques, G. Indoor air quality prediction using optimizers: A comparative study. *J. Intell. Fuzzy Syst.* **2020**, *39*, 7053–7069. [CrossRef]
22. Challoner, A.; Gill, L. Indoor/outdoor air pollution relationships in ten commercial buildings: PM2. 5 and NO$_2$. *Build. Environ.* **2014**, *80*, 159–173. [CrossRef]
23. Challoner, A.; Pilla, F.; Gill, L. Prediction of indoor air exposure from outdoor air quality using an artificial neural network model for inner city commercial buildings. *Int. J. Environ. Res. Public Health* **2015**, *12*, 15233–15253. [CrossRef]
24. Ahn, J.; Shin, D.; Kim, K.; Yang, J. Indoor air quality analysis using deep learning with sensor data. *Sensors* **2017**, *17*, 2476. [CrossRef] [PubMed]
25. Ma, N.; Aviv, D.; Guo, H.; Braham, W.W. Measuring the right factors: A review of variables and models for thermal comfort and indoor air quality. *Renew. Sustain. Energy Rev.* **2021**, *135*, 110436. [CrossRef]
26. Kolarik, J.; Lyng, N.L.; Bossi, R.; Witterseh, T.; Smith, K.M.; Wargocki, P. 3.6 Response of commercially available Metal Oxide Semiconductor Sensors under air polluting activities typical for residences. *Indoor Air Qual. Des. Control. -Low-Energy Resid. Build. (Ebc Annex. 68)* **2020**, 47.
27. Frederickson, L.B.; Petersen-Sonn, E.A.; Shen, Y.; Hertel, O.; Hong, Y.; Schmidt, J.; Johnson, M.S. Low-Cost Sensors for Indoor and Outdoor Pollution. *Air Pollut. Sources Stat. Health Eff.* **2021**, 423–453.
28. Alhasa, K.M.; Mohd Nadzir, M.S.; Olalekan, P.; Latif, M.T.; Yusup, Y.; Iqbal Faruque, M.R.; Ahamad, F.; Abd. Hamid, H.H.; Aiyub, K.; Md Ali, S.H.; et al. Calibration model of a low-cost air quality sensor using an adaptive neuro-fuzzy inference system. *Sensors* **2018**, *18*, 4380. [CrossRef]
29. Topalović, D.B.; Davidović, M.D.; Jovanović, M.; Bartonova, A.; Ristovski, Z.; Jovašević-Stojanović, M. In search of an optimal in-field calibration method of low-cost gas sensors for ambient air pollutants: Comparison of linear, multilinear and artificial neural network approaches. *Atmos. Environ.* **2019**, *213*, 640–658. [CrossRef]
30. Heindel, L.; Hantschke, P.; Kästner, M. A Virtual Sensing approach for approximating nonlinear dynamical systems using LSTM networks. *PAMM* **2021**, *21*, e202100119. [CrossRef]
31. Li, H.; Yu, D.; Braun, J.E. A review of virtual sensing technology and application in building systems. *Hvac&R Res.* **2011**, *17*, 619–645.
32. Yoon, S. Virtual sensing in intelligent buildings and digitalization. *Autom. Constr.* **2022**, *143*, 104578. [CrossRef]
33. Wu, Q.; Cai, W.; Wang, X.; Chakraborty, A. Dehumidifier desiccant concentration soft-sensor for a distributed operating Liquid Desiccant Dehumidification System. *Energy Build.* **2016**, *129*, 215–226. [CrossRef]
34. Hong, Y.; Yoon, S.; Kim, Y.S.; Jang, H. System-level virtual sensing method in building energy systems using autoencoder: Under the limited sensors and operational datasets. *Appl. Energy* **2021**, *301*, 117458. [CrossRef]
35. Li, H.; Hong, T.; Sofos, M. An inverse approach to solving zone air infiltration rate and people count using indoor environmental sensor data. *Energy Build.* **2019**, *198*, 228–242. [CrossRef]
36. Alhashme, M.; Ashgriz, N. A virtual thermostat for local temperature control. *Energy Build.* **2016**, *126*, 323–339. [CrossRef]
37. Zhao, Y.; Zeiler, W.; Boxem, G.; Labeodan, T. Virtual occupancy sensors for real-time occupancy information in buildings. *Build. Environ.* **2015**, *93*, 9–20. [CrossRef]
38. Skön, J.; Johansson, M.; Raatikainen, M.; Leiviskä, K.; Kolehmainen, M. Modelling indoor air carbon dioxide (CO$_2$) concentration using neural network. *Methods* **2012**, *14*, 16.
39. Elbayoumi, M.; Ramli, N.A.; Yusof, N.F.F.M. Development and comparison of regression models and feedforward backpropagation neural network models to predict seasonal indoor PM2. 5–10 and PM2. 5 concentrations in naturally ventilated schools. *Atmos. Pollut. Res.* **2015**, *6*, 1013–1023. [CrossRef]
40. Khazaei, B.; Shiehbeigi, A.; Haji Molla Ali Kani, A. Modeling indoor air carbon dioxide concentration using artificial neural network. *Int. J. Environ. Sci. Technol.* **2019**, *16*, 729–736. [CrossRef]
41. Kusiak, A.; Li, M.; Zheng, H. Virtual models of indoor-air-quality sensors. *Appl. Energy* **2010**, *87*, 2087–2094. [CrossRef]

42. Gabriel, M.; Auer, T. Indoor air pollution estimation using machine learning (ANN and SVR) in smart buildings. In *BauSim Conference 2022, Proceedings of the 9th Conference of IBPSA-Germany and Austria, Weimar, Germany, 20–22 September 2022*; IBPSA-Germany and Austria: Dresden, Germany, 2022; Volume 9. [CrossRef]

43. Leidinger, M.; Sauerwald, T.; Reimringer, W.; Ventura, G.; Schütze, A. Selective detection of hazardous VOCs for indoor air quality applications using a virtual gas sensor array. *J. Sens. Sens. Syst.* **2014**, *3*, 253–263. [CrossRef]

44. Karijadi, I.; Chou, S.Y. A hybrid RF-LSTM based on CEEMDAN for improving the accuracy of building energy consumption prediction. *Energy Build.* **2022**, *259*, 111908. [CrossRef]

45. Jang, J.; Han, J.; Leigh, S.B. Prediction of heating energy consumption with operation pattern variables for non-residential buildings using LSTM networks. *Energy Build.* **2022**, *255*, 111647. [CrossRef]

46. Qolomany, B.; Al-Fuqaha, A.; Benhaddou, D.; Gupta, A. Role of deep LSTM neural networks and Wi-Fi networks in support of occupancy prediction in smart buildings. In Proceedings of the 2017 IEEE 19th International Conference on High Performance Computing and Communications; IEEE 15th International Conference on Smart City; IEEE 3rd International Conference on Data Science and Systems (HPCC/SmartCity/DSS), Bangkok, Thailand, 18–20 December 2017; pp. 50–57.

47. Zhang, L.; Liu, P.; Zhao, L.; Wang, G.; Zhang, W.; Liu, J. Air quality predictions with a semi-supervised bidirectional LSTM neural network. *Atmos. Pollut. Res.* **2021**, *12*, 328–339. [CrossRef]

48. Bai, Y.; Zeng, B.; Li, C.; Zhang, J. An ensemble long short-term memory neural network for hourly PM2. 5 concentration forecasting. *Chemosphere* **2019**, *222*, 286–294. [CrossRef] [PubMed]

49. Demanega, I.; Mujan, I.; Singer, B.C.; Andelkovic, A.S.; Babich, F.; Licina, D. Performance assessment of low-cost environmental monitors and single sensors under variable indoor air quality and thermal conditions. *Build. Environ.* **2021**, *187*, 107415. [CrossRef]

50. Marinov, M.B.; Djermanova, N.; Ganev, B.; Nikolov, G.; Janchevska, E. Performance evaluation of low-cost carbon dioxide sensors. In Proceedings of the 2018 IEEE XXVII International Scientific Conference Electronics-ET, Sozopol, Bulgaria, 3–15 September 2018; pp. 1–4.

51. Hassani, A.; Castell, N.; Watne, Å.K.; Schneider, P. Citizen-operated mobile low-cost sensors for urban PM2. 5 monitoring: field calibration, uncertainty estimation, and application. *Sustain. Cities Soc.* **2023**, *95*, 104607. [CrossRef]

52. Kuula, J.; Friman, M.; Helin, A.; Niemi, J.V.; Aurela, M.; Timonen, H.; Saarikoski, S. Utilization of scattering and absorption-based particulate matter sensors in the environment impacted by residential wood combustion. *J. Aerosol Sci.* **2020**, *150*, 105671. [CrossRef]

53. Alonso, M.J.; Madsen, H.; Liu, P.; Jørgensen, R.B.; Jørgensen, T.B.; Christiansen, E.J.; Myrvang, O.A.; Bastien, D.; Mathisen, H.M. Evaluation of low-cost formaldehyde sensors calibration. *Build. Environ.* **2022**, *222*, 109380. [CrossRef]

54. Arsiwala, A.; Elghaish, F.; Zoher, M. Digital twin with Machine learning for predictive monitoring of CO_2 equivalent from existing buildings. *Energy Build.* **2023**, *284*, 112851. [CrossRef]

55. Trilles, S.; Juan, P.; Chaudhuri, S.; Fortea, A.B.V. Data on CO_2, temperature and air humidity records in Spanish classrooms during the reopening of schools in the COVID-19 pandemic. *Data Brief* **2021**, *39*, 107489. [CrossRef]

56. Toschke, Y.; Lusmoeller, J.; Otte, L.; Schmidt, J.; Meyer, S.; Tessmer, A.; Brockmann, C.; Ahuis, M.; Hüer, E.; Kirberger, C.; et al. Distributed LoRa based CO_2 monitoring network–A standalone open source system for contagion prevention by controlled ventilation. *HardwareX* **2022**, *11*, e00261. [CrossRef]

57. Willmott, C.J.; Matsuura, K. Advantages of the mean absolute error (MAE) over the root mean square error (RMSE) in assessing average model performance. *Clim. Res.* **2005**, *30*, 79–82. [CrossRef]

 *buildings*

Article

Applicability of Deep Learning Algorithms for Predicting Indoor Temperatures: Towards the Development of Digital Twin HVAC Systems

Pooria Norouzi, Sirine Maalej * and Rodrigo Mora *

British Columbia Institute of Technology, Burnaby, BC V5G 3H2, Canada
* Correspondence: smaalej@bcit.ca (S.M.); rodrigo_mora@bcit.ca (R.M.)

Abstract: The development of digital twins leads to the pathway toward intelligent buildings. Today, the overwhelming rate of data in buildings carries a high amount of information that can provide an opportunity for a digital representation of the buildings and energy optimization strategies in the Heating, Ventilation, and Air Conditioning (HVAC) systems. To implement a successful energy management strategy in a building, a data-driven approach should accurately forecast the HVAC features, in particular the indoor temperatures. Accurate predictions not only increase thermal comfort levels, but also play a crucial role in saving energy consumption. This study aims to investigate the capabilities of data-driven approaches and the development of a model for predicting indoor temperatures. A case study of an educational building is considered to forecast indoor temperatures using machine learning and deep learning algorithms. The algorithms' performance is evaluated and compared. The important model parameters are sorted out before choosing the best architecture. Considering real data, prediction models are created for indoor temperatures. The results reveal that all the investigated models are successful in predicting indoor temperatures. Hence, the proposed deep neural network model obtained the highest accuracy with an average RMSE of 0.16 °C, which renders it the best candidate for the development of a digital twin.

Keywords: indoor temperature; HVAC; machine learning; deep learning; educational building

Citation: Norouzi, P.; Maalej, S.; Mora, R. Applicability of Deep Learning Algorithms for Predicting Indoor Temperatures: Towards the Development of Digital Twin HVAC Systems. *Buildings* **2023**, *13*, 1542. https://doi.org/10.3390/buildings13061542

Academic Editor: Etienne Saloux

Received: 25 April 2023
Revised: 31 May 2023
Accepted: 2 June 2023
Published: 16 June 2023

1. Introduction

The world's energy consumption has significantly increased during the past few decades. According to the estimates from the International Energy Agency, over two decades, primary energy demand has grown by 49%, and CO_2 emissions have seen a 43% increase, with an average increase of 2% and 1.8%, respectively, each year [1]. A relationship between global warming, climate change, rising pollution, and the expansion in global energy consumption has revealed that reducing energy consumption can impact all the mentioned issues positively. Reducing global energy consumption is a subject worthy of more research and analysis because global population growth is anticipated to increase in the upcoming years, and global energy demands are expected to keep growing.

One of the critical energy consumers in today's world is the building sector. Building energy consumption has risen to the level of transportation and industry over the last decades as a result of population growth, improved building amenities, and comfort levels, plus increasing time spent in buildings and the spread of building services, particularly heating, ventilation, and air conditioning (HVAC) systems [2].

In Canada, buildings account for 30% of the total energy consumption, and HVAC systems consist of a large portion of it in residential and non-residential buildings [3]. HVAC systems with 59% of total energy consumption account for the largest share of energy use in commercial and institutional buildings in Canada [3].

Optimizing the HVAC energy efficiency benefits the environment and the economy, and implementing proper operational and management strategies is crucial for achieving the best level of energy consumption in buildings. Improving control strategies in HVAC systems through data-driven modeling and prediction techniques has the potential to reduce the overall energy consumption in buildings. Since today people spend most of their time inside buildings, their comfort level must be maintained. Therefore, a key concern in the field of energy management is how to optimize energy consumption without sacrificing thermal comfort level. Indoor temperature predictions can be an efficient measure and effective strategy to optimize the HVAC system [3] while maintaining the occupants' comfort level.

Incorporating temperature predictions within a Digital Twin HVAC system offers significant advantages in terms of optimizing HVAC system operation. By integrating forecasted temperature conditions, the system gains the ability to proactively adapt its cooling and heating operations. This proactive approach is made possible by leveraging temperature forecasting models that consider variables such as weather patterns and building HVAC characteristics. The system can then make informed decisions regarding the optimal allocation of resources to maintain the desired indoor temperature, ensuring occupant comfort while minimizing energy consumption.

Furthermore, the integration of temperature predictions facilitates the optimization of scheduling within the Digital Twin HVAC system. By considering the anticipated future temperature conditions, the system can strategically activate and deactivate HVAC operations. This approach prevents unnecessary energy consumption during the periods of low occupancy or when temperature adjustments are not required. By aligning the timing of HVAC system operations with forecasted temperature needs, the Digital Twin HVAC system could achieve a higher level of energy efficiency and reduce energy consumption.

Traditionally physical or semi-physical models have been employed to model and predict indoor temperature, but these models' input parameters are often based on particular building attributes and occupant activity, which are not always easily accessible [4–7].

Nowadays, modern buildings can negotiate settings that are continually changing because of the interactive and adaptive nature of the buildings, their components, and the surrounding environment. Advanced sensor technologies in buildings are increasingly more capable of inferring valuable information about the building's properties, outdoor weather conditions (such as temperature, wind, and solar radiation), and occupant behavior (e.g., temperature set points, CO_2 level). Massive and diverse datasets regarding every aspect of building operations are produced by the continuous communications between smart hubs, sensors within buildings, smart meters, and equipment. For data processing and real-time decision-making, these massive datasets require increasingly automated and adaptable methods to optimize building operations. Therefore, many researchers over the last decade have studied data-driven models to predict indoor temperatures, heating or cooling load, and energy consumption levels.

The majority of data-driven models in the early 2000s were statistical models such as ARIMA (Auto-Regressive Integrated Moving Average) [8]. Common machine learning (ML) algorithms, such as SVR (Support Vector Regression) [9], Random Forest [10], XGBoost [11], etc., have also been widely investigated. Due to the simplicity of the nonlinear parameters included, the minimal demands on feature engineering, and the capacity to handle interactions among nonlinear features, neural networks (NN) have also been increasingly popular in the indoor temperature predictions and energy optimization [12,13].

Both academic scholars and building industry professionals have generally acknowledged predictive modeling as a useful technique to support building design and control. Predictive modeling extracts patterns seen in historical datasets through mathematical methods seeking to predict future events. The building industry has a chance to employ predictive algorithms to lower failure rates, maintenance expenses, and energy consumption while enhancing occupant thermal comfort.

This research investigates suitable data-driven models to predict indoor temperatures toward building a Digital Twin HVAC system. The accuracy of the models was compared through the prediction of the indoor temperature of seven different zones—six classrooms and one office area. The preprocessing and modeling procedure is presented through the use of five-year real operational data collected from the NE01 building at the British Columbia Institute of Technology, located in the city of Burnaby, BC, Canada. Finally, the results were evaluated to propose the most efficient and accurate model for the prediction of indoor temperatures.

2. Related Studies

Indoor temperature prediction is one of the important methods for determining thermal comfort levels in buildings and identifying energy-saving potential, particularly when it is used to develop HVAC system control strategies. Over the last decade, different studies have been carried out to model HVAC systems and simulate indoor temperatures.

Classically, to model and optimize HVAC systems, physics-based models have been widely developed to express the complex laws of physics, energy, heat transfer, and thermodynamics in buildings. Most of these models were built for the purpose of systems control and energy efficiency. In this regard, Nassif et al. [14–16] developed a supervisory control strategy based on a simplified physics-based model for a Variable Air Volume (VAV) system to optimize the multi-zone HVAC system. As HVAC systems become more complex, non-linear, and large-scale involving numerous constraints and variables, developing physics-based models for buildings energy management becomes more challenging [17]. In order to generate accurate physics-based models, high-order complex models are often used. The latter are computationally expensive while reducing the model order leads to an increase in prediction error [18]. Furthermore, high computational time makes complex models difficult to apply and implement in real-time applications [19]. Thus, the produced physics-based models are usually deterministic, requiring multiple assumptions and simplifications on their parameters, rendering models less applicable to represent and interfere with the buildings' daily operations.

To overcome the shortfalls of classical physics-based models, many data-driven approaches have been developed in recent years, and many studies have concentrated on developing predictive models based on data mining techniques. Data-driven models also referred to as "black box" models in artificial intelligence are built straight from data by an algorithm, which cannot be easily comprehended nor interpreted to explain the integration of variables to produce predictions.

Therefore, in recent years, researchers working on modeling HVAC systems have focused on indoor temperature predictions using artificial intelligence algorithms. These algorithms employ various approaches to build models, such as machine learning (ML) tree-based models and deep learning (DL) neural networks.

Tree-based machine learning models are constructed by recursively dividing the considered observations following specific criteria. These criteria are created by comparing all possible splits in the data and selecting the one that provides the highest mean squared error (MSE) reduction in the variance of child nodes. Tree-based ensemble methods enhance performance and create stronger predictive models by combining multiple decision tree predictors. Several studies [20–22] have examined the effectiveness of ensemble methods, including Random Forest and Extra Trees, in forecasting time series datasets and specifically HVAC systems.

On the other hand, a deep learning strategy is based on artificial neural networks (ANN). It involves scanning the data with an algorithm to find features that correlate, then combining those features to facilitate rapid learning [23]. These algorithms have the capacity to learn on their own and generate outputs that are independent of their inputs. They are also capable of accomplishing several tasks in parallel, without impacting the system's performance. Studies show that systems with the ANN modeling approach can handle the internal [24] and external [25] disturbances that impact the modeling process. In addition, researchers have developed strong numerical foundations that enables ANN to handle real-time events because they can learn from examples and apply them when a comparable situation occurs [26,27]. In order to model indoor temperatures, different studies [28–33] developed several simple to advanced models with different scenarios, and all discovered that ANN models provided an acceptable accuracy in temperature predictions. However, the literature on artificial neural network (ANN) modeling for buildings primarily focuses on residential buildings and dedicated labs. Researchers in this field often utilize a limited amount of data, typically centered around a single zone within the building, which can impact the comprehensiveness of their findings. Furthermore, the selection of input variables and the preprocessing phase are commonly overlooked, leaving gaps in understanding the methodology employed.

Thomas et al. [28] investigated two types of buildings to develop ANN models for indoor temperature prediction. The first one was a small experimental building, with a total data collection period of 624 h. The second building was a factory building, and the studied area was a laboratory room, with a total data collection of 140 h. Mirzaei et al. [29] presented a simplified model that uses artificial neural network (ANN) techniques to establish a correlation between weather parameters and indoor thermal conditions in residential buildings. The study included a measurement campaign in 55 residential buildings in Montreal. The model incorporates neighborhood-specific parameters, building characteristics, and occupant behavior. Fewer works considered educational buildings to develop ANN models and forecast indoor temperatures. Temperature management and optimization of the thermal conditions in educational buildings is a crucial aspect that significantly impacts the learning environment and students' well-being. Educational buildings have unique specifications and requirements such as the existence of different areas (e.g., classrooms, offices), the occupancy density and the variety of HVAC features involved in providing a comfortable and productive learning environment. As an example of work carried out in an educational building, Atoue et al. [30] studied office areas of an old school building. The model was developed based on indoor and outdoor temperatures and humidity as well as solar radiation; in addition, collected data from two summer months were used for this study.

In buildings modeled using data-driven techniques, investigations usually consider a single zone area [28,31] and rarely include multiple zones. Afroz et al. [32] evaluated the performance of single-zone and multi-zone indoor temperature prediction models using different combinations of training datasets and inputs. The studied area was concentrated on the second floor of a library building and the data were colected for a few seasons, from summer to winter.

The collected data for the ANN modeling found in the literature were frequently extremely brief, ranging from a few days [10] to a few months [34]. It is well known that ANN performs better when trained on large datasets and that small dataset can significantly limit ANN performance [35]. In fact, small training and test datasets can lead to generalization problems because daily fluctuation of the indoor temperature is insignificant. Therefore, it is essential to look closely at "low" prediction errors; for example, an average RMSE of 0.8 °C might already be too high in comparison to the average daily variance.

The selection of input variables is a crucial subject of investigation in the process of creating models. The input variables and the performance of artificial neural network (ANN) models from different studies [28,29,36] show that researchers often employ input variables without a clear explanation of systematic rationale behind their selection. Moreover, the majority of the studies employing historical data for the purpose of forecasting future interior temperatures have generally failed to provide explicit explanations regarding the preprocessing steps involved [31,32,37,38].

To conclude, it was noticed that there is a lack of comprehensive research exploring the use of deep learning methods for predicting indoor temperatures in educational buildings while considered multiple zones, expanded data collection over many years and detailed data preparation and processing. The research gaps of previous studies can be summarized in the following aspects:

- Limited research exists on the application of deep learning for indoor temperature prediction in educational buildings.

- Consideration of single-zone predictions versus multiple-zone ones.
- Lack of explicit explanations of data preprocessing.
- Extremely brief training periods for ANN models.

The overarching goal of this research is to create a model that would be used to build a digital twin for optimization of the HVAC system operation. The accurate prediction of indoor temperatures is required to achieve this goal. The primary objective of this paper is to select a suitable algorithm to accurately predict indoor temperatures. The secondary objectives are the following:

- To compare the performance and scalability potential of machine learning versus deep learning algorithms to predict indoor temperatures.
- To investigate the system parameters and select the best set of input parameters for model development.

3. Methodology

To achieve the objectives listed above, this research uses a case study HVAC system serving a group of classrooms and an office at the British Columbia Institute of Technology (BCIT). The research consists of four main parts to develop prediction models: data collection and preparation, feature selection and extraction, model development, and performance evaluation.

3.1. Data Collection and Preparation

The data for this study are provided by the Building Management System (BMS) of the British Columbia Institute of Technology. The BMS centralizes building management operations and collects real-time sensor and equipment parameters data.

The raw data that have been collected and archived over the last few years are available and accessible for the institute researchers. Building meaningful features and datasets involves several phases, including data cleaning and transformation.

Several parameters require preprocessing to increase the quality of the data. The database in this study is based on a time series. To reduce the error produced by time delay and system error, the original data were aggregated to 15 min interval data by resampling and filling in missing parameters' data, well-thought-out in the time intervals.

3.2. Feature Selection and Extraction

The intended purpose of feature selection and extraction is to explore options to improve performance by selecting or extracting features from the existing datasets.

The selection of parameters is based on two approaches. The first approach is based on domain knowledge, and the second approach relies on statistical and data analysis approaches.

The feature extraction process is crucial in many artificial intelligence applications because it helps the prediction model generate useful information that will enable accurate predictions [39]. As a feature extraction strategy, the timestamps feature is simplified by separating time, day, month, and year. For example, months are treated like dummy variables, days are divided into working days and weekends, and time is categorized by daytime. Feature extraction could greatly improve the quality of the data by transforming the data to have better distribution, removing linear dependencies, etc., and is deemed crucial to increase the data quality.

3.3. Model Development

A critical comparison of machine learning and deep learning algorithms is conducted to select the most suitable algorithms for indoor temperature prediction. The indoor temperatures are a time series problem. As a result, the algorithms applied in this research were chosen based on their shown effectiveness in time series contexts.

The room temperatures are known to be impacted by historical HVAC features. As a result, a lookback sliding window approach was applied for the resampling procedure to effectively train the considered models.

The first training sample for the input variable "*x*" is built as a matrix with the dimensions $W \times F$, where W is the window length ($W = n \times 15$ min, $if\ n = 4$, *then* W = 1 h), and F is the number of features. As shown in Figure 1, the window then slides forward by one step (15 min), the data from time step i + 1 create the second sample, and so on. The size of this sliding window controls how much historical data influences the prediction (e.g., 30 min, 60 min, 120 min, etc.).

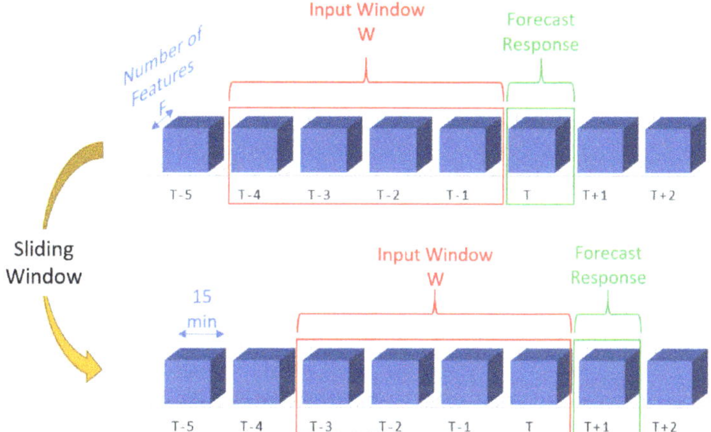

Figure 1. Sliding Window Sampling Procedure.

A reading from a single time step can be included in numerous sliding windows and overlaps when considering the 15 min step. The final training dataset has the shape of a matrix $[N, n, F]$; N is the number of rows of data in total, n is the window size (length of window) and F is the number of features. The model predicts the room temperatures for $T + i$ (output) for each sample (window length W), where i is the horizon response in the future. Each model is created to accurately predict the room temperatures at one or more alternative forecast responses, such as T + 1, T + 2, T + 3, or T + 4 (15, 30, 45, or 60 min in the future).

3.4. Performance Evaluation

To assess the model performance, three metrics were used for each test sample: mean squared error (*MSE*), root mean squared error (*RMSE*), and mean absolute error (*MAE*). The final score was calculated by averaging the metric scores across all samples.

$$MSE = \frac{\sum_{i=1}^{n}(y_i - Y_i)^2}{n}, \tag{1}$$

$$RMSE = \sqrt{\frac{\sum_{i=1}^{n}(y_i - Y_i)^2}{n}}, \tag{2}$$

$$MAE = \frac{\sum_{i=1}^{n}|y_i - Y_i|}{n}, \tag{3}$$

where n is the total number of observations, Y_i and y_i are the actual and predicted values, respectively. *MSE*, *MAE*, and *RMSE* are frequently employed with time series forecasts [33,10,11]. Since the prediction error is expressed in the same units as the predicted variables, they offer a simple method for determining the prediction error. While *MSE* is calculated as the average of the squared differences between the predicted and actual values, *RMSE* gives relatively significant weight to the large errors since they are squared before being averaged and *MAE* evaluates the average magnitude of the errors in the prediction set. A higher accuracy of the model under consideration is implied by lower *MSE*, *RMSE*, and *MAE* values.

4. Machine Learning and Deep Learning Algorithms

Machine learning (ML) and deep learning (DL) are two subfields of artificial intelligence (AI) that involve the development of algorithms capable of learning from data and making predictions or decisions. While both ML and DL algorithms aim to extract patterns and insights from data, they differ in their approach and level of complexity.

Deep learning focuses on developing algorithms inspired by the structure and function of the human brain's neural networks. DL algorithms are designed to automatically learn hierarchical representations of data by using multiple layers of interconnected artificial neurons. Machine learning algorithms often provide more interpretability, while deep learning algorithms, with their complex architectures and a large number of parameters, can be considered as "black boxes" with less interpretability. In addition, deep learning algorithms, due to their complex architectures, often require significant computational resources, while machine learning algorithms are generally less computationally demanding. In this study, we focused on deep learning algorithms as they are expected to be more suitable for the application. However, we did explore the potential of the less complex tree-based machine learning algorithms.

Five data mining algorithms, namely Extra Trees and Random Forest (RF) as tree-based algorithms and the Multilayer Perceptron (MLP), Long-Short Term Memory (LSTM), and convolutional neural networks (CNN) as deep learning algorithms have been applied to the studied HVAC system based on their proven potential in related studies. The Extra Trees and Random Forest are supervised tree-based machine learning models that solve classification or regression problems by building a tree-like structure to generate predictions. Recursively dividing the observations under consideration according to some criteria results in the construction of tree-based models. However, the MLP neural network as a deep learning algorithm is a widely used feed-forward neural network with several neurons and multiple layers. The MLP can recognize and learn patterns based on input datasets and the corresponding target values by adaptively modifying the weights in supervised learning. The LSTM is a type of neural network architecture that was designed with memory cells and gates to selectively remember or forget information in time series. On the other hand, the key innovation of CNNs lies in their use of convolutional layers that can effectively extract spatial features from raw data.

By selecting these five algorithms, a comprehensive and multi-faceted approach to indoor temperature prediction was examined. Each algorithm contributes unique capabilities that address different aspects of the problem, such as ensemble learning (Extra Trees and Random Forest algorithms) for improved accuracy, MLP for non-linear relationship modeling, LSTM for handling temporal dependencies and CNN for capturing spatial patterns. These algorithms are amongst the most popular algorithms in artificial intelligence and HVAC system modeling.

4.1. Tree-Based Algorithms

Multiple decision tree predictors are combined in ensemble methods (such as bagging and boosting) which are known as tree-based algorithms to improve performance and create more robust prediction models. The bagging method makes use of average forecasts from various trees that were built using various data subsets; meanwhile, boosting is an iterative method that fits a series of trees made from random samples. Simple models are fitted to the data at each phase, and the data are assessed for errors. In the past, several researchers have examined the efficiency of ensemble-based models for forecasting time series datasets and HVAC systems in particular [20,21,42].

4.1.1. Extra Trees

According to Alawadi et al. [20], the Extra Trees algorithm is the most precise and effective model for HVAC system temperature prediction. Therefore, the Extra Trees algorithm with randomly selected decision rules is one of the selected algorithms in this study for modeling purposes.

The sampling for each tree in the Extra Trees algorithm is random and without replacement. By sampling the complete dataset, Extra Trees avoids the biases that different subsets of the data may impose on the results. Additionally, each tree receives a specific number of features, chosen at random from the entire set of features. The random selection of a splitting value for a feature is also the most significant and distinctive aspect of Extra Trees. As a result, the trees become more diverse and uncorrelated, which reduces variance and lessens the influence of specific features or patterns in the data.

4.1.2. Random Forest

The performance of a single decision tree in regression models is unstable because the final regions are influenced by the data's properties, and even little changes in the data might produce drastically different outcomes. Therefore, Breiman [43] suggested using random forest algorithms to produce more stable models with improved prediction accuracy. The Random Forest algorithm also showed reliable performance in dealing with time series datasets [44]. In addition, it has been frequently employed in HVAC systems for modeling purposes [26,45,46].

The Random Forest (RF) method subsamples the input data with replacement. The RF technique is based on the idea that numerous uncorrelated models work significantly better together than they do separately. With random sampling from robust decision trees, the RF algorithm decreases the risk of overfitting, while decision trees generally tend to tightly fit all the samples throughout training, which exposes them to overfitting.

4.2. Deep Learning Algorithms

A group of machine learning algorithms known as "deep learning" go beyond simple learning and focus on learning from experience [23]. These algorithms undergo many linear or non-linear transformations of the input data before obtaining an output.

Deep learning (DL) is made up of several hidden layers of neural networks that perform complex operations on massive amounts of data. Due to the superior predictive modeling applications of DL, it has advanced predictive modeling by enabling more accurate predictions and the ability to handle vast amounts of complex data, thus opening up new possibilities for data-driven insights and decision-making in a number of sectors.

DL techniques have become more popular among researchers in recent years as an alternative to manual feature extraction. This strategy has gained popularity with ANN architectures and has produced groundbreaking solutions for time series problems [47].

In this study, three deep learning algorithms are investigated: the multilayer perceptron (MLP), long short-term memory (LSTM), and convolutional neural networks (CNN). These algorithms were chosen because they are pertinent to both the application under consideration and the studied case study.

4.2.1. Multilayer Perceptron

Multilayer perceptron (MLP) is a deep learning algorithm that is one of the selected algorithms in this study because it has already demonstrated a good performance in time series prediction problems [48]. In addition, it proved its potential in HVAC systems as time-dependent systems [46,49,50]. Figure 2 shows the structure of the deep neural network model.

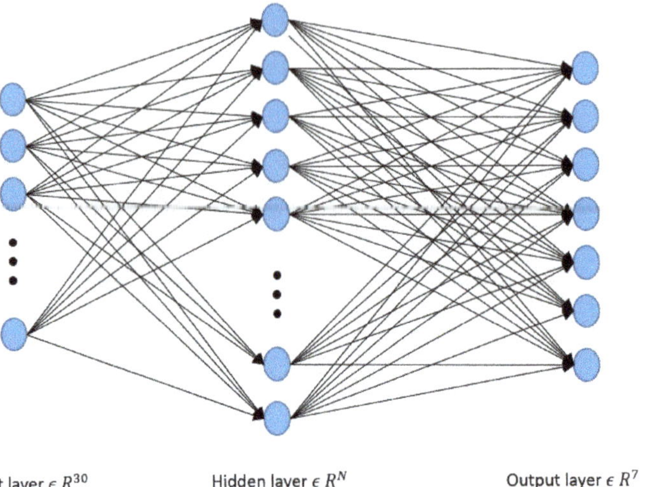

Figure 2. Structure of the deep neural network.

By updating the weights of the neurons, the MLP algorithm has a significant capability for mapping the link between input parameters and output parameters. Three layers made up the MLP model in this study: the input layer, the hidden layer, and the output layer. First, the forward propagation process from the input layer to the output layer is created when the input parameters are added to the MLP model. Then, using a backpropagation method, the weights and thresholds between the input layer and hidden layer as well as between the hidden layer and output layer are tuned. The first and second stages are repeated until the training error can meet the desired settings.

The perceptron/neuron is the most basic learnable artificial neural network with input visible units $\{v_i\}_{i=1}^{D}$, trainable connection weights $\{w_i\}_{i=0}^{D}$, a bias, and an output unit y as shown in Equation (4). The perceptron model is also known as a single-layer neural network since it only contains one layer of output units, excluding the visible input layer. Given an input $v \in R^D$, the value of the output unit y is derived from an activation function $f(.)$ by taking the weighted total of the inputs as follows:

$$y(v;\theta) = f\left(\sum_{i=1}^{D} v_i w_i + w_0\right) = f\left(w^T v + w_0\right), \tag{4}$$

where $\theta = \{w, w_0\}$ stands for a parameter set, $w = \{w_i\}_{i=1}^{D} \in R^D$ is a connection weight vector, and w_0 is a bias. In this study, a "relu" function is utilized as the activation function $f(.)$ and the activation variable z is defined by the weighted sum of the inputs, i.e., $z = w^T v + w_0$.

As the studied HVAC system involves multiple zones, it needs to be extended to a multi-output model, so multiple output parameter vector $\{y_k\}_{k=1}^{K}$ is added to each output. The outputs' respective connection weights $\{w_{ki}\}_{i=1,\dots,D;k=1,\dots,K}$ are as follows:

$$y_k(v;\theta) = f\left(\sum_{i=1}^{D} v_i w_{ki} + w_{k0}\right) = f\left(w_k^T v + w_{k0}\right), \tag{5}$$

where $\theta = \{w \in R^{K \times D}\}$, W_{ki} denotes a connection weight from v_i to y_k.

During the training phase, neural networks learn by iteratively adjusting the parameters (weights and biases). The biases are initially set to zero and the weights are created at random to establish the parameters. The data are then forwarded across the network to provide model output. The process of back-propagation is the last one. Several iterations of a forward pass, back propagation, and parameter update are commonly included in the model training process until the model reaches the predefined setting values, and the best parameters of the network are obtained, which could link the inputs to the outputs with a high accuracy or low error, even in highly non-linear and sophisticated systems.

4.2.2. Long Short-Term Memory (LSTM)

LSTM topology has been presented as a way to directly integrate time dependence. The fundamental concept is to dynamically integrate a sequential structure in order to enhance the functionality of conventional ANN approaches [51]. LSTM has demonstrated exceptional performance because it can learn the short- and long-term dependencies. The simultaneous inclusion of slow- and fast-moving phenomena makes LSTM a good option for indoor temperature prediction problems [52]. Time series data processing and prediction are all very well suited to LSTM networks [34,53]. The LSTM algorithm was considered in this research since it can capture features and remember them over time. A typical illustration of an LSTM cell is shown in Figure 3.

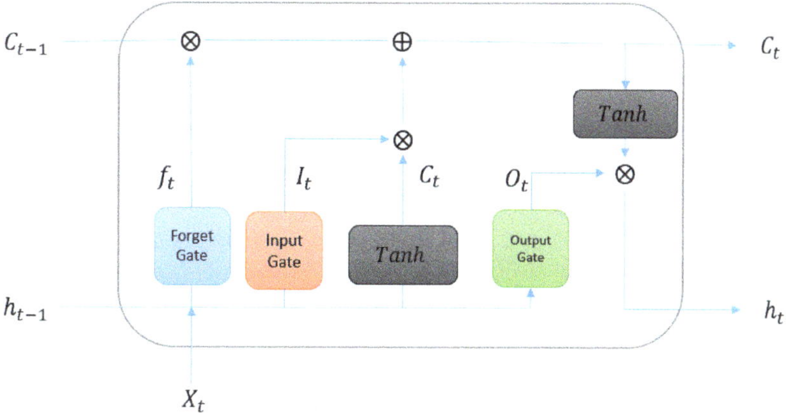

Figure 3. A typical LSTM neuron structure.

When compared to cutting-edge black-box modeling techniques used for indoor temperature prediction, ANN-based algorithms have high predictive power. Moreover, recurrent neural network-based methods, particularly LSTM, have an outstanding capacity for "learning" the dynamics of non-linear problems with time dependency.

4.2.3. Convolutional Neural Networks (CNN)

Convolutional neural networks, generally known as CNNs, are powerful types of artificial neural networks and deep learning algorithms. CNNs have been demonstrated to be very successful in a variety of computer vision tasks such as image classification, object recognition, and segmentation. According to their design, CNNs can automatically and adaptively learn spatial feature hierarchies from basic to complex patterns. The feature extraction is carried out by the convolution and pooling layers, and then it is mapped by the fully connected layer into a final output [54].

In general, CNNs are effective tools for processing and interpreting complicated data because of their hierarchical structure and learnable filters. Due to the proven capability and success of CNNs in solving time series problems [55], CNN is also considered in this study.

5. Case Study

Numerous initiatives have been implemented on the Burnaby campus of the British Columbia Institute of Technology to reduce environmental impacts of its educational buildings. One of these initiatives aims to reduce BCIT's greenhouse gas emissions by 33% by 2023 [56]. Therefore, the primary objective of this study is to develop a model to predict indoor temperatures which can provide a meaningful way to reduce the energy consumption of the system and meet the requirement of greenhouse gas emissions reduction at BCIT.

The British Columbia Institute of Technology (BCIT) is located in the city of Burnaby, BC, Canada. It is a post-secondary institution with 62 buildings and five campuses. The HVAC system under study is owned and operated by BCIT at the Burnaby Campus. The investigated HVAC system is installed at the NE01 building (NE stands for North East). The building is 20,076.88 m^2 of gross floor area consisting of offices, classrooms, restrooms, and mechanical rooms. NE01 is a 4-floor educational building dedicated to the Construction and Environmental Department at the Burnaby campus; see Figure 4.

Figure 4. NE01 building overview.

The selection of the NE01 building for this research was based on several key factors, namely its high energy consumption compared to other buildings in the BCIT campus, the alignment with BCIT's sustainability goals, the availability of archived data over many years for multiple zones served by the same system, and the potential for implementation of the proposed solution to prove significant energy savings and improved occupant experience.

The temperatures in the NE01 building zones are controlled by ten independent Air Handling Units (AHUs). In this study, the data and parameters of one AHU (AHU7) are investigated. AHU7 delivers air to seven Variable Air Volume (VAV) units, serving seven interior zones on the fourth floor (top floor) of building NE01. The zones consist of six classrooms and an office area as shown on the floor plan of Figure 5. The BMS is used to collect the data from sensors and virtual meters of the air handling system.

Figure 6 shows a schematic diagram of AHU7 with the available monitored parameters. Airflow is controlled via the supply and return fans. Recirculated air from the seven service zones is mixed with outdoor air and reconditioned through the AHU. The mixed-air conditions are controlled by three dampers regulating the percentages of air exhausted from the system, entering the system, and recirculating in the system. The mixed air is pulled through the filter via the supply fan and then cooled or heated when passing through the cooling and heating coils. The heating coil uses hot water supplied by the boiler. Heating is controlled by adjusting the flow of water through the coil, which is controlled by an electronic valve.

AHU7 serves seven VAV units to further condition the supplied air to meet the required temperature by each zone. The seven reheat systems have the same design. Figure 7 represents a sample VAV box and the reheat system for each zone. Air is supplied by the AHU to the VAV boxes, which control the amount of air that is supplied to the zone by a VAV damper. At the VAV, the air also passes through a reheat coil to be warmed based on the temperature required in the zone. Similar to the heating coil at the AHU level, the reheating coil uses hot water supplied by the boiler, and it is controlled by an electronic valve to adjust the flow of water in the coil.

Figure 5. Schematic of the zones supplied by AHU7.

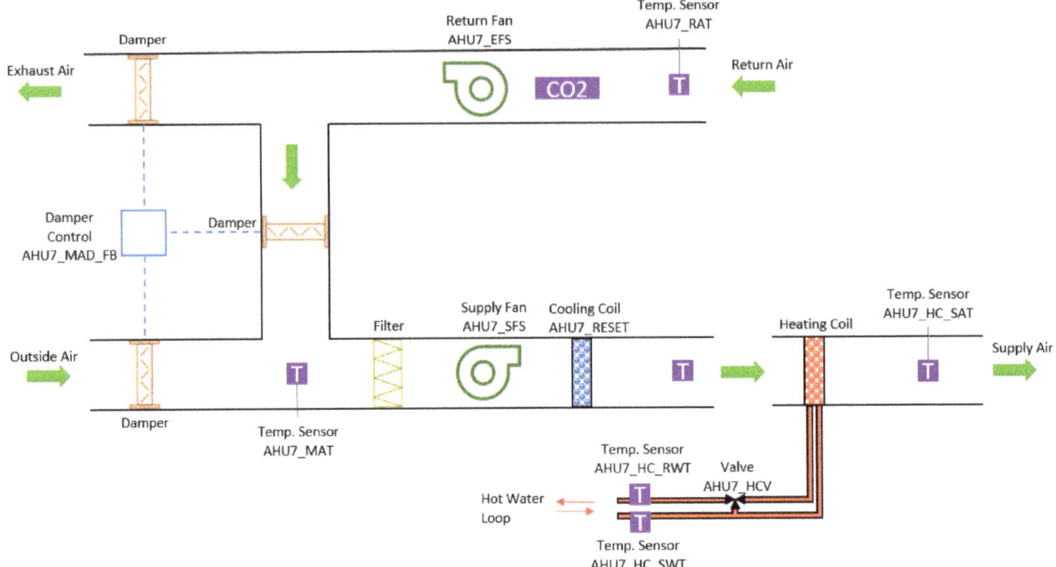

Figure 6. Schematic diagram of AHU7.

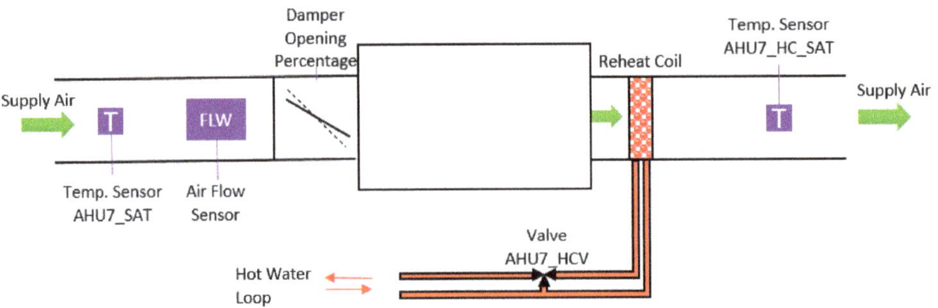

Figure 7. VAV box and reheat coil subsystem for each zone.

6. Description of Available Data

The facilities department at BCIT has adopted an advanced BMS over the past few years. The data from many of the system sensors has been recorded and archived since 2016. Due to the COVID-19 pandemic, there was very little activity on campus between March 2020 and September 2021. Therefore, the data from this period were not considered for this application of predictive modeling. The data used for this study are for two periods. The first period is from February 2016 to December 2019, and the second period is from September 2021 to November 2022.

The first period represents almost four years and about 140,000 data points per system component, and the second period represents 14 months and about 40,000 data points; in total, about 180,000 data points were considered. Each row of these data corresponds directly to the state of the system at 15 min time intervals. Among the studied dataset, last year's data were assigned as the test dataset, and the remaining data were divided into two parts: 80 percent was used for training models, and 20 percent for validation purposes.

7. Parameter Selection

Since it is too complicated to establish the mapping between inputs and outputs using all features, parameter selection is a crucial step before building a data-driven model. A typical HVAC system is a complex, non-linear network with hundreds of parameters reflecting every particular element. While some of the data obtained are closely related to the result, others are unnecessary or redundant for the modeling process. In data mining, the inclusion of redundant or unnecessary parameters may hide primary trends. Additionally, redundant parameters substantially or entirely replicate the data in one or more other parameters, making the model considerably more challenging than it has to be. The accuracy, scalability, and understandability of the resulting models may be enhanced by removing irrelevant or related characteristics. In addition, by reducing the dimensionality, a model's complexity may also be significantly reduced [57].

For a preliminary analysis, a dataset with 75 parameters was selected based on the domain knowledge. To determine how closely parameters were coupled, a correlation analysis was applied. The range of the correlation between two parameters is from -1 to 1. The two features are positively associated if the correlation coefficient is greater than 0, and the higher the number, the stronger the correlation. For example, in this study, the correlation analyzes illustrated that Return Fan Speed and Supply Fan Speed were highly correlated with correlation equal to 1. It means Supply and Return Fan Speeds were changing at the same rate, and both were also correlated with the Status of the Return Fan with a 0.98 correlation. These are expected because the blowing air into the zones must be exhausted, and if the building is made tight enough, return fans are good controllable options to control the exhausted air. Similarly, other highly correlated parameters were diagnosed and eliminated from the list of input parameters in order to reduce the model complexity and also to reduce the risk of overfitting. With this method, the number of parameters was reduced from 75 to 57 parameters.

To increase the computational efficiency and lower the generalization error, a sequential backward feature selection approach was used to reduce the dimensionality of the initial feature subspace from N- to K-features with a minimum reduction in model performance [58]. Sequential backward feature selection was initialized by considering a feature set containing available predictors in the dataset. The regression model was trained using these features and its performance was evaluated using mean squared error (MSE). Iteratively, one feature at a time was removed from the current feature set. After each removal, the regression model was retrained, and its performance was assessed using the chosen evaluation metric. Then, the performance of the regression model with the reduced feature subset was compared to the performance achieved with the previous feature set. If the performance dropped beyond a predefined threshold or significantly deteriorated, the process was stopped, and the previous feature set was retained as the final selection. However, if the performance did not decrease significantly, the next iteration proceeded. The removal and evaluation steps were repeated until the stopping criterion was satisfied. Features with the highest impact on error rate indicated a larger contribution to the predicted output parameters. Based on the domain knowledge, correlation consideration of parameters and the sequential backward feature selection results, the final set of variables reached 24 features.

In addition, as the studied zones are on the top floor of the building and through the roof they are exposed to solar radiation, the solar radiation and outdoor air temperature data were also added to the dataset. The data were recorded by the sensors installed on the nearby weather station on the campus, and because of their stochastic nature, solar radiation and outdoor air temperature were considered as external disturbances. The selected 26 parameters are shown in Table 1.

Table 1. Parameters Description.

ID	Parameter Name	Description	Unit	No.
01	OAT	Outdoor weather temp.	°C	1
02	AHU7_SAT	AHU7 Supply air temp. after heating coil	°C	1
03	AHU7_SAT_SP	AHU7 supply air temp. set point	°C	1
04	AHU7_EF_SPD	AHU7 Return Fan speed	%	1
05–11	VAV_x^1_SAT_SP	VAV_x^1 supply air temp. set point	°C	7
12–18	VAV_x^1__SAT	VAV_x^1 supply air temp.	°C	7
19–25	VAV_x^1_RT	Room temp. in each zone supplied by VAVs	°C	7
26	horiz_solar_rad	Roof horizontal solar radiation	W/M^2	1

x^1 represent the specific VAV number.

Furthermore, the HVAC system under consideration operates on set schedules. Based on the operational specifications, schedules change depending on the day of the week (weekday/weekend), the hour of the day, as well as holidays. The RTs were affected by this operation schedule. Therefore, the characteristics that reflect seasonality such as the hour of the day and the day of the week (weekday/weekend) were also used as inputs in an effort to capture the temporal dependency in the model. The schedule-based variables that were specified as input features for the prediction models are listed in Table 2. Therefore, the total number of input parameters was 30.

Table 2. Schedule-based features.

ID	Variables	Parameters Description	No.
27	Day of Week	Weekdays vs Weekends	1
28–30	24 h	Day time, evening, night	3

8. Results and Discussion

The five algorithms detailed in Section 4 were trained to predict indoor temperatures up to a 60 min forecast horizon with 15 min intervals and 30 min, 60 min, and 120 min sliding window sizes. Since the forecast horizon (T + 1, T + 2, etc.) and the time delay of input variables (sliding window width) have a major impact on the prediction model accuracy, various combinations of the two parameters are established in order to examine their interaction. The width of sliding windows of 2 (30 min), 4 (60 min), and 8 (120 min) time steps was investigated. As the system is notably changing fast, a lookback window beyond 120 min appears unnecessary and could potentially introduce noise to the prediction results without conferring any benefits. In addition, the RTs for the 15 to 60 min forecast horizon were predicted using the model since the equipment shows a rapid response to control commands.

First of all, the model architecture was trained to predict indoor thermal temperatures in each zone supplied by VAVs, while the evaluation metrics were calculated for different window lengths and prediction horizons.

Figure 8 illustrates the RMSE obtained on the test dataset for various lookback sliding windows on the MLP model. The metrics for the error analysis evaluation are plotted against the window length and forecast horizon to show that the model performance degrades when the forecast horizon increases and the window length goes beyond 30 min. For a prediction of 15 min ahead, the lowest errors were observed for a sliding window of 60 min. The size of the lookback window has an insignificant effect on the prediction when it is one hour or beyond.

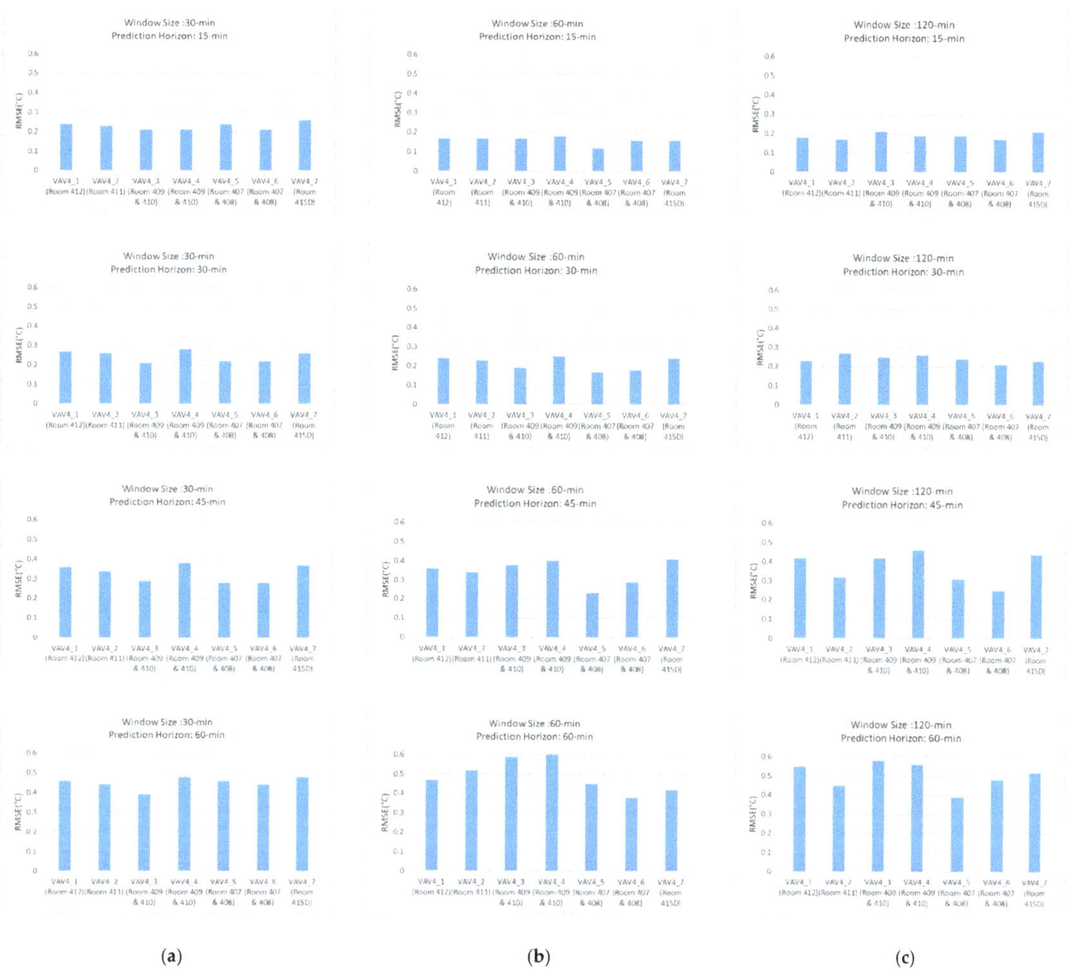

Figure 8. Error analysis evaluation on window length and forecast horizon of the proposed MLP model. (**a**) 30 min lookback window; (**b**) 60 min lookback window; (**c**) 120 min lookback window.

The investigation of the sliding window and forecast horizon effect on the RMSE was applied to all studied algorithms, and for all of them, similar results were seen. Thus, the best window length for the sliding window sampling method was selected as 60 min with 15 min forecast horizon as it resulted in the lowest RMSE in the seven zones; therefore, it was applied for training the final developed model.

The MSE, RMSE, and MAE for the two tree-based machine learning algorithms and the deep learning models are shown in Tables 3 and 4 for the 15 min forecast horizon. The performance of the different architectures was assessed. It was noted that all models exhibit comparable overall prediction accuracy. Although the ML and DL models under consideration can predict the RTs with an acceptable error, the rate of their errors was analyzed in detail. The model sensitivity is crucial since this analysis is a critical part of a larger study that aims to represent a digital twin of the studied HVAC system for long-term predictions and application of energy optimization strategies.

Table 3. The 15 min prediction horizon error [°C] of validation and test datasets of the trained tree-based algorithm models for the seven zones supplied by AHU7.

Zones	Dataset	Extra Trees			Random Forest		
		MSE	RMSE	MAE	MSE	RMSE	MAE
Room 412	Validation	0.02	0.15	0.10	0.03	0.17	0.11
	Test	0.11	0.33	0.24	0.10	0.32	0.22
Room 411	Validation	0.02	0.14	0.10	0.02	0.16	0.11
	Test	0.10	0.32	0.21	0.09	0.30	0.20
Room 410	Validation	0.01	0.12	0.07	0.02	0.14	0.09
	Test	0.20	0.44	0.29	0.22	0.47	0.30
Room 409	Validation	0.01	0.13	0.08	0.02	0.15	0.09
	Test	0.18	0.42	0.28	0.20	0.45	0.29
Room 408	Validation	0.01	0.11	0.07	0.01	0.13	0.09
	Test	0.13	0.36	0.22	0.15	0.39	0.23
Room 407	Validation	0.01	0.11	0.07	0.01	0.13	0.09
	Test	0.04	0.20	0.13	0.05	0.23	0.15
Room 415D	Validation	0.02	0.15	0.10	0.02	0.17	0.11
	Test	0.12	0.35	0.20	0.13	0.37	0.21

Table 4. The 15 min prediction horizon error [°C] of validation and test datasets of the trained deep learning algorithm models for the seven zones supplied by AHU7.

Zones	Dataset	MLP			LSTM			CNN		
		MSE	RMSE	MAE	MSE	RMSE	MAE	MSE	RMSE	MAE
Room 412	Validation	0.02	0.14	0.09	0.01	0.13	0.09	0.02	0.16	0.12
	Test	0.03	0.17	0.12	0.02	0.16	0.11	0.03	0.17	0.13
Room 411	Validation	0.01	0.13	0.09	0.02	0.15	0.11	0.02	0.16	0.13
	Test	0.03	0.17	0.13	0.03	0.18	0.13	0.02	0.17	0.13
Room 410	Validation	0.01	0.12	0.08	0.01	0.12	0.08	0.01	0.13	0.10
	Test	0.03	0.17	0.12	0.02	0.17	0.11	0.02	0.14	0.10
Room 409	Validation	0.01	0.13	0.09	0.01	0.12	0.08	0.02	0.15	0.11
	Test	0.03	0.18	0.12	0.02	0.17	0.11	0.03	0.18	0.13
Room 408	Validation	0.01	0.10	0.07	0.01	0.10	0.07	0.02	0.16	0.13
	Test	0.01	0.12	0.09	0.01	0.13	0.09	0.02	0.15	0.12
Room 407	Validation	0.01	0.11	0.08	0.01	0.11	0.08	0.01	0.13	0.10
	Test	0.02	0.16	0.11	0.01	0.11	0.09	0.01	0.13	0.10
Room 415D	Validation	0.01	0.13	0.09	0.01	0.13	0.09	0.02	0.14	0.10
	Test	0.02	0.16	0.11	0.03	0.19	0.12	0.02	0.15	0.10

According to Tables 3 and 4, the Extra Trees and Random Forest algorithms are the least accurate compared to the deep learning algorithms. This demonstrates the necessity of more complex ML and DL algorithms that learn time dependency and patterns of room temperature changes over time. It was noticed that the average RMSE rate on the test dataset for all seven room temperatures was 0.36 °C for Random Forest and 0.35 °C for Extra Trees algorithms. All the deep learning algorithms outperformed tree-based algorithms and showed a lower RMSE rate on test datasets with an average RMSE of 0.16 °C. In addition, the difference between the MSE results of the deep learning algorithms in validation and test datasets with an average of 0.01 is insignificant, which means a low variance in the proposed models. In contrast, the MSE error difference between validation and test datasets for tree-based algorithms is higher. It shows the higher capability of the deep learning algorithms in modeling indoor temperatures compared to the Extra Trees and the Random Forest algorithms.

Additionally, it was noted that the RMSE values of both the validation and test sets for each of the investigated algorithms are higher than the MSE and MAE. This is a reference to the samples in the dataset having significant error levels. A detailed examination of the significant errors revealed that the models perform poorly in forecasting abrupt changes in room temperatures. Therefore, the larger error instances were checked and analyzed and it was noticed that at harsh temperature changes, the prediction error sometimes can surpass ±0.8 °C. However, the number of samples with this error rate is below 0.2% on the whole samples in the dataset. Samples with high errors can be treated as outliers in the modeling process, hence they are considered as a nature of the current system in this study which is part of the existing HVAC systems. In addition, investigations showed that high prediction errors are recorded in the early mornings and at the end of the day when the HVAC system starts and stops based on schedules. During these transitions, considerably different set points are assigned to the system abruptly, which leads to harsh temperature changes in a short time period. Appendix A illustrates a snapshot of the predicted room temperatures versus actual room temperatures in each zone.

It should be noted that direct comparison of the results presented in this paper with those of other published studies is not feasible due to variations in factors such as the buildings considered, input parameter sets employed, and quantities of collected data utilized. However, the modeling quality achieved in terms of RMSE, MSE, and MAE in this study is comparable to the outcomes reported in other published researches. Table 5 lists a concise summary of the deep learning model results from previous studies. The predictions of room temperatures obtained in this study through the utilization of deep learning algorithms show superior performance when compared to the outcomes of previous comparable relevant studies.

Table 5. Performance of ANN models for indoor temperature predictions in previous works.

Previous Studies	Forecast Horizon	MSE
B. Thomas et al. [28]	15 min	0.1069
P.A. Mirzaei et al. [29]	1 h	3.16
T.G. Ozbalta et al. [38]	daily	0.1521
Ch. Xu et al. [34]	5 min	0.1285

Figure 9 shows the distribution of the average MAE per month across the year in detail for the studied zones. It was found that all deep learning models can predict all room temperatures with high accuracy all over the year. The high errors are captured in the prediction of the transition period (April to June) during spring and then these decrease during the summer months and again they start to gradually increase from September to November almost in all algorithms. It means that the models are less accurate in predicting indoor temperatures during these periods, for example, in the spring when heating is on and cooling starts to take over the control of the HVAC system.

In addition, Figure 9 shows the MAE of the MLP algorithm for all seven zones is more consistent compared to the other algorithms. The LSTM and CNN algorithms predict each individual zones at monthly highly different error rates, while the MLP algorithm is producing more persistent results for all zones. In other words, LSTM and CNN algorithms can predict some of the zones with high accuracy, while prediction errors of some other zones can be twice (or more) as prevalent.

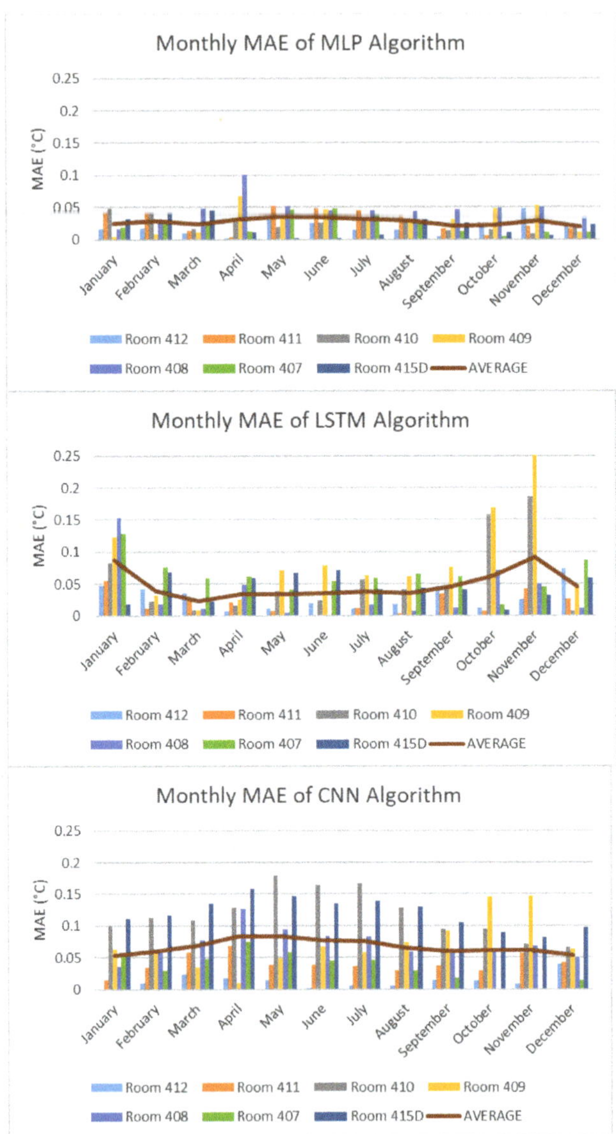

Figure 9. Distribution of monthly mean absolute errors.

The purpose of this study was first to determine how well algorithms predict HVAC parameters and then to select the best model settings for predicting room temperatures in each zone. Based on the performance evaluation detailed above, it was found that the deep learning models outperform the tree-based models. Specifically, the MLP model showed a more consistent and accurate prediction among the investigated deep learning models. This is a promising result leading to the potential employment of the MLP model and more generally the LSTM and CNN models to adjust the equipment setting of the HVAC in real-time, without human involvement and scheduled rules.

9. Conclusions

Accurate indoor temperature prediction is necessary to develop HVAC system digital twin that can eventually provide opportunities to optimize building energy consumption and maintain indoor thermal comfort. The main goal of this study was to suggest the most accurate data-driven model for HVAC systems. In this regard, this study investigated five suitable algorithms to predict indoor temperatures of a multi-zone HVAC system of an educational building. Seven room temperatures supplied by an AHU and seven VAVs at the NE01 building of the British Columbia Institute of Technology were predicted using five algorithms and collected real data.

Firstly, feature selection techniques were applied to determine the most critical features. The effectiveness of the proposed feature selection methods was successfully demonstrated as they proved capable of identifying significant and independent input parameters without compromising the overall prediction performance. This highlights the robustness and reliability of the feature selection methods in determining the most relevant features while ensuring that the prediction accuracy is high.

This study investigated machine learning and deep learning algorithms and proved that the latter produced better outcomes throughout the analysis with an average RMSE of 0.16 °C. The MLP algorithm as a deep learning model produced the best results compared to the other investigated deep learning algorithms with its consistent and stable prediction throughout the months of the year. It was demonstrated that room temperatures could be precisely predicted within a 15 min forecast horizon in advance using the developed MLP deep learning algorithm. As a result, the study offers the most accurate algorithm for development of a digital twin that includes various HVAC systems and the rooms they service in a specific educational building. Regarding building a digital twin model, the proposed model will be utilized and addressed in future work to simulate and produce the best settings for the studied HVAC system, which would result in lower energy consumption without sacrificing the occupant's comfort level.

In this study, the test dataset spans a whole year. Therefore, a more thorough analysis of the models' performance during this time is necessary. Using whole-year data could lead to less accurate transitional period predictions because no input variables have been provided to address transitions over the months in this study. In addition, occupancy level parameters are not included as a predictor variable, so their effect is not considered during the modeling process. A qualitative consideration revealed that some significant errors are the result of unexpected indoor activities, such as high occupancy levels during particular periods. Future research could further increase prediction accuracy by incorporating the scheduled occupancy data and analysis of the influence of heat and humidity produced by occupants, where the "full" effect of occupants can be considered.

Author Contributions: Conceptualization, S.M.; Formal analysis, P.N.; Investigation, P.N.; Resources, R.M.; Data curation, P.N.; Writing—original draft, P.N.; Supervision, S.M. and R.M.; Project administration, R.M.; Funding acquisition, S.M. All authors have read and agreed to the published version of the manuscript.

Funding: This work was supported by the Institute Research Fund granted by the British Columbia Institute of Technology.

Data Availability Statement: Data sharing not applicable. No new data were created or analyzed in this study. Data sharing is not applicable to this article.

Acknowledgments: The authors would like to express their sincere gratitude to the British Columbia Institute of Technology for supporting this project. The facilities department and the Energy Team are also acknowledged for their invaluable support, guidance, and assistance throughout the processes of case study selection and data collection.

Conflicts of Interest: The authors declare no conflict of interest.

Appendix A

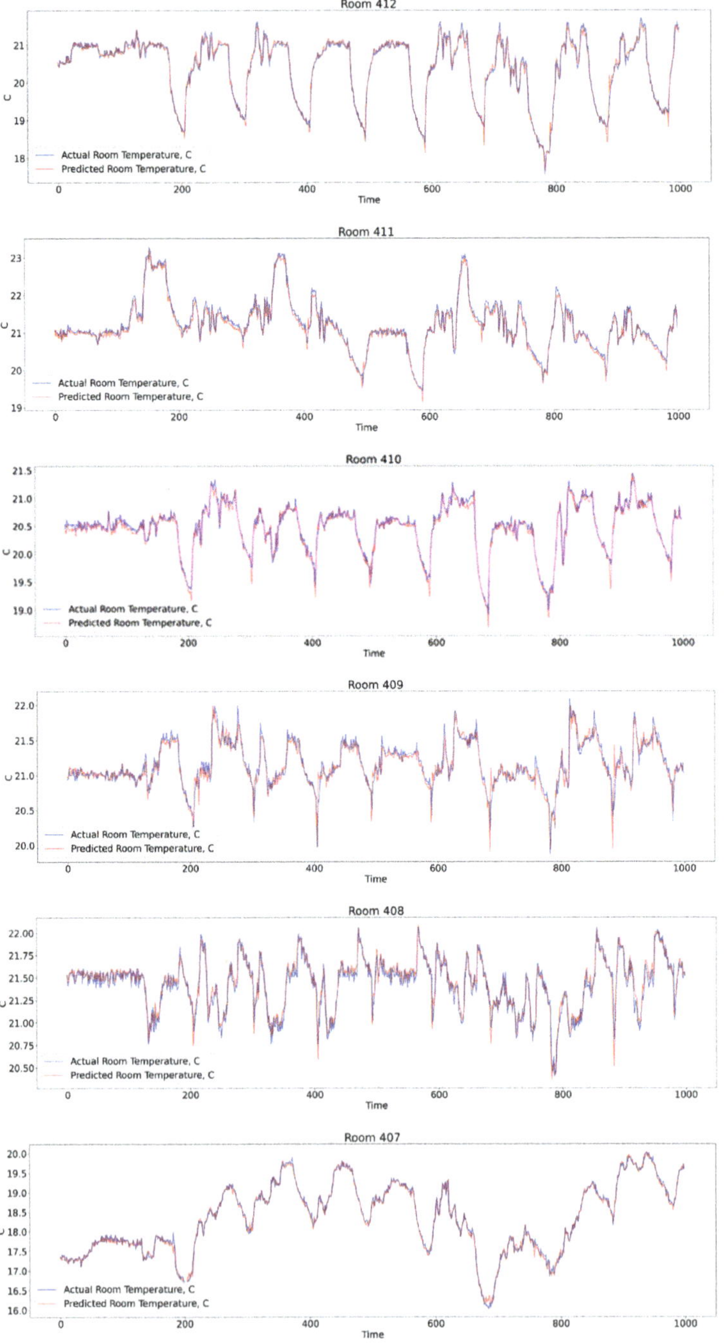

Figure A1. *Cont.*

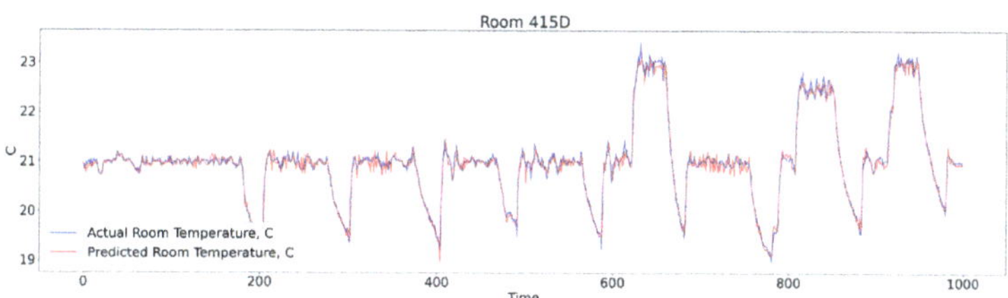

Figure A1. The 15 min ahead and 1 h sliding window predicted indoor temperature for the MLP model.

References

1. International Energy Agency. International Energy Outlook 2006, June 2006. Available online: https://www.iea.org/reports/world-energy-outlook-2006 (accessed on 1 June 2023).
2. Perez-Lombard, L.; Ortiz, J.; Pout, C. A Review on Buildings Energy Consumption Information. *Energy Build.* **2008**, *40*, 394–398. [CrossRef]
3. Natural Resources Canada. *Report Energy Fact Book*; Natural Resources Canada: Ottawa, ON, Canada, 2021.
4. Bourdeau, M.; Zhai, X.Q.; Nefzaoui, E.; Guo, X.; Chatellier, P. Modeling and forecasting building energy consumption: A review of data-driven techniques. *Sustain. Cities Soc.* **2019**, *48*, 101533. [CrossRef]
5. Braun, J.E.; Chaturvedi, N. An inverse gray-box model for transient building load prediction. *HVAC R Res.* **2002**, *8*, 73–99. [CrossRef]
6. Maasoumy, M.; Razmara, M.; Shahbakhti, M.; Sangiovanni-Vincentelli, A. Handling Model Uncertainty in Model Predictive Control for Energy Efficient Buildings. *Energy Build.* **2014**, *77*, 377–392. [CrossRef]
7. Hao, H.; Kowli, A.; Lin, Y.; Barooah, P.; Meyn, S. Ancillary Service for the Grid via Control of Commercial Building HVAC Systems. In Proceedings of the American Control Conference, Washington, DC, USA, 17–19 June 2013.
8. Jeong, K.; Koo, C.; Hong, T. An estimation model for determining the annual energy cost budget in educational facilities using SARIMA, (seasonal autoregressive integrated moving average) and ANN (artificial neural network). *Energy* **2014**, *71*, 71–79. [CrossRef]
9. Zhang, F.; Deb, C.; Lee, S.E.; Yang, J.; Shah, K.W. Time series forecasting for building energy consumption using weighted support vector regression with differential evolution optimization technique. *Energy Build.* **2016**, *126*, 94–103. [CrossRef]
10. Ahmad, M.W.; Mourshed, M.; Rezgui, Y. Trees vs neurons: Comparison between random forest and ann for high-resolution prediction of building energy consumption. *Energy Build.* **2017**, *147*, 77–89. [CrossRef]
11. Huang, Y.; Miles, H.; Zhang, P. A sequential modelling approach for indoor temperature prediction and heating control in smart buildings. *arXiv* **2020**, arXiv:2009.09847.
12. Ogunmolu, O.; Gu, X.; Jiang, S.; Gans, N. Nonlinear systems identification using deep dynamic neural networks. *arXiv* **2016**, arXiv:1610.01439.
13. Ferreira, P.; Ruano, A.; Silva, S.; Conceição, E. Neural networks based predictive control for thermal comfort and energy savings in public buildings. *Energy Build.* **2012**, *55*, 238–251. [CrossRef]
14. Nassif, N.; Moujaes, S. A cost-effective operating strategy to reduce energy consumption in a HVAC system. *Int. J. Energy Res.* **2008**, *32*, 543–558. [CrossRef]
15. Nassif, N.; Kajl, S.; Sabourin, R. Evolutionary algorithms for multi-objective optimization in HVAC system control strategy. *Fuzzy Inf.* **2004**, *1*, 51–56.
16. Nassif, N.; Kajl, S.; Sabourin, R. Optimization of HVAC control system strategy using two objective genetic algorithm. *HVAC R Res.* **2005**, *11*, 459–486. [CrossRef]
17. Platt, G.; Li, J.; Li, R.; Poulton, G.; James, G.; Wall, J. Adaptive HVAC zone modeling for sustainable buildings. *Energy Build.* **2010**, *42*, 412–421. [CrossRef]
18. Goyal, S.; Barooah, P. A Method for Model-Reduction of Nonlinear Thermal Dynamics of Multi-Zone Buildings. *Energy Build.* **2012**, *47*, 332–340. [CrossRef]
19. Wang, S.; Ma, Z. Supervisory and optimal control of building HVAC systems: A Review. *HVAC R Res.* **2008**, *14*, 3–32. [CrossRef]
20. Alawadi, S.; Mera, D.; Fernández-Delgado, M.; Alkhabbas, F.; Olsson, C.M.; Davidsson, P. A comparison of machine learning algorithms for forecasting indoor temperature in smart buildings. *Energy Syst.* **2020**, *13*, 689–705. [CrossRef]
21. Ampomah, E.K.; Qin, Z.; Nyame, G. Evaluation of Tree-Based Ensemble Machine Learning Models in Predicting Stock Price Direction of Movement. *Information* **2020**, *11*, 332. [CrossRef]
22. Manivannan, M.; Behzad, N.; Rinaldi, F. Machine Learning-Based Short-Term Prediction of Air-Conditioning Load through Smart Meter Analytics. *Energies* **2017**, *10*, 1905. [CrossRef]
23. Fan, C.; Xiao, F.; Zhao, Y. A short-term building cooling load prediction method using deep learning algorithms. *Appl. Energy* **2017**, *195*, 222–233. [CrossRef]

24. Kusiak, A.; Xu, G.; Zhang, Z. Minimization of energy consumption in HVAC systems with data-driven models and an interior-point method. *Energy Convers. Manag.* **2014**, *85*, 146–153. [CrossRef]
25. Ben-Nakhi, A.E.; Mahmoud, M.A. Energy conversation in buildings through efficient A/C control using neural networks. *Appl. Energy* **2002**, *73*, 5–23. [CrossRef]
26. Wei, X.; Kusiak, A.; Li, M.; Tang, F.; Zeng, Y. Multi-objective optimization of the HVAC (heating, ventilation, and air conditioning) systems performance. *Energy* **2015**, *83*, 294–306. [CrossRef]
27. Kusiak, A.; Tang, F.; Xu, G. Multi-objective optimization of HVAC system with an evolutionary computation algorithm. *Energy* **2011**, *36*, 2440–2449. [CrossRef]
28. Thomas, B.; Soleimani-Mohseni, M. Artificial neural network models for indoor temperature prediction: Investigations in two buildings. *Neural Comput. Appl.* **2007**, *16*, 81–89. [CrossRef]
29. Mirzaei, P.A.; Haghighat, F.; Nakhaie, A.A.; Yagouti, A.; Giguere, M.; Keusseyan, R.; Coman, A. Indoor thermal condition in urban heat Island—Development of a predictive tool. *Build. Environ.* **2012**, *57*, 7–17. [CrossRef]
30. Attoue, N.; Shahrour, I.; Younes, R. Smart building: Use of the artificial neural network approach for indoor temperature forecasting. *Energies* **2018**, *11*, 395. [CrossRef]
31. Mba, L.; Meukam, P.; Kemajou, A. Application of artificial neural network for predicting hourly indoor air temperature and relative humidity in modern building in humid region. *Energy Build.* **2016**, *121*, 32–42. [CrossRef]
32. Afroz, Z.; Urmee, T.; Shafiullah, G.M.; Higgins, G. Real-time prediction model for indoor temperature in a commercial building. *Appl. Energy* **2018**, *231*, 29–53. [CrossRef]
33. Lu, T.; Viljanen, M. Prediction of indoor temperature and relative humidity using neural network models: Model comparison. *Neural Comput. Appl.* **2009**, *18*, 345–357. [CrossRef]
34. Xu, C.; Chen, H.; Wang, J.; Guo, Y.; Yuan, Y. Improving prediction performance for indoor temperature in public buildings based on a novel deep learning method. *Build. Environ.* **2019**, *148*, 128–135. [CrossRef]
35. Jiang, B.; Gong, H.; Qin, H.; Zhu, M. Attention-LSTM architecture combined with Bayesian hyperparameter optimization for indoor temperature prediction. *Build. Environ.* **2022**, *224*, 109536. [CrossRef]
36. Makridakis, S.; Spiliotis, E.; Assimakopoulos, V. Statistical and machine learning forecasting methods: Concerns and ways forward. *PLoS ONE* **2018**, *13*, e0194889. [CrossRef]
37. Weron, R. Electricity price forecasting: A review of the state-of-the-art with a look into the future. *Int. J. Forecast.* **2014**, *30*, 1030–1081. [CrossRef]
38. Ozbalta, T.G.; Sezer, A.; Yildiz, Y. Models for prediction of daily mean indoor temperature and relative humidity: Education building in Izmir, Turkey. *Indoor Built Environ.* **2012**, *21*, 772–781. [CrossRef]
39. Khalid, S.; Khalil, T.; Nasreen, S. A survey of feature selection and feature extraction techniques in machine learning. In Proceedings of the Science and Information Conference, London, UK, 27–29 August 2014; IEEE: Piscataway, NJ, USA, 2014; pp. 372–378.
40. Galicia, A.; Talavera-Llames, R.; Troncoso, A.; Koprinska, I.; Martínez-Álvarez, F. Multi-step forecasting for big data time series based on ensemble learning. *Knowl.-Based Syst.* **2019**, *163*, 830–841. [CrossRef]
41. Fan, C.; Sun, Y.; Zhao, Y.; Song, M.; Wang, J. Deep learning-based feature engineering methods for improved building energy prediction. *Appl. Energy* **2019**, *240*, 35–45. [CrossRef]
42. Maalej, S.; Lafhaj, Z.; Yim, J.; Yim, P.; Noort, C. Prediction of HVAC System Parameters Using Deep Learning. In Proceedings of the 12th Conference of IBPSA, Ottawa, ON, Canada, 22–23 June 2022.
43. Breiman, L. Random forests. *MLear* **2001**, *45*, 5–32.
44. Qiu, X.; Zhang, L.; Suganthan, N.; Amaratunga, G.A.J. Oblique random forest ensemble via Least Square Estimation for time series forecasting. *Inf. Sci.* **2017**, *420*, 249–262. [CrossRef]
45. Tang, F.; Kusiak, A.; Wei, X. Modeling and short-term prediction of HVAC system with a clustering algorithm. *Energy Build.* **2014**, *82*, 310–321. [CrossRef]
46. Kusiak, A.; Li, M. Reheat optimization of the variable-air-volume box. *Energy* **2010**, *35*, 1997–2005. [CrossRef]
47. Li, D.; Zhang, J.; Zhang, Q.; Wei, X. Classification of ECG signals based on 1D convolution neural network. In Proceedings of the 19th IEEE International Conference on E-Health Networking, Applications and Services, Dalian, China, 12–15 October 2017; pp. 1–6.
48. Wong, F.S. Time series forcasting using backpropagation neural networks. *Neurocumputing* **1990**, *91*, 147–159.
49. Kusiak, A.; Li, M. Cooling output optimization of an air handling unit. *Appl. Energy* **2010**, *87*, 901–909. [CrossRef]
50. Kusiak, A.; Li, M.; Tang, F. Modeling and optimization of HVAC energy consumption. *Appl. Energy* **2010**, *87*, 3092–3102. [CrossRef]
51. Mandic, D.; Chambers, J. Recurrent Neural Networks for Prediction: Learning Algorithms. In *Architectures and Stability*; Wiley: Chichester, UK, 2001.
52. Afram, A.; Janabi-Sharifi, F. Theory and applications of HVAC control systems—A review of model predictive control (MPC). *Build. Environ.* **2014**, *72*, 343–355. [CrossRef]
53. Sülo, I.; Keskin, S.R.; Dogan, G.; Brown, T. Energy efficient smart buildings: LSTM neural networks for time series prediction. In Proceedings of the 2019 International Conference on Deep Learning and Machine Learning in Emerging Applications (Deep-ML), Istanbul, Turkey, 26–28 August 2019; IEEE: Piscataway, NJ, USA, 2019.

54. Yamashita, R.; Nishio, M.; Gian Do Kinh, R.; Togashi, K. Convolutional Neural Networks: An Overview and Application in Radiology. *Insights Imaging* **2018**, *9*, 611–629. [CrossRef] [PubMed]
55. Krizhevsky, A.; Sutskever, I.; Hinton, G.E. ImageNet Classification with Deep Convolutional Neural Networks. *Adv. Neural Inf. Process. Syst.* **2012**, *25*, 1097–1105. [CrossRef]
56. Available online: https://www.bcit.ca/facilities/facilities-services/energy-greenhouse-gas-management/ (accessed on 1 October 2022).
57. Wang, J. *Data Mining: Opportunities and Challenges*; Idea Group Pub.: Hershey, PA, USA, 2003.
58. Aha, D.W.; Bankert, R.L. A Comparative Evaluation of Sequential Feature Selection Algorithms. In *Learning from Data*; Springer: Berlin/Heidelberg, Germany, 1996. [CrossRef]

Article

Buildings' Heating and Cooling Load Prediction for Hot Arid Climates: A Novel Intelligent Data-Driven Approach

Kashif Irshad [1,2,*], Md. Hasan Zahir [1], Mahaboob Sharief Shaik [3] and Amjad Ali [1]

[1] Interdisciplinary Research Center for Renewable Energy and Power Systems (IRC-REPS), Research Institute, King Fahd University of Petroleum & Minerals, Dhahran 31261, Saudi Arabia

[2] Researcher at K.A. CARE Energy Research & Innovation Center, Dhahran 31261, Saudi Arabia

[3] Computer Science & Engineering Department, Ellenki College of Engineering and Technology, Hyderabad 502319, India

* Correspondence: kashif.irshad@kfupm.edu.sa

Abstract: An important aspect in improving the energy efficiency of buildings is the effective use of building heating and cooling load prediction models. A lot of studies have been undertaken in recent years to anticipate cooling and heating loads. Choosing the most effective input parameters as well as developing a high-accuracy forecasting model are the most difficult and important aspects of prediction. The goal of this research is to create an intelligent data-driven load forecast model for residential construction heating and cooling load intensities. In this paper, the shuffled shepherd red deer optimization linked self-systematized intelligent fuzzy reasoning-based neural network (SSRD-SsIF-NN) is introduced as a novel intelligent data-driven load prediction method. To test the suggested approaches, a simulated dataset based on the climate of Dhahran, Saudi Arabia will be employed, with building system parameters as input factors and heating and cooling loads as output results for each system. The simulation of this research is executed using MATLAB software. Finally, the theoretical and experimental results demonstrate the efficacy of the presented techniques. In terms of Mean Square Error (MSE), Root Mean Square Error (RMSE), Regression (R) values, Mean Absolute Error (MAE), coefficient of determination (R2), and other metrics, their prediction performance is compared to that of other conventional methods. It shows that the proposed method has achieved the finest performance of load prediction compared with the conventional methods.

Keywords: energy consumption; data-driven; prediction; building; heating load; cooling load; optimization

Citation: Irshad, K.; Zahir, M.H.; Shaik, M.S.; Ali, A. Buildings' Heating and Cooling Load Prediction for Hot Arid Climates: A Novel Intelligent Data-Driven Approach. *Buildings* **2022**, *12*, 1677. https://doi.org/10.3390/buildings12101677

Academic Editor: Tomasz Sadowski

Received: 16 August 2022
Accepted: 5 October 2022
Published: 12 October 2022

Publisher's Note: MDPI stays neutral with regard to jurisdictional claims in published maps and institutional affiliations.

1. Introduction

The proportion of residence structures has grown during the last ten years of global concern about climate change, worldwide carbon emissions, global warming, urbanization, and rapid construction development [1]. Many procedures and technologies in residential and commercial buildings serve to keep the environment at a pleasant and favorable level, but they cost energy, which adds to the heating and cooling burden [2]. A lot of studies have been conducted on the energy profile of buildings, as well as many elements of efficient building development [3,4]. In Saudi Arabia, numerous residential buildings are attached or semidetached, which require more cooling and heating than ordinary flat residences [5]. Temperature, humidity, the operations of sunlight devices, and the construction and design elements of buildings all have a role in the heating and cooling of structures [6].

The material used in wall surfaces, the relative compactness of building structures, the glazed windows region, the ceiling dimensions, the outer layer and density of the building, the outer layer and density of the wall, the roof height, the number of wall surfaces and their region, the orientation of the halls and the building, and the stand over height are all involved in construction and relate to the environment. However, several aspects of the building design and layout have a significant effect with regard to the building's warming and chilling load, which has a direct impact on the building's overall performance [7]. Thus, by alleviating the computational load for optimum design, which includes multiple building feature subsets, an accurate and rapid forecast of space heating as well as cooling loads improves energy saving and carbon emission mitigation [8]. Developing machine learning techniques for predicting heating and cooling needs can assist in improving the effectiveness and precision in real time [9]. Conventional heating and cooling predictive modeling algorithms include the minimax probability machine regression (MPMR) [10], deep neural network (DNN) [11], Gaussian process regression (GPR) [12], and gradient boosted machine (GBM) [13]. Moreover, artificial neural networks (ANN) [14], categorization and regression trees (CART) [15], general linear regressions (GLR) [16], and chi-squared automated interaction detectors (CHAID) [17] were used to forecast the cooling and heating requirements of the building.

As a result, various hybrid methodologies based on ANN and meta-heuristic schemes have been developed for forecasting a building's heating and cooling demand efficiency, including the imperialist competitive algorithm (ICA) [18], artificial bee colony (ABC) [19], genetic algorithm (GA) [20], whale optimization [21], bat optimization [22], and particle swarm optimization (PSO) [23]. However, such factors have been utilized in certain studies with little benefit. Meteorological parameters were employed as an indication and input for estimating apartment building cooling/heating demands in the majority of the preceding academic studies [24]. Environmental and climatic conditions do not affect the cooling/heating loads of residential construction; this is indisputable. However, sudden weather changes might cause sustainable models to be disrupted, lowering the reliability factor and enhancing the error in the process [25].

This research focuses on the construction and design characteristics of the building, as well as their effects on heating/cooling loads. Besides constructing supervised classification forecasting models, the research applied in-depth testing on structural features for building energy. The quantity of the cooling/heating load was regarded as an outcome parameter, although a collection of information on the structural attributes of the structure was regarded as an input parameter. The following are the main research contributions:

- A dataset was produced in the Dhahran area of Saudi Arabia for estimating power requirements based on building attributes in a dry climate.
- After collecting the data, the preprocessing and feature extraction function is applied for improving the prediction model using a knowledge-based approach.
- Then, to predict building heating/cooling demands, the SSRD-SsIF-NN technique is presented with various parameter tunings.
- Building energy demand simulations are conducted to anticipate heating and cooling demands in dry climates. In contrast to various studies, the prediction methods depend on the characteristics of the study instead of on previous results on energy usage.
- A simulation analysis is conducted by varying the input parameters.
- The actual performance and the theoretical load prediction of the existing structure are compared.

The rest of the essay is organized as follows. The definition of investigation gaps is given in Section 2, along with a short assessment of the relevant publications. Section 3 describes how the issue is stated. Section 4 includes detailed explanations of the suggested technique. Section 5 discusses the experimental outcomes as well as the efficiency comparison with state-of-the-art frameworks. Section 6 is the paper's conclusion.

2. Related Work

The heating/cooling loads in buildings are closely connected to energy performance, and various studies have been undertaken in this area. Because cooling/heating loads are considered key factors for examining building energy efficiency, the necessity to anticipate and assess them for residential structures appears to be unavoidable. As a result, Xu, Yuanjin, Fei Li, and Armin Asgari [26] sought to optimize the multi-layer perceptron-based neural network utilizing a variety of optimization techniques in order to anticipate the heating/cooling of energy-efficient architecture. The database used for this investigation is made up of eight different variables, such as total area, space limitations, wall region, and so on. Optimizing has the highest accuracy in both the learning and test data for cooling/heating loads. The normed RMSD, RMSD, and MAE have the lowest values, as well as adjusted R2, as per the study findings.

The multi-target forecasting of heating/cooling loads through HVAC systems uses hybrid intelligent methodologies: the wind-driven-based optimization (WDO), grasshopper optimization algorithm (GOA), and biogeography based optimization (BBO) were employed by [27]. Some swarm-based rounds are carried out to optimize the applicable methods, and the optimum design for each simulation is provided. In regard to the heating load, the suggested WDO-ANN offered an accurate forecast, while in terms of cooling capacity, it provided the finest forecast.

While earlier research has focused on point forecasts, Rana and Mashud [28] focused on predicting prediction pauses for the building cooling/heating load in this work. A data-driven technique for predicting prediction periods for building cooling demand is provided here, which initially employs a machine learning subset of feature techniques to find a limited but useful collection of factors. The findings demonstrate that the suggested method may yield a narrow and trustworthy forecasting period while fulfilling the penetration probabilities that have been defined.

To determine the energy requirement of the structures for heating and cooling, Li, Xinyi, and Runming Yao [29] combined the physical analytical model with the data-driven technique. The severity of heating/cooling energy consumption was then predicted using a variety of machine learning algorithms (EUI). The findings reveal that machine learning methods can accurately estimate building heating and cooling EUI. At the single-cell level, the most accurate method is quadratic kernel-based supported vectors extraction, while the Gaussian perceptron support-vector training has the highest accuracy at the inventory levels. Kim, Daeung Danny, and Hye Soo Suh [30] used the statistical technique to design a forecast model for energy usage in residential structures. The links between the design elements and heating/cooling load of energy usage in residential structures were detailed utilizing the response surface approach. To establish a prediction model for the heating/cooling load of energy usage, the connection has validated the dependencies of the energy consumption on key design factors of exterior technologies in residential structures. With only a few design factors, the created model can provide a quick energy estimate for apartment structures. Additionally, it may quickly determine the most significant design component for creating a more efficient energy residential building layout.

Tran et al. [31] developed an evolutionary Neural Machine Inference Model (ENMIM) for predicting energy usage using actual data from residential structures. Their novel ensembles model combines the Radial Basis Function Neural Network and the Least Squares Support Vector Regression (LSSVR), two separate supervised learning devices (RBFNN). For forecasting resource usage, the created model is more accurate than previous comparable artificial intelligence systems.

In the study that was conducted by Zhou et al. [32], the artificial bee colony (ABC) and particle swarm optimization (PSO) metaheuristic algorithms were used to optimize the MLP neural network. This was done in order to make accurate predictions regarding the heating and cooling loads of energy-efficient buildings that are used for residential purposes. In order to do this, they made use of a dataset that had eight independent variables. According to the findings of their study, making use of the ABC and PSO algorithms makes the MLP perform better. In addition to this, they came to the conclusion that, in terms of MLP performance improvement, PSO was superior to ABC.

In order to study how well machine learning can estimate the heating and cooling loads of buildings, Seyedzadeh et al. [33] created two datasets using two distinct kinds of modelling software. In order to investigate the different permutations of the model parameters, a gridsearch and a cross-validation approach were used. The findings revealed that, among the five models that were investigated, the Gradient Boosted Regression Trees

(GBRT) deliver the most accurate forecast predictions depending on the RMSE. On the other hand, NNs were shown to be the most effective at dealing with complicated datasets.

Sharif and Hammad [34] centered on the creation of an ANN model with the goal of predicting energy consumption using a big and difficult dataset supplied by the SBMO model. According to the results of this research, the ANN models that were recommended were able to yield accurate predictions. These scenarios included the building envelope, HVAC, and lighting systems.

In a work by Singaravel et al. [35], 201 design scenarios were used to evaluate the deep learning model against a simulation of a building's functionality. With an R2 of 0.983, the deep learning model demonstrated remarkable accuracy for cooling predictions. With additional heating data from a more accurate sample model, the model's R2 of 0.848 inaccuracy for heating predictions may be eliminated. According to the research, deep learning simulation results may be obtained in 0.9 s, which is regarded as a high calculation speed for simulating structure efficiency.

The cooling demand of big industrial structures was predicted by Gao et al. [36] using a hybrid forecasting model based on the random forest-improvement parallel whale optimizing-extreme learning machine neural network (RF-IPWOA-ELM). The experimental findings demonstrate that the RMSE and MAPE of the RF-IPWOA-ELM model accurately estimate the cooling demand for these two structures. The suggested hybrid model may be used as a trustworthy tool for cooling load prediction in the administration and energy saving of air conditioning systems.

Wei et al. [37] used seven common machine learning techniques to determine the best prediction method in a local heating load sample from Shanghai, China. To evaluate the model's effectiveness, data from a power transmission sensor, a heat sensor, and the current weather are merged into several input options. The findings demonstrate that SVR outperforms all others in MAPE. Further analysis reveals that the continual lengthening of previous datasets does not affect performance. The preceding literature analysis demonstrates the successful application of data-driven algorithms to handle building heating/cooling load forecast issues. Nevertheless, subsequent research deficiencies have been identified:

- The majority of the research is being undertaken in non-arid regions such as Canada, Greece, the United States, and China.
- The benchmark dataset of the conventional technique is used in many studies that employ building attributes as inputs to the forecasting model. Some self-generated statistics are kept secret and cannot be replicated experimentally.
- There is some advice on how to utilize deep learning techniques and how to modify them for the highest predicted accuracy and completeness for the assigned task. While most research includes machine learning techniques, deep learning, as well as optimization, are infrequently employed.
- As a result, utilizing a bigger, accessible self-generated database in Dhahran, a typical desert climatic zone, this research sets out an applicable strategy for adapting smart data-driven frameworks to building energy performance data.

3. Problem Statement

Predictive cooling/heating load is an effective method for ensuring future energy use. A significant number of academics are investigating the strategies and models for predicting cooling/heating demand in green buildings using machine learning and artificial intelligence techniques. Due to numerous issues that the researchers encountered, linked to building features, weather conditions, and data produced by the process control itself, the proposed methodologies differ in their ability to produce precise and reliable outcomes. Because linear regression is often utilized, it is more difficult to describe the function that connects to the aforementioned issue than the cooling/heating load. Furthermore, the non-linear structure of building systems complicates the connection.

The forecasting models require a lot of specific details from the building features, which might be difficult to assess and calculate. The operations and conduct of the residents within the building are predictable, since their behavior does not follow a specific order and varies irregularly. Thus, this research proposed an intelligent data-driven approach to anticipate the cooling and heating load in Dhahran buildings in response to the aforementioned issues.

4. Proposed Framework

This research is focused on thermal load characteristics for residential structures, as their real-world activities are heavily influenced by building design requirements. Building design rules attempt to decrease energy usage by taking into account two main terms of heating/cooling load. The proposed framework of a predictive model is illustrated in Figure 1. Some procedures were followed in the current investigation. The factors of the residential building layout were first discovered. The buildings in Dhahran, Saudi Arabia, were chosen for the data collection on design characteristics. The initial step in data preparation is to filter the data. Furthermore, by obtaining relevant features for validation, feature extraction methods may be utilized to improve the prediction performance of the proposed model. A prediction model for the cooling and heating load of the development of a domestic structure's energy usage uses SSRD-SsIF-NN. Following that, the proposed models anticipate the study's outputs of heating and cooling demand. In the last phase, the suggested models' error efficiency is measured using the leftover 30% of the test data depending on the discrepancies between the true calculated data and the anticipated values derived from the developed model. Then, the performance analysis is performed for the effective measurements.

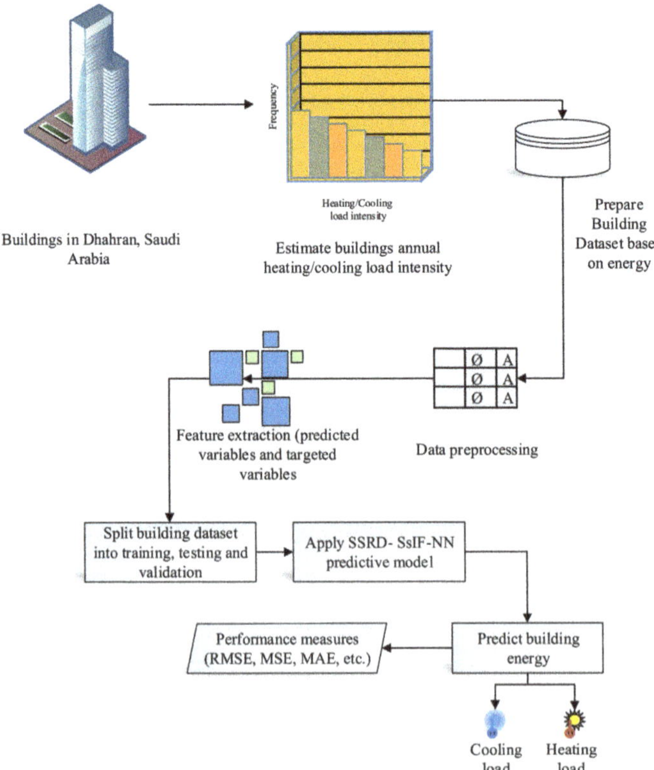

Figure 1. Proposed framework of the predictive model.

4.1. Pre-Processing and Feature Extraction

Filtering the data is the initial step in data preparation. The information is organized by the date-time index in order of decreasing importance. Data that are erroneous, anomalous, or duplicated are found and deleted. The missing data are again filled using knowledge-based interpolation. The method of translating raw information into information features that can be examined while keeping the information from the source dataset is carried out by feature extraction. It produces better outcomes than applying prediction techniques to raw information automatically. By splitting the inputs and predicting the aim, a singular sample is produced after the feature extraction procedure.

The acquired dataset of the heating and cooling loads is arbitrarily split into two different portions—the training data and the testing data—in this stage. Additionally, 70% of the entire data are utilized for building the model in order to develop a good prediction model and to develop the link across both the heating and cooling objectives and their significant components, according to a well-delivered train/test database selection. In order to test and validate the model, the other 30% of the information would be employed.

4.2. SSRD-SsIF-NN-Based Prediction

A quantitative performance indicator is assessed for validity. SSRD-SsIF-NN is the combination of a self-systematized intelligent fuzzy reasoning-based neural network and a hybrid meta-heuristic optimization approach. The parameters of SsIF-NN are tuned by the SSRD optimization technique.

4.2.1. SsIF-NN Methodology

These tools are frequently used to represent difficult engineering problems. By generating non-linear relationships, this artificial intelligence (AI)-based approach will attempt to build a link between a sequence of given input layers and one or more output neurons. A fuzzy inference system layer, four hidden layers, and a defuzzification layer make up the SsIF-NN structure. Figure 2 depicts the suggested predictive model. The fuzzification layer transforms the feature selection's sharp input into a fuzzy collection of values. Floors Area, Number of flats, Gross Area, Roof Area (m^2), Study Area, Module Orientation, Parapet Wall Height (m), Annual Consumption (kWh), and Utilization Factor are the inputs of the SsIF-NN structure. The heating and cooling load is the output of the proposed approach. The input activation function and layer result were both specified in Equations (1) and (2), correspondingly.

$$Input = a\left[z_1^{(s)}, z_2^{(s)}, \ldots z_n^{(s)}; b_1^{(s)}, b_2^{(s)}, \ldots .b_n^{(s)}\right] \tag{1}$$

$$Output = F_o^{(s)} = f_a(input) = f_a^{(a)} \tag{2}$$

where the inputs to this unit are $z_1^{(s)}, z_2^{(s)}, \ldots \ldots z_n^{(s)}$ and the link weights are $b_1^{(s)}, b_2^{(s)}, \ldots \ldots b_n^{(s)}$. The layer number is denoted by $f_a^{(\cdot)}$ and the superscript in the overhead equation is denoted by (s). The activation function is described as follows: each node's second function is to create an activation value based on its primary input.

The following six stages of the prediction model are explained. There are no computations performed by this layer. This layer's terminals, each of which corresponds to a certain input factor, only transmit data to the following layer. This is accurate, and the first layer connection weight factor is $\left[b_i^{(1)}\right]$ one, according to Equation (3).

$$a = z_1^{(s)} \ and \ f_a^{(1)} = aa = z_1^{(s)} \ and \ f_a^{(1)} = u \tag{3}$$

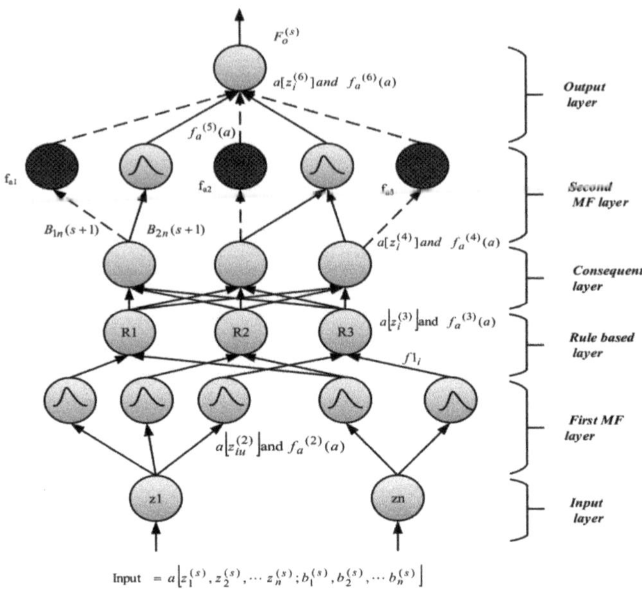

Figure 2. The proposed model of the SsIF-NN method.

Fuzzification is achieved in the second layer by finding the membership function parameters of an input to a group of Gaussian MFs. Each component in this provided a good one to one of the linguistic values of the input variables in the first layer (medium, small, large, etc.). This research makes use of a Gaussian membership characteristic, which has been proved to be a global prediction technique of any dynamic function on the basis of Equation (4).

$$a\left[z_{iu}^{(2)}\right] = -\frac{\left[z_i^{(2)} - f_{iu}\right]^2}{\sigma_{iu}^2} \ and \ f_a^{(2)}(a) = t^a \tag{4}$$

where the Gaussian MF of the u_{th} element of the i_{th} input factor has a mean and variance of f_{iu} and σ_{iu}, respectively. As a solution, the weight of a second-layer link may be expressed as f_{iu}. Equation (5) may be used to compute the normalized fuzzy closeness between a fresh fuzzy sample z_{1f} and u_{th} the stored characteristic $L_1(u)$,

$$N_u = \frac{\|z_{1f} - L_1(u)\|_g}{\sum_{u=1}^n \|z_{1f} - L_1(u)\|_g} \tag{5}$$

Here, g-norm is the abbreviation for $\|.\|_g$. The g-norm $\|d\|_{g+z} \leq \|d\|_g$ for $d \in \Re^n$, $u \geq 1$, $g \geq 0$ of each given vector $\|d\|_g$ does not expand with g; all other norms are lower-bounded by the 1-norm. As a consequence, the Euclidean system was implemented $g = 2$. Radial basis models may also be used to determine rule neuron activation thresholds. Equation (6) is used in this section.

$$f1_i = 1 - N_u \tag{6}$$

where $f1_i$, $N_u \in [0,1]$. This threshold controls the model's sensitivity to the generation or modification of rule neurons $q \in [0,1]$. With higher scores, larger numbers of hidden neurons and attributes are feasible. Assume that the threshold value is set at 0.3 by default, which was created in each of the iterations. If the number of neurons develops at a faster pace than the set rate, the number of neurons grows at a slower rate $F > \varepsilon$, and the number of neurons increases at $F > \varepsilon$

$$\text{If } F > \varepsilon \text{ , } q \text{ is reduced, } q(n+s) = \left[1 + \frac{F - \varepsilon}{s}\right] q(n) \tag{7}$$

$$\text{If } w < \beta, \ q \text{ is increased, } q(n+s) = \left[1 + \frac{\varepsilon}{s}\right] q(n) \tag{8}$$

This layer's nodes also each have one fuzzy inference system rule and execute prerequisite testing. For the third layer part, the AND function was utilized, as seen below

$$a\left[z_i^{(3)}\right] = \Pi_i z_i^{(3)} = t^{-[R_i(z-f_i)]q[R_i(z-f_i)]} \text{ and } f_a^{(3)}(a) = t^a \tag{9}$$

The number of second layers is stated as being engaged in the IF component of the fuzzy rule, and the diagonals are written as

$$R_i = d\left(\frac{1}{\sigma_{i1}}, \frac{1}{\sigma_{i2}}, \ldots \ldots \frac{1}{\sigma_{in}}\right) \text{ and } f_i = d(f_{i1}, f_{i2}, \ldots \ldots f_{in})^T$$

In the third layer, there is just one weight connection, $\left[b_i^{(3)}\right]$. The firing intensity of the connected fuzzy rule is reflected in the third layer consequences. The subsequent layer has the same number of components as the third layer, the firing strength estimated in the third layer is normalized in this layer by Equation (10), and the weighted link in the fourth layer is also one $\left[b_i^{(4)}\right]$.

$$a\left[z_i^{(4)}\right] = \sum_i z_i^{(4)} \text{ and } f_a^{(4)}(a) = \frac{z_i^{(4)}}{t^a} \tag{10}$$

A discriminative method that estimates the probability and a sample from training data rather than the prediction model delivers superior results that accurately depict the data distribution. The purpose of generative training is to reduce the chances of people making poor decisions. The second MF layer has two distinct modes, which are shown in Figure 2 as blank and shaded circles, respectively. The basic node, denoted by empty circles, is a fuzzy set specified by the Gaussian membership degree of the outcome variable. In the local mean of maximum (LMOM)-based defuzzification approach, the center of each Gaussian membership value is simply relayed to the next layer, while the width is just employed for output grouping. The activation of the winning neuron is propagated by Equation (11), using a saturated scaling factor of the type

$$G_{max} = \begin{cases} 0 & if\, G(F_{max})B_2 < 0 \\ 1 & if\, G(F_{max})B_2 > 1 \\ G(F_{max})B_2 & otherwise \end{cases} \tag{11}$$

Furthermore, G_{max} is the neuron with the highest membership value and $G(F_{max})$ is the activation. The error e^* among the actual fuzzy outcome vector z_{1f} and $G(F_{max})$ is contrasted to a threshold q. If the mistake exceeds the threshold, a rule neuron is formed. Meanwhile, the leading neuron weight parameters w_1 and w_2 are generated from Equation (12) for the shaded and blanked portion,

$$B_{1n}(s+1) = B_{1n}(s) + \mu_1(z_i - B_{1n}) \tag{12}$$

$$B_{2n}(s+1) = B_{1n}(s) + \mu_2 G_{max} c^* \tag{13}$$

Here, μ_1 and μ_2 are the constant learning values, and z_i is the ith input vector. The same fuzzy numbers can be given for various rules if many fourth-layer terminals are linked to the same fifth-layer empty element. By integrating these two components in the fifth layer, the whole function provided by this layer can be explained.

$$f_a^{(5)}(a) = \left(\sum_i z_{iu} z_u + f_{a0j} + B_{1n}(s+1) + B_{2n}(s+1) \right) z_j^5 \tag{14}$$

The mean of the Gaussian MF is expressed. Only when the shaded element is necessary is it produced. The summing is over the important phrases associated with the darkened node alone, and q_{xj} is the relevant variable. This layer's nodes each relate to a single output variable. The node collects all fifth-layer ideas and functions as a defuzzifier, predicting the proper outcomes.

$$a\left[z_i^{(6)} \right] = \sum_i z_i^{(6)} \text{ and } f_a^{(6)}(a) = a \tag{15}$$

The output of the proposed algorithm provided the consequences of the heating and cooling load in residential buildings.

4.2.2. SSRD of Parameter Tuning

The SSRD optimization algorithm is a combination of the shuffled shepherd and red deer optimization algorithms. The fitness of both algorithms is considered for the parameter tuning of the proposed SsIF-NN predictive model in the residential building energy load. The purpose of optimization is to create a global solution that takes into account all of the problem's factors. Figure 3 also displays the flowchart for the suggested prediction system. The values of the fuzzy variable and t^a parameters are to be optimized in this case.

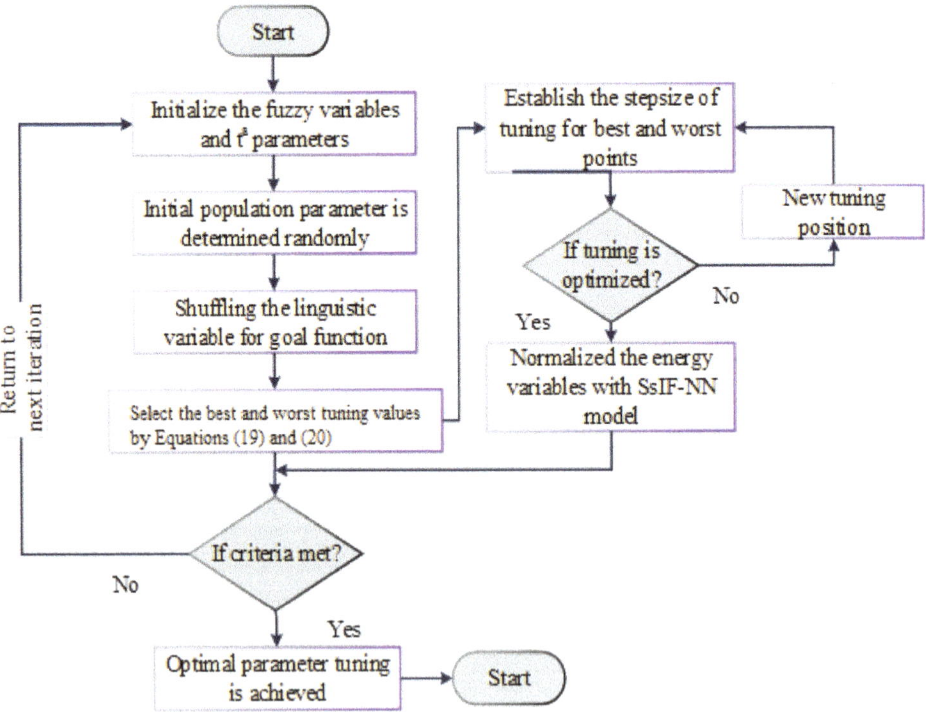

Figure 3. The flowchart of the proposed SSRD for SsIF-NN parameter optimization.

Initialization: The algorithm is initialized; the parameters in the array form Equation (16):

$$p(r) = f(t_1, t_2, \ldots \ldots t_n) \tag{16}$$

In the solution space, the mathematical analysis initiates SSRD with a randomly determined beginning population parameter:

$$T_{q,w}^0 = T_{min} + ran \times (T_{max} - T_{min}); \quad q = 1,2,\ldots\ldots x \text{ and } w = 1,2,\ldots\ldots y \tag{17}$$

where T_{min} and T_{max} are the lowest and maximum fuzzy model parameter bounds, respectively; ran is a random variable formed between 0 and 1 for each element; x is the number of persons in each parameter group; and y is the total number of parameters in the groups.

Shuffling: According to their goal features, the initial point x of each population is randomly placed in the first column of the cross conditions (Equation (18)), as between members of each population. The subsequent members x are chosen in the same way as the previous step and are organized in a random order in the section to create the second column of the multi-community parameter. This procedure is continued y until the following multi-community matrix is created:

$$T_p = \begin{bmatrix} T_{1,1} & T_{1,2} & T_{1,y} & T_{1,y} \\ T_{2,1} & T_{2,2} & T_{2,y} & T_{2,y} \\ T_{q,1} & T_{q,2} & T_{q,w} & T_{q,y} \\ T_{x,1} & T_{x,2} & T_{x,w} & T_{x,y} \end{bmatrix} \tag{18}$$

It is worth noting that each row of the multi-community parameter reflects an individual from each group, with the top column being the best values from each group. Furthermore, the persons in the last segment are the weakest variables in the group.

Optimal value selection: After shuffling the variables, the best and worst values are selected for the finest tuning of the SsIF-NN model. Equations (18) and (19) are used to calculate the worst and best functions of the step size for adjusting the parameter

$$S_q^w, w = \alpha \times ran_1 \times (T_{q,w} - T_{q,w}) \tag{19}$$

$$S_q^f, w = \beta \times ran_2 \times (T_{q,f} - T_{q,w}) \tag{20}$$

Compared to $T_{q,w}$, ran_1 and ran_2 are random variables, with each component formed between 0 and 1, respectively; $T_{q,f}$ and $T_{q,w}$ are the superior and worse variables in terms of optimal value. To establish a specific step size for each member of the group, two factors are utilized. The functional form for the step size is as follows:

$$S_{q,w=S_{q,w}^w + S_{q,w}^f} \ldots q = 1,2,\ldots\ldots x \text{ and } w = 1,2,\ldots\ldots y \tag{21}$$

The potential to explore more areas of the solution space is shown by the first variable S_q^w, w. The capacity to explore the surroundings of previously visited prospective solution space portions of the intensification approach is the second variable S_q^f, w.

It is worth noting that the xth community's initial parameter $T_{q,1}$ lacks an affiliate that is superior to it. As a result, S_q^f, w has the same value as 0. As a result of the xth group's final parameters, $T_{q,y}$ does not have a worse parameter than itself. As a result, S_q^w, w is also zero. Furthermore, α and β are the variables that have an impact on both exploration and exploitation.

New Tuning Position: If the neighbors' objective functions are better than the attained fuzzy values, the fuzzy values are replaced with the preceding ones. Allow all fuzzy values to modify their positions in reality. The following equation is presented to update the location of the fuzzy value:

$$S_{new} = \begin{cases} S_{old} + t_1 \times ((U - L) * t_2) + L) & \text{if } t_3 \geq 0.5 \\ S_{old} - t_1 \times ((U - L) * t_2) + L) & \text{if } t_3 < 0.5 \end{cases} \tag{22}$$

Limit the search field to where and when older value neighborhood responses are appropriate. As a result, they are the upper and lower boundaries of a random search, U and L. S_{old} denotes the current fuzzy scenario, whereas S_{new} denotes the modified position. Homogeneity between 0 and 1 is employed to develop t_1, t_2 and t_3 for the randomization of nature's tuning mechanism.

Normalized Energy variables: The following expression can be used to calculate the general normalized power.

$$E_k = \left| \frac{P_k}{\sum_{i=C_1}^{N} P_i} \right| \tag{23}$$

where p_k is the energy of the kth main node and N_c is the number of variables. To find the nearest hind, divide the distance among an MF, which is estimated with the ith dimension.

Termination: After a chosen maximum iteration, the optimization procedure will be completed. If it is not, it goes back to step one for another round of repetitions.

5. Results and Discussion

The proposed prediction models were developed using the MATLAB 2019b software program and settings on a desktop PC with an Intel Core i-7 9700K processor, 16 GB RAM, and a 3.6 GHz clock speed, as well as the Windows 10 64-bit operating platform.

5.1. Case Study

The district meteorological station has been established in Al-Dhahran. The research employed the use of climatic variables from Al-Dhahran. In the Saudi Arabian metropolis of Al-Dhahran, the study area covers more than 100 km^2 and includes 33,000 residential properties. For more than a 38-year span, the average global temperature data were obtained. The hottest month is July, with the maxima reaching 49 °C and a mean high temperature of 43 °C. The coldest month is January, with a mean low temperature of 11 °C. Summers in Al-Dhahran are hot and muggy, with an estimated average of 100% relative humidity (RH) ranging between 61 and 90 percent and the daily total minimum RH ranging between 15 and 46 percent over the year. During the year, the area features clear skies with rare sandstorms that reduce sun irradiation. With a total surface area of 254 m^2, a 1.7 m-high parapet wall, and a PV utilization ratio of 0.13, the roof is rectangular. In addition, site inspections were made to further understand the roof's qualities and the nearby regions of 70 model buildings. To understand the differences in building features, specimens from several residential districts around the city were tested. The database includes 70 samples, each with nine characteristics, namely, z1, z2, z3, z4, z5, z6, z7, and z8, along with b1 and b2 as objective functions (see Table 1). Using the preceding properties as objective functions, this study tries to predict y1 as the heating load and y2 as the cooling load.

Table 1. Description of the case study's input and output data.

Variables	Symbols	Values
Floors Area	z1	504 m^2
Number of Flats	z2	14
Gross Area	z3	254 m^2
Roof Area (m^2)	z4	254
Study Area	z5	100 m^2
Module Orientation	z6	Main elevation facing east
Parapet Wall Height (m)	z7	1.7
Annual Consumption (kWh)	z8	188,740
Utilization Factor, UF	z9	0.3
Cooling Load	b1	-
Heating Load	b2	-

5.2. Performance Analysis

The NMAE and NRMSE are context-independent and may be used to compare the performance of the model on building heating and cooling load strength with various input ranges. Smaller numbers for RMSE, like MSE, imply a better performance of the model. The correlation between the actual and predicted variables is measured by the R value. The nearer the R value is to 1, the greater the association is and the better the model's effectiveness is. R2 measures how much variance in the relying factor can be anticipated from the independent factors. The nearer the R2 number is to 1, the greater the relative value is and the better the model's performance is. The preceding formulae are used to compute each of the categories referenced, where a_p and b_p are the actual and anticipated values for sample p, respectively. Furthermore, $\overline{a}$ and $\overline{b}$ show the average of the actual and desired heating/cooling load intensity for a building, where M is the representative sample and $\overline{A}$ is a simulation run in MATLAB that represents the average of the building's initial heating and cooling load strength.

$$R = \frac{\sum_{p=1}^{M} \left(a_p - \overline{a}\right)\left(b_p - \overline{b}\right)}{\sqrt{\sum_{p=1}^{M}\left(a_p - \overline{a}\right)^2 \sum_{p=1}^{M}(b_p - \overline{b})^2}} \tag{24}$$

$$R^2 = 1 - \frac{\sum_{p=1}^{M}\left(a_p - b_p\right)^2}{\sqrt{\sum_{p=1}^{M}(a_p - \overline{b})^2}} \tag{25}$$

$$\text{RMSE} = \sqrt{\frac{1}{M}\sum_{p=1}^{M}\left(a_p - b_p\right)^2} \tag{26}$$

$$\text{MSE} = \frac{1}{M}\sum_{p=1}^{M}\left(a_p - b_p\right)^2 \tag{27}$$

$$\text{MAE} = \frac{1}{M}\sum_{p=1}^{M}\left|a_p - b_p\right| \tag{28}$$

$$\text{NRMSE} = \frac{RMSE}{\overline{A}} \tag{29}$$

$$\text{NMAE} = \frac{MAE}{\overline{A}} \tag{30}$$

During the training stage, Figure 4 shows a good correlation coefficient between the actual values and the projected value for the presented approach. Considering the massive correlation between the goal data and each channel's outcome, it is evident that all of these systems survived the training phase with excellent grades. Great training implies that the system can recognize statistical properties in the types of information and forecast new data using the learned structures, allowing each system to learn how much cooling and heating load is necessary for every building with unique features. Each model can anticipate the quantity of cooling and heating loads based on the test stage's data input with some of this training.

Each model is verified by early test data once it has been trained (30 percent of 100 percent). This is a form of a practice run for the training stage, which is handled entirely by the system. Figure 5 shows the forecast error in the testing or validation stage, which is one of the foremost essential metrics in assessing the outcomes, in a histogram manner, for the SSRD-SsIF-NN. The minimal and largest prediction errors are indicated by the error histogram framework's error. This indicates that the number of inaccuracies that all of the trained systems can have in forecasting the cooling and heating loads for such a testing set is equivalent to the quantity stated in the statistics.

It can be inferred by examining and analyzing each of the above statistics, which reflect the effectiveness of each system throughout the first training stage as well as the testing stage, that the suggested techniques' training has been well verified using the needed data. It is worth mentioning that, whenever a model is trained with extreme accuracy, it is well built, and the number of failures in the validation as well as the first testing procedure is more dependent on the data quality. It also indicates that the system will be able to examine and forecast new and untested data with accuracy. During training, each system is preserved as a black box. During the training stage, the system was able to recognize trends inside this black box. New and untested data must first be utilized to evaluate these systems and forecast cooling and heating loads for buildings. To accomplish this, 15% (five samples) of the information, which was preserved as unidentified and unique data, was employed. Figure 6a,b demonstrate the outcomes of predicting heating and cooling loads for updated information by trained SSRD-SsIF-NN models.

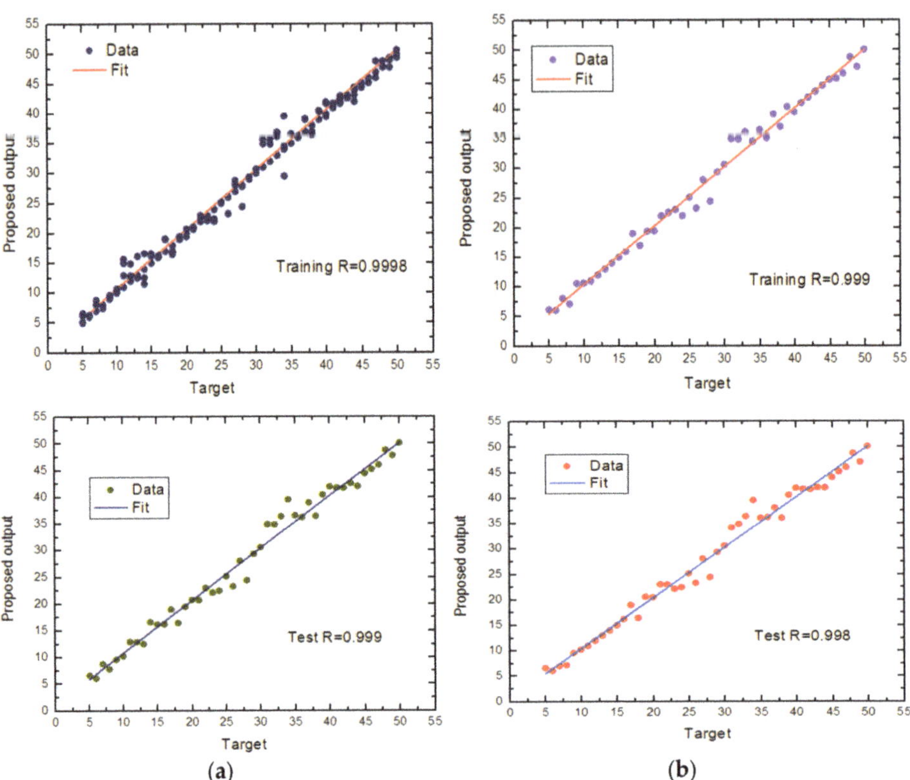

(a)   (b)

Figure 4. The training and testing phases of the proposed model's correlation coefficient. (**a**) Heating load, (**b**) cooling load.

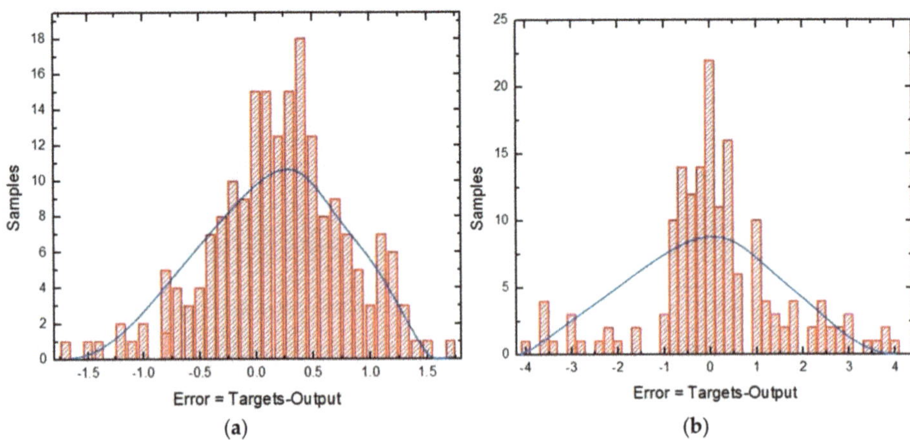

(a)   (b)

Figure 5. Heating load and cooling load histogram testing error. (**a**) Heating load, (**b**) cooling load.

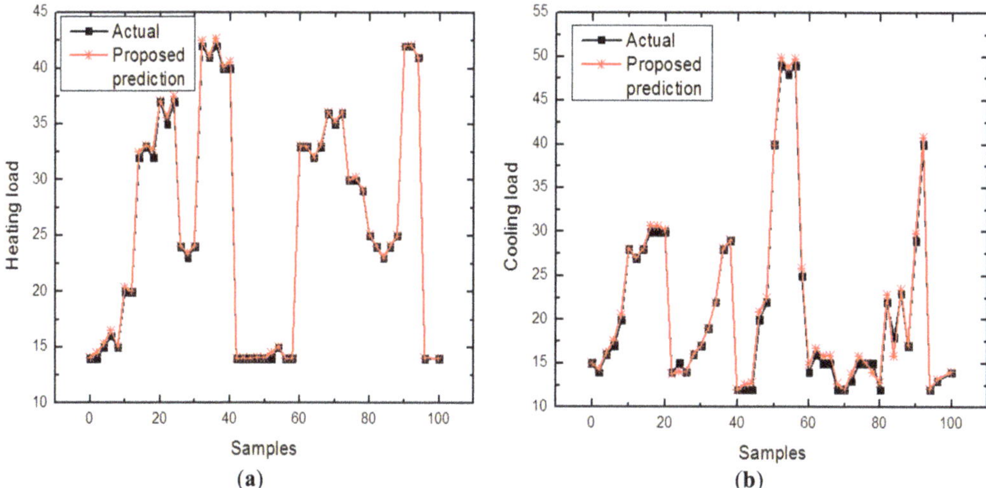

Figure 6. Forecasting heat load and cooling load using the proposed method. (**a**) Heating load, (**b**) cooling load.

In the log scale, Figure 7 shows the MSE effectiveness of the proposed models for the training, validation, and test datasets. The best performing network on the validation dataset is the completed system. As a result of training, the system can now predict simultaneous heating and cooling needs. The MSE of the developed framework reduced quickly, resulting in lower error levels. Using the validation sample, the suggested framework predicted the heating load with an MSE of 0.01530 at epoch 115 (Figure 7a) and the cooling load with an MSE of 0.148 at epoch 110 (Figure 7b).

Figure 8 depicts the forecasting accuracy of models tuned using training/testing sets with various sample sizes. When the sample size is increased from 50 to 250, the criteria for evaluating prediction performance drop considerably. When the sample size exceeds 250, meanwhile, the process of diminishing prediction standard evaluation metrics decreases or, indeed, reverses.

Table 2 shows the R, R2, RMSE, MSE, MAE, NRMSE, and NMAE performance evaluations of the proposed approaches. The best prediction was connected to the forecast of the heating load by the suggested technique, which had the greatest value of R (0.9998) and the lowest errors of MSE (0.01530), RMSE (0.21), MAE (0.2), NRMSE (1.5), and NMAE (0.2).

The suggested strategy in the prediction of the cooling load was likewise linked to the best ratings of MSE & RMSE prediction errors. The kind of data input has a significant impact on the application of proposed algorithms and the outcomes. There is indeed a difference in the outcomes of each platform's prediction of cooling and heating loads, and the heating load is anticipated with a high degree of accuracy. This disparity arises from a lack of connection between the data input and the degree of cooling load in comparison to the number of the heating load.

5.3. Comparative Analysis

It is vital to compare the findings acquired with the findings of earlier research to assess the usefulness of the offered approaches in this study. Using identical datasets, comparisons must be conducted with caution. For that purpose, some studies with similar findings for estimating cooling and heating loads were chosen for comparison. The findings of numerous tests conducted to estimate cooling and heating loads by relevant metrics were compared with the experimental results obtained in this paper to represent the efficacy of the data structure in the accuracy of the results. This comparison is made in Table 3.

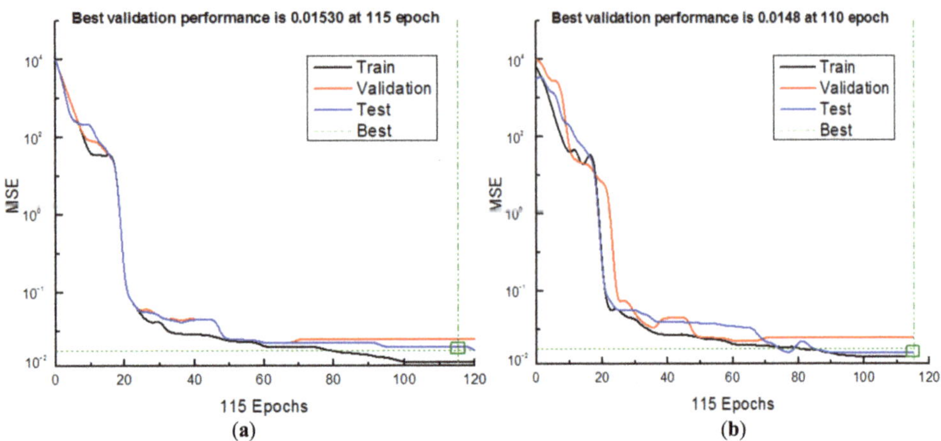

Figure 7. Analysis of the best performing model in terms of prediction using the MSE metric. (**a**) Heating load, (**b**) cooling load.

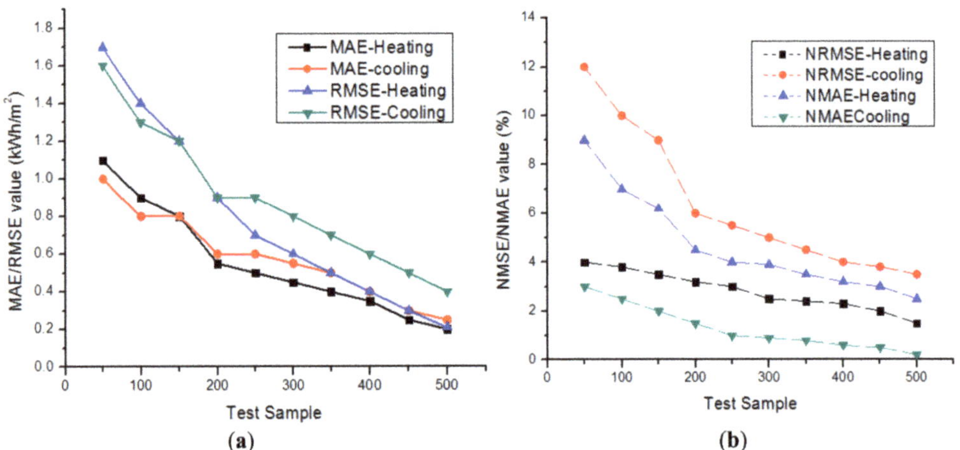

Figure 8. (**a**,**b**) Developed model performance using varying sample sizes in the training and testing sets.

Table 2. The suggested approach results in terms of predicting heating and cooling loads.

Loads	Performance Metrics						
	R	R2	RMSE (kWh/m²)	MSE (kWh/m²)	MAE (kWh/m²)	NRMSE (%)	NMAE (%)
Heating load	0.9998	0.9987	0.21	0.01530	0.2	1.5	2.5
Cooling load	0.999	0.9978	0.4	0.148	0.25	3.5	0.2

The comparison in Table 3 demonstrates the precision and robustness of the proposed approaches in this research for projecting a building's cooling and heating demands. In residential structures, the use of the proposed SSRD-SsIF-NN methods and the choice of the most appropriate approach for energy prediction and energy-efficient technologies are highly beneficial in reducing energy consumption. With their great accuracy, the chosen approaches were able to accomplish the objective of the study and achieve this key goal. Finally, it is worth noting that the presented techniques may be applied to real-world data as well. The SSRD-SsIF-NN approach was used to forecast the yearly spatial heating/cooling of load intensities in individual groups in this study. Meanwhile, the database of annual residential heating and cooling load intensities created in this study will be a useful resource of household energy statistics for traditional research on WDO-ANN [27], RF-IPWOA-ELM [31], and SVR [32] approaches. Compared to conventional methods, the developed model has achieved much fewer errors of MSE, RMSE, and other significant metrics. The MATLAB simulation is used to create the information on the building structures' heating and cooling of load intensities for this study.

Table 3. Comparative analysis of the proposed method and conventional approaches.

Loads	Performance Evaluation	Model			
		WDO-ANN [27]	RF-IPWOA-ELM [31]	SVR [32]	Proposed SSRD-SsIF-NN
Heating load	R	0.99	0.9978	0.997	0.9998
	R2	0.97	0.987	0.991	0.9987
	RMSE (kWh/m^2)	0.2476	0.146	0.3458	0.21
	MSE (kWh/m^2)	0.1459	0.01597	0.26	0.01530
	MAE (kWh/m^2)	0.28	0.39	0.85	0.2
	NRMSE (%)	1.16	2.04	1.89	1.5
	NMAE (%)	2.92	3.18	2.987	2.5
Cooling load	R	0.9987	0.99	0.9912	0.999
	R2	0.8931	0.928	0.948	0.9978
	RMSE (kWh/m^2)	0.599	0.492	0.643	0.4
	MSE (kWh/m^2)	0.234	0.285	0.68	0.148
	MAE (kWh/m^2)	0.396	0.46	0.26	0.25
	NRMSE (%)	4.2	4.6	3.9	3.5
	NMAE (%)	0.39	0.52	0.4	0.2

6. Conclusions

The necessity of energy protection and sustainability has created several obstacles in predicting a building's heating and cooling needs. Numerous strategies and techniques for estimating heating and cooling loads are offered by most experts in this subject to improve predictive performance. In this research, SSRD-SsIF-NN is offered as a method for predicting a residential structure's cooling and heating demands. In this work, considerable improvements can be made by including additional values in constructing structural features and switching from a shallow to a profound design-based forecasting model. During the training stage, after developing each of the proposed frameworks, the essential features of a residence were utilized as sources, and the heating/cooling loads were utilized as the output results of each system. To validate the trained networks and anticipate the heating and cooling needs, unique and unidentified information was employed. In forecasting the heating load, this proposed model had an MSE of 0.01530, an MAE of 0.2, an RMSE of 0.21, and an R and R2 both as great as 0.998, and in forecasting the cooling load, it had an MSE of 0.148, an MAE of 0.25, an RMSE of 0.4, and an R and R2 both as great as 0.99. Because the generated prediction methods were dependent on the building attributes, the findings of the study may be useful for developers during the pre-design phase of the energy-efficient heating/cooling of residential buildings.

Author Contributions: Conceptualization, K.I.; methodology, K.I. and M.S.S.; validation, K.I., and M.H.Z.; formal analysis, M.S.S. and M.H.Z.; data curation, K.I., M.H.Z. and M.S.S.; writing—review and editing, K.I. and A.A. All authors have read and agreed to the published version of the manuscript.

Funding: This research was funded by Interdisciplinary Research Center for Renewable Energy and Power Systems (IRC-REPS), King Fahd University of Petroleum & Minerals (KFUPM), Saudi Arabia under Project No. INRE2113.

Data Availability Statement: The data presented in this study are available on request from the corresponding authors.

Acknowledgments: The authors gratefully acknowledge the funding (Project No. INRE2113) from the Interdisciplinary Research Center for Renewable Energy and Power Systems (IRC-REPS), King Fahd University of Petroleum & Minerals (KFUPM), Saudi Arabia. Kashif Irshad acknowledges the funding support provided by the King Abdullah City for Atomic and Renewable Energy (K. A. CARE).

Conflicts of Interest: The authors declare no conflict of interest.

References

1. Felimban, A.; Prieto, A.; Knaack, U.; Klein, T.; Qaffas, Y. Assessment of Current Energy Consumption in Residential Buildings in Jeddah, Saudi Arabia. *Buildings* **2019**, *9*, 163. [CrossRef]
2. Krarti, M.; Dubey, K.; Howarth, N. Evaluation of Building Energy Efficiency Investment Options for the Kingdom of Saudi Arabia. *Energy* **2017**, *134*, 595–610. [CrossRef]
3. Irshad, K.; Algarni, S.; Islam, N.; Rehman, S.; Zahir, M.H.; Pasha, A.A.; Pillai, S.N. Parametric Analysis and Optimization of a Novel Photovoltaic Trombe Wall System with Venetian Blinds: Experimental and Computational Study. *Case Stud. Therm. Eng.* **2022**, *34*, 101958. [CrossRef]
4. Alassery, F.; Alzahrani, A.; Khan, A.I.; Irshad, K.; Islam, S. An Artificial Intelligence-Based Solar Radiation Prophesy Model for Green Energy Utilization in Energy Management System. *Sustain. Energy Technol. Assess.* **2022**, *52*, 102060. [CrossRef]
5. Alshamrani, O.; Alshibani, A.; Mohammed, A. Operational Energy and Carbon Cost Assessment Model for Family Houses in Saudi Arabia. *Sustainability* **2022**, *14*, 1278. [CrossRef]
6. Soni, A.; Das, P.K.; Yusuf, M.; Pasha, A.A.; Irshad, K.; Bourchak, M. Synergy of RHA and silica sand on physico-mechanical and tribological properties of waste plastic–reinforced thermoplastic composites as floor tiles. *Environ. Sci. Pollut. Res.* **2022**, 1–19. [CrossRef]
7. Esmaeil, K.K.; Alshitawi, M.S.; Almasri, R.A. Analysis of energy consumption pattern in Saudi Arabia's residential buildings with specific reference to qassim region. *Energy Effic.* **2019**, *12*, 2123–2145. [CrossRef]
8. Islam, N.; Irshad, K.; Zahir, M.H.; Islam, S. Numerical and experimental study on the performance of a photovoltaic Trombe wall system with Venetian blinds. *Energy* **2021**, *218*, 119542. [CrossRef]
9. Feng, Y.; Duan, Q.; Chen, X.; Yakkali, S.S.; Wang, J. Space cooling energy usage prediction based on utility data for residential buildings using machine learning methods. *Appl. Energy* **2021**, *291*, 116814. [CrossRef]
10. Sajjad, M.; Khan, S.U.; Khan, N.; Haq, I.U.; Ullah, A.; Lee, M.Y.; Baik, S.W. Towards efficient building designing: Heating and cooling load prediction via multi-output model. *Sensors* **2020**, *20*, 6419. [CrossRef]
11. Irshad, K.; Khan, A.I.; Irfan, S.A.; Alam, M.M.; Almalawi, A.; Zahir, M.H. Utilizing artificial neural network for prediction of Occupants thermal comfort: A case study of a test room fitted with a thermoelectric air-conditioning system. *IEEE Access* **2020**, *8*, 99709–99728. [CrossRef]
12. Zeng, A.; Ho, H.; Yu, Y. Prediction of building electricity usage using gaussian process regression. *J. Build. Eng.* **2020**, *28*, 101054. [CrossRef]
13. Malik, A.; Saggi, M.K.; Rehman, S.; Sajjad, H.; Inyurt, S.; Bhatia, A.S.; Farooque, A.A.; Oudah, A.Y.; Yaseen, Z.M. Deep learning versus gradient boosting machine for Pan evaporation prediction. *Eng. Appl. Comput. Fluid Mech.* **2022**, *16*, 570–587. [CrossRef]
14. Bagheri-Esfeh, H.; Safikhani, H.; Motahar, S. Multi-objective optimization of cooling and heating loads in residential buildings integrated with phase change materials using the artificial neural network and genetic algorithm. *J. Energy Storage* **2020**, *32*, 101772. [CrossRef]
15. Amasyali, K.; El-Gohary, N. Machine learning for occupant-behavior-sensitive cooling energy consumption prediction in office buildings. *Renew. Sustain. Energy Rev.* **2021**, *142*, 110714. [CrossRef]
16. Tsoka, S.; Tolika, K.; Theodosiou, T.; Tsikaloudaki, K.; Bikas, D. A method to account for the urban microclimate on the creation of 'typical weather year' datasets for building energy simulation, using stochastically generated data. *Energy Build.* **2018**, *165*, 270–283. [CrossRef]
17. Lin, Y.; Zhou, S.; Yang, W.; Shi, L.; Li, C.-Q. Development of building thermal load and discomfort degree hour prediction models using data mining approaches. *Energies* **2018**, *11*, 1570. [CrossRef]
18. Moayedi, H.; Gör, M.; Kok Foong, L.; Bahiraei, M. Imperialist competitive algorithm hybridized with multilayer perceptron to predict the load-settlement of square footing on layered soils. *Measurement* **2021**, *172*, 108837. [CrossRef]
19. Wang, W.; Tian, G.; Chen, M.; Tao, F.; Zhang, C.; AI-Ahmari, A.; Li, Z.; Jiang, Z. Dual-objective program and improved artificial bee colony for the optimization of energy-conscious milling parameters subject to multiple constraints. *J. Clean. Prod.* **2020**, *245*, 118714. [CrossRef]
20. Green, C.; Garimella, S. Residential microgrid optimization using grey-box and black-box modeling methods. *Energy Build.* **2021**, *235*, 110705. [CrossRef]
21. Cui, X.; E, S.; Niu, D.; Chen, B.; Feng, J. Forecasting of carbon emission in China based on gradient boosting decision tree optimized by modified whale optimization algorithm. *Sustainability* **2021**, *13*, 12302. [CrossRef]

22. Mokhtara, C.; Negrou, B.; Settou, N.; Settou, B.; Samy, M.M. Design optimization of off-grid hybrid renewable energy systems considering the effects of building energy performance and climate change: Case study of Algeria. *Energy* **2021**, *219*, 119605. [CrossRef]
23. Chaturvedi, S.; Bhatt, N.; Gujar, R.; Patel, D. Application of PSO and GA stochastic algorithms to select optimum building envelope and air conditioner size—A case of a residential building prototype. *Mater. Today: Proc.* **2022**, *57*, 49–56. [CrossRef]
24. Almalawi, A.; Khan, A.I.; Alqurashi, F.; Abushark, Y.B.; Alam, M.M.; Qaiyum, S. Modeling of remora optimization with deep learning enabled heavy metal sorption efficiency prediction onto biochar. *Chemosphere* **2022**, *303*, 135065. [CrossRef]
25. Usman, M.; Frey, G. Multi-objective techno-economic optimization of design parameters for residential buildings in different climate zones. *Sustainability* **2021**, *14*, 65. [CrossRef]
26. Xu, Y.; Li, F.; Asgari, A. Prediction and optimization of heating and cooling loads in a residential building based on multi-layer Perceptron neural network and different optimization algorithms. *Energy* **2022**, *240*, 122692. [CrossRef]
27. Moayedi, H.; Mosavi, A. Double-target based neural networks in predicting energy consumption in residential buildings. *Energies* **2021**, *14*, 1331. [CrossRef]
28. Rana, M.; Sethuvenkatraman, S.; Goldsworthy, M. A data-driven approach based on quantile regression forest to forecast cooling load for commercial buildings. *Sustain. Cities Soc.* **2022**, *76*, 103511. [CrossRef]
29. Li, X.; Yao, R. Modelling heating and cooling energy demand for building stock using a hybrid approach. *Energy Build.* **2021**, *235*, 110740. [CrossRef]
30. Kim, D.D.; Suh, H.S. Heating and cooling energy consumption prediction model for high-rise apartment buildings considering design parameters. *Energy Sustain. Dev.* **2021**, *61*, 1–14. [CrossRef]
31. Tran, D.H.; Luong, D.L.; Chou, J.S. Nature-inspired metaheuristic ensemble model for forecasting energy consumption in residential buildings. *Energy* **2020**, *191*, 116552. [CrossRef]
32. Zhou, G.; Moayedi, H.; Bahiraei, M.; Lyu, Z. Employing artificial bee colony and particle swarm techniques for optimizing a neural network in prediction of heating and cooling loads of residential buildings. *J. Clean. Prod.* **2020**, *254*, 120082. [CrossRef]
33. Seyedzadeh, S.; Rahimian, F.P.; Rastogi, P.; Glesk, I. Tuning machine learning models for prediction of building energy loads. *Sustain. Cities Soc.* **2019**, *47*, 101484. [CrossRef]
34. Sharif, S.A.; Hammad, A. Developing surrogate ANN for selecting near-optimal building energy renovation methods considering energy consumption, LCC and LCA. *J. Build. Eng.* **2019**, *25*, 100790. [CrossRef]
35. Singaravel, S.; Suykens, J.; Geyer, P. Deep-learning neural-network architectures and methods: Using component-based models in building-design energy prediction. *Adv. Eng. Inform.* **2018**, *38*, 81–90. [CrossRef]
36. Gao, Z.; Yu, J.; Zhao, A.; Hu, Q.; Yang, S. A hybrid method of cooling load forecasting for large commercial building based on Extreme Learning Machine. *Energy* **2022**, *238*, 122073. [CrossRef]
37. Wei, Z.; Zhang, T.; Yue, B.; Ding, Y.; Xiao, R.; Wang, R.; Zhai, X. Prediction of residential district heating load based on machine learning: A case study. *Energy* **2021**, *231*, 120950. [CrossRef]

Article

Cascaded Control for Building HVAC Systems in Practice

Chris Price [1], Deokgeun Park [2] and Bryan P. Rasmussen [2,*]

1   Oak Ridge National Laboratory, Oak Ridge, TN 37830, USA
2   J. Mike Walker '66 Department of Mechanical Engineering, Texas A&M University, College Station, TX 77840, USA
*   Correspondence: brasmussen@tamu.edu

Abstract: Actuator hunting is a widespread and often neglected problem in the HVAC field. Hunting is typically characterized by sustained or intermittent oscillations, and can result in decreased efficiency, increased actuator wear, and poor setpoint tracking. Cascaded control loops have been shown to effectively linearize system dynamics and reduce the prevalence of hunting. This paper details the implementation of cascaded control architectures for Air Handling Unit chilled water valves at three university campus buildings. A framework for implementation the control in existing Building Automation software is developed that requires only a single line of additional code. Results gathered for more than a year show that cascaded control not only eliminates hunting in control loops with documented hunting issues, but provides better tracking and more consistent performance during all seasons. A discussion of efficiency losses due to hunting behavior is presented and illustrated with comparative data. Furthermore, an analysis of cost savings from implementing cascaded chilled water valve control is presented. Field tests show 2.2–4.4% energy savings, with additional potential savings from reduced operational costs (i.e., maintenance and controller retuning).

Keywords: building; HVAC; control; energy efficiency; faults

Citation: Price, C.; Park, D.; Rasmussen, B.P. Cascaded Control for Building HVAC Systems in Practice. *Buildings* **2022**, *12*, 1814. https://doi.org/10.3390/buildings12111814

Academic Editor: Etienne Saloux

Received: 1 September 2022
Accepted: 10 October 2022
Published: 28 October 2022

Publisher's Note: MDPI stays neutral with regard to jurisdictional claims in published maps and institutional affiliations.

1. Introduction

Actuator hunting is a known and well documented problem in the HVAC field affecting a wide range of systems from vapor compression systems to Variable Air Volume (VAV) terminal units. Hunting is an undesired oscillation in a system's control input due only to the interaction of the controller with the system dynamics, in contrast to oscillations due to a changing external input. The phenomenon is the result of nonlinear and time varying dynamics associated with HVAC systems. For example, VAV units have steady-state input/output gains that can vary by more than an order of magnitude over the full range of operating conditions [1]. Fixed controllers will struggle to provide consistent performance when a system operates far from its tuning conditions. Hunting can also be spread to upstream and downstream components in an HVAC system, making identification of the root cause difficult. A survey at Texas A&M University showed campus Air Handling Units displayed high levels of hunting, with chilled water valves hunting 70% and supply fans hunting nearly 25% of operating time [2].

While hunting is often easily identifiable by visual inspection of a measured signal, there are several automated methods to detect the behavior. These methods are able to distinguish between daily disturbances such as outdoor air temperature and the high frequency oscillations that stem from the controller. The time between consecutive zero crossings of the Integrated Absolute Error signal is used to detect the presence of oscillations [3]. Hunting is detected when two or more crossings occur far enough apart and with sufficient magnitude. This paper uses the simple method proposed by [2] to identify and measure hunting times. This method only requires the input signal and uses the magnitude and time between consecutive sign changes to identify hunting. For details on the algorithm parameters and implementation, see [4].

Most control loops utilized in the HVAC field are Proportional-Integral-Derivative (PID) controllers. As shown in Equation (1), PID is a low order controller consisting of three parts that respond to instantaneous error (P), steady-state error (I), and changes in error (D). Despite its simplicity, PID has proven versatility and robustness for controlling a wide range of systems in many fields [5–7]. In many HVAC applications, the time derivative 'D' term is not used due to its sensitivity to noise and additional implementation complexity. For effective operation, the control gains, k_p, k_i, and k_d must to be tuned for the equipment and operating conditions. There are numerous methods for tuning PID controllers each with their own unique goals and procedures [8]. The most common of which are the methods of Ziegler and Nichols. Building Automation System (BAS) software often have built-in PID functions that implement the controller with native anti-windup and saturation solutions. Such commands typically have recommended initial gains from which the tuning process can begin. Further background on PID control can be found in [7,8].

$$\text{PID}(t) = k_p e(t) + k_i \int_0^\infty e(t)\,dt + k_d \frac{d}{dt} e(t), \tag{1}$$

The tendency for HVAC systems to have large, load-dependent nonlinearities often causes difficulties for PID controllers; see [1] for examples. Nonlinearities are the result of fundamental heat transfer processes and system actuators that typically do not have linear flow profiles. PID controllers tuned in high system gain conditions will have very slow response times when operating conditions change. Conversely, a controller tuned in low gain conditions can easily develop hunting behavior as system gains increase. These variations in performance lead to hunting behavior and decreased system efficiency. To address such issues, some have focused on assessing poor control and designed control quality factors (CQF) to analyze control performance based on measured data [9].

Improving the control of HVAC subsystems is important for two main reasons. First, the nonlinear power profiles of actuators such as fans or pumps (Equation (2)) causes overall energy use and operational costs to increase with oscillatory (hunting) behavior. Second, hunting behavior increases actuator wear and, finally, system-level coordinating controllers, such as Model Predictive Control (MPC), are increasingly used to optimize system setpoints. These supervisory controllers depend on subsystem controllers that can consistently track setpoints. Hunting can also interfere with model predictions, reducing the effectiveness of advanced control techniques and resulting in lost efficiency.

$$\frac{P_1}{P_2} = \left(\frac{\omega_1}{\omega_2}\right)^3, \tag{2}$$

This paper details the implementation of cascaded control architectures for building Air Handing Unit (AHU) temperature control. Research has shown that cascaded control loops are an effective strategy to reduce common HVAC issues stemming from nonlinear dynamics and input/output coupling, including the elimination of hunting behavior. The proposed architecture uses nested PID control loops to improve system performance by isolating and linearizing system dynamics. Cascaded loops are inherently low order and easily implementable in existing Building Energy Management (BEM) software. As will be shown, this approach requires no special software commands and can be implemented with a single line of additional code, facilitating adoption in HVAC applications.

The simplest embodiment of a cascaded control architecture is the addition of a proportional control loop inside of a standard Proportional-Integral (PI) controller. An example of this approach, would be the representation of a cascaded control loop applied to nonlinear system, as seen in Figure 1. The nonlinear plant is represented as a Hammerstein model consisting of a nonlinear gain function dependent on operating condition 'σ' and some dynamic of unitary gain. In this model, inner and outer loop signals (y_i and y_o) and nonlinearities (ψ_i and ψ_o) can be equal or unique. Note that the plant nonlinearity is contained inside the inner loop control where it is effectively linearized by proportional feedback with gain k_L. This affect can be seen in Equation (3), where the inner and outer loop nonlinearities are placed in the numerator and denominator of the inner loop transfer function. This structure allows nonlinearities to counteract themselves over all operating conditions and thereby reducing their overall effect. Additionally, the inner loop process will essentially become a static gain as the inner loop gain becomes large. If the nonlinearities are equal or multiplicatively related, all dependence on operating condition is eliminated. This behavior is guaranteed if both $\psi_i(\sigma)$ and $\psi_o(\sigma)$ are monotonic and share the same trends. For more details on cascaded control and properties of the inner loop gain, see [10].

$$L(s, k_L, \sigma) = \frac{k_L \psi_o(\sigma) G_o(s)}{1 + k_L \psi_i(\sigma) G_i(s)},\tag{3}$$

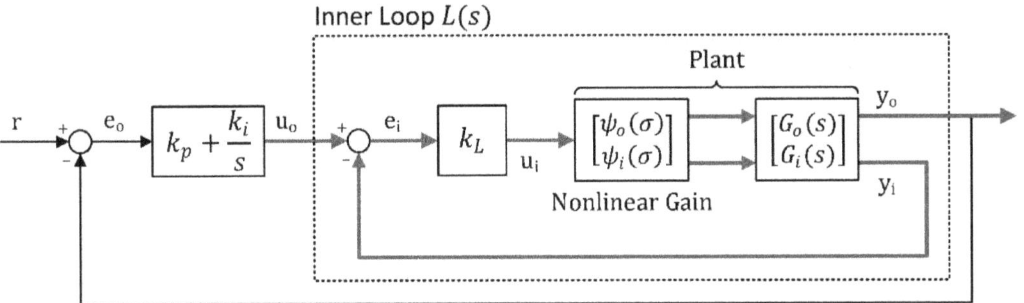

Figure 1. Block diagram of a generalized cascaded control loop. Subscripts 'i' and 'o' denote the inner and outer loops, respectively.

Recent work with cascaded control architectures has shown it to be an effective strategy for control of a wide range of HVAC systems. Simulations have improved control of and eliminated hunting behavior in VAV units, hydronic radiator systems, and AHUs [1]. The architecture was also able to decouple the dynamics of a multi-evaporator refrigeration systems and improve individual tracking performance [11]. Although cascaded loops require an additional control loop and a more complex tuning process, simple metrics have been developed to quantify the benefits of cascaded control and tune inner and outer loop gains accordingly along with optimal frameworks [10] and simple tuning rules [12].

The linearization and decoupling effects of cascaded control are particularly important for Model Predictive Control algorithms that rely on consistent and linear sub-system behavior. The simplicity of cascaded controllers can enable MPC algorithms to better optimize building HVAC performance while still guaranteeing control stability. Neither MPC nor cascaded control, however, has seen widescale testing in real building systems. Numerous studies have implemented cascaded controllers in simulation with simple models of HVAC systems or with experimental test rigs [13]. In [14], a Hybrid Expansion Valve (HEV) was used to linearize the response of a small laboratory vapor compression cycle system. The HEV used a combination of physical and digital feedback to implement the cascaded control loop and eliminate differences in high flow and low flow conditions. Cascaded control has also appeared in trade manuals [15] that adjust equipment exit temperatures based on room temperature setpoint errors. These manuals, however, do not focus or test the linearization behavior of the architecture. Testing of MPC controllers in buildings has similarly been mostly in simulation [16] with some studies beginning to test on real building systems [17].

This paper presents results of a widescale implementation of cascaded control loops for AHU discharge air temperature control in three university campus buildings (9 AHUs). These controllers regulated the position of chilled water supply valves and utilized standard building automation software. The implementation required the addition of only a single line of code to existing control routines. Data was collected over multiple years, comparing building operation under existing control algorithms, and the proposed cascaded control approach. A comparative analysis of this data is presented, including the primary source of wasted energy and costs for discharge air temperature control. The results lay the groundwork for future work testing the implementing of an MPC algorithm across a whole building HVAC system.

2. Materials and Methods

Most building automation software can implement PID control loops using a built-in command. Consider, for example, a LOOP command given below that has 11 usable inputs. These inputs define the direction of control (type = 0 direct control, type = 1 indirect), the regulated signal (pv), the control signal (cv), the setpoint signal (sp), PID control gains (pg, ig, and dg), sample time (st), loop bias, and the saturation limits of the controller (lo and hi). Loop gains have a divisor (usually 1000) and have recommend values such as pg = 1000 and ig = 20 that provide good control for a wide range of systems. A sampling of AHUs at Texas A&M reveals that most PI control loops have these standard values, which strongly indicates that many loops operate with factory defaults and may never receive additional tuning unless problems are detected [4]. The relationship between LOOP gains and a standard PI control formulation is given in Equation (4) where it is important to note the multiplication of the sampling time and integral gain.

$$\text{LOOP(type, pv, cv, sp, pg, ig, dg, st, bias, lo, hi, 0)}$$
$$u = k_p \cdot e + k_i t_s \cdot \sum e = \tfrac{\text{pg}}{1000} \cdot e + \tfrac{\text{ig}}{1000} t_s \cdot \sum e, \tag{4}$$

A first pass implementation of a cascaded control loop in building software is given by Algorithm 1 for a discharge air temperature controller. Note that the inner and outer loops are implemented using two LOOP commands and an intermediate virtual point named 'AHU.DATLOOP1.ILSP' (Discharge Air Temperature Loop 1, Inner Loop Set Point) that stores the inner loop setpoint signal (i.e., the outer loop output). Although most LOOP commands in building software will have built-in saturation and anti-windup solutions, the interaction of the two loops must be considered. When the inner loop output (i.e., valve position) becomes saturated, the outer loop controller must also be disabled to avoid windup while the inner loop is disabled. Lines 6 through 7 deal with this issue by checking if the inner loop signal is saturated and then dynamically enabling/disabling the outer LOOP command on Line 11 accordingly. Although intuitive, this code is somewhat lengthy, requires the creation of intermediate virtual points, and has seven tunable variables.

Algorithm 1 Cascaded control implementation with two LOOPs

```
 1:   C Point Name Abbreviations
 2:   DEFINE(X,"AH01.")
 3:   DEFINE(Y,"DATLOOP1.")
 4:   DEFINE(Y,"DATLOOP2.")
 5:   C Outer Loop Anti-Windup
 6:   IF("%X%CCV" .GT. 1 .AND. "%X%CCV" .LT. 99) THEN SET(0,SECND2)
 7:   IF(SECND2 .GT. "DISABLE.TIMER") THEN DISABL(110) ELSE ENABLE(110)
 8:   C Inner Loop Control
 9:   LOOP(128,"%X%DAT","%Y%ILSP","%X%DAT.S","%Y%P","%Y%I",0,"%Y%TIME","%Y%BIAS",50,70,0)
10:   C Outer Loop Control
11:   LOOP(0,"%X%DAT","%X%CCV","%Y%ILSP","%Z%P",0,0,"%Z%TIME","%Z%BIAS",50,70,0)
```

An alternative approach to cascaded control can shorten the code required and simplify implementation. Consider the inner loop control signal given in Equation (5) where $e_1 = r - y_1$, $e_2 = u_1 - y_2$, B_1, and B_2 are outer and inner loop errors and biases, respectively. Note that the first two terms resemble the output of a PI controller with PI gains of $k_L k_p$ and $k_L k_i$ while the final terms are a combination of loop biases and inner loop feedback.

$$
\begin{aligned}
u_2 &= k_L e_2 + B_2 \\
&= k_L(u_1 - y_2) + B_2 \\
&= k_L\left[\left(k_p e_1 + k_i \sum e_1 + B_1\right) - y_2\right] + B_2 \\
&= k_L k_p(r - y_1) + k_L k_i \sum(r - y_1) + B_2 + k_L B_1 - k_L y_2,
\end{aligned}
\tag{5}
$$

Expressed in this form, the cascaded controller can clearly be implemented as a *single* LOOP command without the need for the extra intermediate virtual point as before. This is important because inner/outer loop anti-windup issues are avoided as the new algorithm takes advantage of built-in saturation features. Code based on this implementation for AHU control is given by Algorithm 2 taking into account that the outer and inner loops are reverse and direct acting, respectively. The bias term is calculated and stored in a local variable $LOC1 on Line 4 because some software does not allow for calculations inside of function calls. Note that the simplified code eliminates fives lines and reduces the number of tuning variables to five. One disadvantage of this implementation is the loss of ability to have different sampling times for the inner and outer loops. Despite this, all benefits of cascaded control can still be realized even through the two loops operate at the same sampling rate.

Algorithm 2 Simplified cascaded control implementation with one LOOP

```
1:   C Point Name Abbreviation
2:   DEFINE(X,"AH01.")
3:   C Bias Term Calculation
4:   $LOC1 = "%X%BIAS" + "%X%KL"*"%X%DAT"
5:   C Cascaded Control
6:   LOOP(0,"%X%DAT","%X%CCV","%X%DAT.S","%X%P","%X%I",0,"%X%TIME",$LOC1,0,100,0)
```

The final sections of this paper detail results of applying cascaded control within three campus buildings. Details about the size, layout, and location of each building will be provided as well as comparisons between original PI and cascaded control. Finally, a discussion of the cost of poor AHU control is presented with a savings estimate based on observed performance improvements.

3. Results

Working with the staff at the Utilities and Energy Services, limited access to the HVAC control systems of Building 1497, 0474 and 1600 was established.

3.1. Building 1497 Results

This building is a single-story, rectangular building with an area of 12,040 ft^2 (1119 m^2) and consisting of ten temperature-controlled zones and one unconditioned server room with the general floor plan shown in Figure 2. The building is serviced by a 14 ton single rooftop AHU consisting of a chilled water coil with valve, return/outdoor air dampers, and variable speed fan capable of suppling 6425 CFM of air. The unit has two sensors for discharge air temperature and end static pressure. Zones 1–10 have VAV terminal boxes equipped with a hot water reheat coil and an air damper. The hot and cold water needs of the building are serviced by two dedicated loops that provide access to the university's centralized heating and cooling water supply.

Building 1497 uses a complex, nested PI-based architecture for its HVAC control (Figure 3). During normal operation, PI controller (1) modulates the speed of the supply fan to maintain static pressure in the air ducts. The End Static Pressure (ESP) setpoint is the output of another PI controller (2) that compares the damper demand given by Equation (6) to a design setpoint $D_{set} = 60$. Room air temperature is regulated by a cascaded damper control architecture similar to the one discussed in [1]. An outer loop PI controller (3) uses room temperature error to calculate a flow demand $F_i \in [0, 100]$ that determines the flow rate required for each room. Flow demand is converted to a flow rate though linear interpolation between minimum ventilation requirements and the maximum system output. Inner loop control (4) uses local control and a flow rate sensor to match the outer loop flow setpoint. Similar to ESP control, the AHU discharge air temperature setpoint is generated by a PI controller (5) using the cooling demand calculation of Equation (7) and the design setpoint $C_{set} = 60$. PI controllers (6–7) modulate hot and cold-water supply valves to match the exit/supply air temperature setpoint.

$$D = \frac{3}{5}\max(\theta_i) + \frac{2}{5}\left(\frac{1}{n}\sum_{i=1}^{n}\theta_i\right), \tag{6}$$

$$C = \frac{3}{5}\max(F_i) + \frac{2}{5}\left(\frac{1}{n}\sum_{i=1}^{n}F_i\right), \tag{7}$$

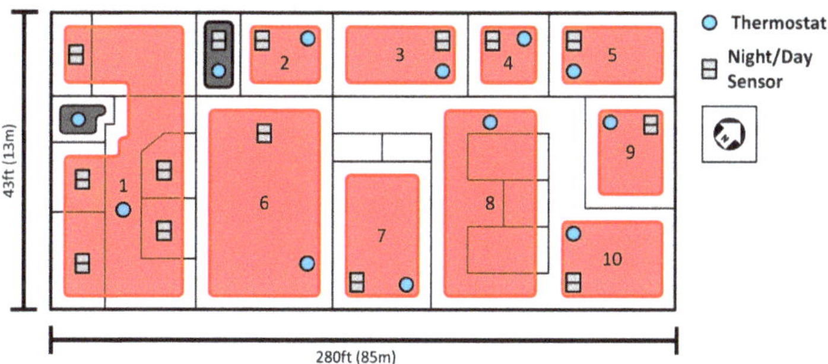

Figure 2. Layout of HVAC zones for Building 1497.

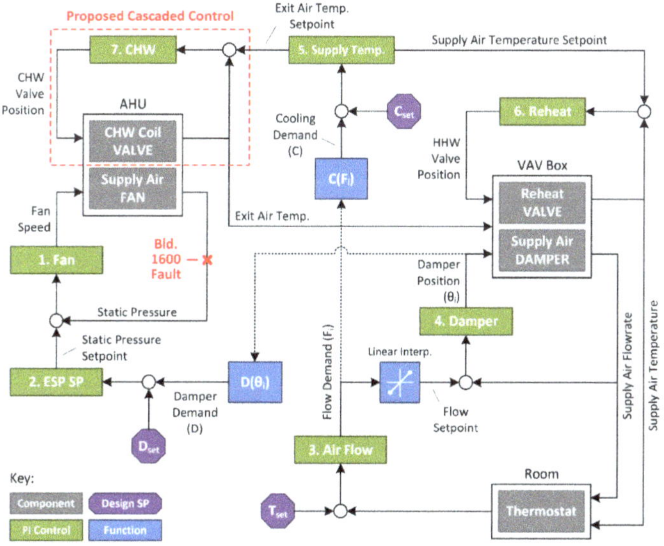

Figure 3. HVAC control system diagram for Building 1497.

Chilled water valve control used to regulate AHU exit air temperature for Building 1497 has documented issues with actuator hunting. Oscillations are most pronounced during low load conditions such as early morning or during cool winter weather. For example, the valve hunted 57% of its operating time during the three-month period of 1 November 2013 to 1 February 2014 while the valve hunted only 14% from 1 May to 1 August 2016. PI valve control has three distinct hunting behaviors as seen in Figure 4.

Under high load, valve control typically does not hunt (19 March 2016). In early spring, temperatures are usually warm in the afternoon but cool in the evening resulting in hunting late in the day (23 March 2016). On other spring days, there is never enough load to prevent hunting behavior (30 March 2016). This behavior indicates that control performance is strongly tied to the operating conditions of the system.

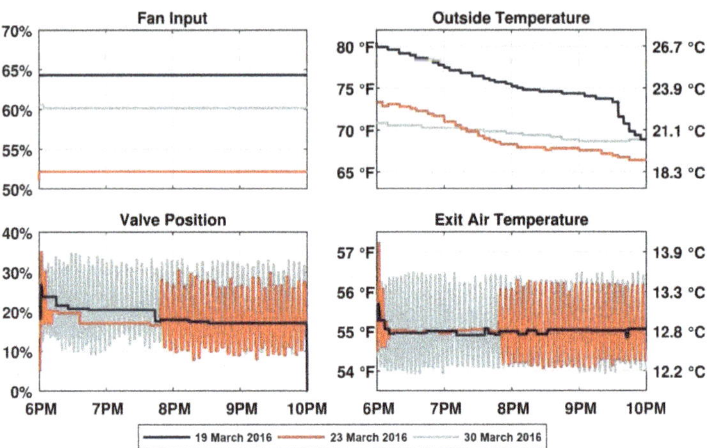

Figure 4. Building 1497 PI control performance displays three hunting behaviors depending on system load (outside temperature). The chilled water valve will not hunt, hunt late in the day, or hunt continuously.

Cascaded control was applied to Building 1497 chilled water valve control from approximately October through December of 2015. Testing utilized Algorithm 1 with gains tuned using step identification tests for a range of supply fan speeds (system loads) from 20–90%. Cascaded gains of $k_L = 4$, $k_{pc} = 1.25$, and $k_{ic} = 0.2$ were chosen using the analysis and the tuning procedure from [10]. Figure 5 shows that valve hunting modes seen with the original PI control for a range of system loads has been eliminated without sacrificing performance.

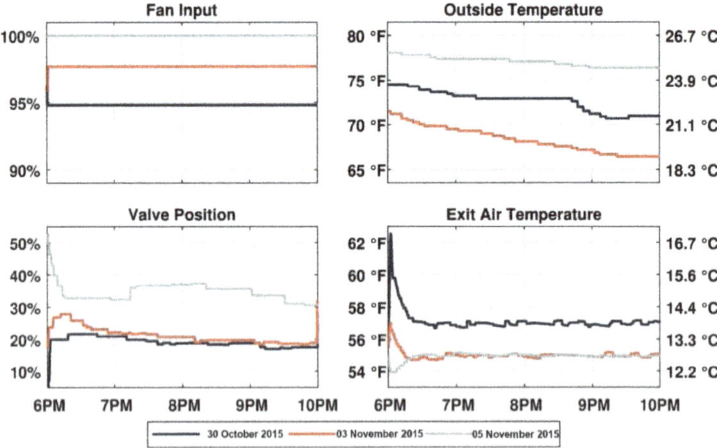

Figure 5. Building 1497 cascaded control performance displays no hunting behavior over a range of system loads.

3.2. Building 0474 (Philosophy Department) Results

Building 0474 is a four-story building originally completed in 1914 that houses the university Philosophy Department. A total renovation in 2012 included upgrading the entire HVAC system and controls. Each floor has a dedicated AHU for the floor and is numbered for the floor it covers. AHU1, AHU2, AHU3 and AHU4 have capacities of 35, 27, 21 and 22 tons and total air flow capacity of 12,000, 10,000, 8000, and 8000 CFM, correspondingly. Building 0474 has approximately 54000 ft^2 (5017 m^2) of office space with approximately 20 heating and cooling zones per floor, controlled by the AHU in the middle of the floor area (Figure 6). Each floor has its own AHU where return and outside air are mixed and conditioned. Zones have a parallel fan powered VAV terminal box with hot water reheat coil and return air ducting that draws warm air from the ceiling plenum for 'free' reheat. The heating coil can be used for substitute reheat when at the minimum supply air flow rate. The building control system has a wide array of sensors including relative humidity, CO_2, and outside air flow rate (ventilation). The overall temperature control structure is the same as at Building 1497 (Figure 3) with the exception of additional complexity due to the upgraded terminal boxes and ventilation sensors.

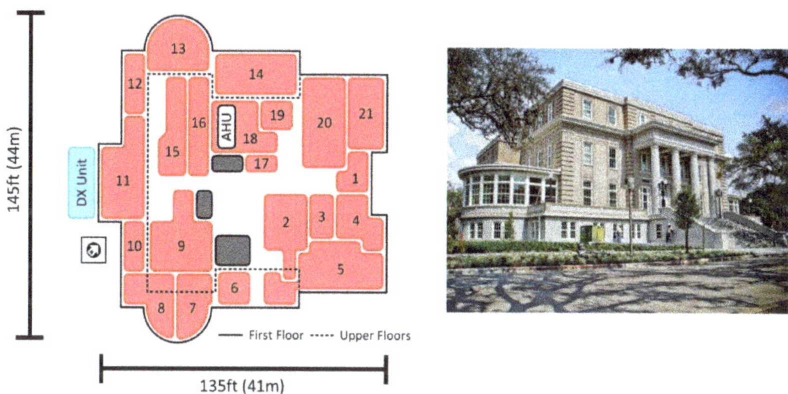

Figure 6. Layout of HVAC zones and exterior of Building 0474.

Cascaded control Algorithm 2 was initially tested on the fourth floor AHU chilled water valve and later applied to the other three floors. All four original PI controllers had gains of pg $= 1000$ and ig $= 20$ with sampling times of $t_s = 1$ s, which are the recommended LOOP gains from [18]. As a starting point, the inner loop gain was set at a conservative value $k_L = 0.5$ and the outer loop gains at $k_{pc} = 1$ and $k_{ic} = 0.04$. When converted to nominal gains using the relationships of Equation (5), the resulting LOOP gains are equal to the original PI LOOP gains. This choice should provide similar transient performance to the original control, but with the added linearization benefits of the inner loop control. The resulting control gains (pg$_c$ and ig$_c$) used in Algorithm 2 are calculated using Equation (8). The inner loop bias is the average of the minimum and maximum valve position (i.e., $B_2 = 50\%$). The outer loop bias is the average of the minimum and maximum allowable exit/discharge air temperatures, 52 °F and 65 °F, respectively. The overall bias term B for the PPCL code is therefore given by Equation (9), where DAT is discharge air temperature. Note that the bias term of the LOOP command has no scaling factor. Inner loop gains for all units were later increased to $k_L = 1$ starting in March 2018 to increase the level of cascaded linearization.

$$\text{pg}_c = 1000 k_L k_{pc} = 500 \quad \& \quad \text{ig}_c = 1000 k_L k_{ic} = 20, \tag{8}$$

$$\begin{aligned}
B &= B_2 + k_L(\text{DAT} - B_1) \\
&= 50\% + \left(0.5\tfrac{\%}{°\text{F}}\right)\left(\text{DAT} - \tfrac{65°\text{F}+52°\text{F}}{2}\right) \\
&= 20.75\% + \left(0.5\tfrac{\%}{°\text{F}}\right)\text{DAT},
\end{aligned} \tag{9}$$

Building 0474 operations were transferred to a new server in the spring of 2017 with full historical trending of relevant HVAC operating points beginning approximately 1 August at 5 min intervals. Table 1 gives the results of analyzing each floor's AHU operation for fan and chilled water valve hunting with PI control through 31 December 2017. Overall, CHW (Chilled Water) control in Building 0474 displays very little hunting behavior except for the third floor where the valve hunts just over 10% of its operating time. Observations of building performance show that identified hunting in AHU3 occurs almost entirely in low cooling conditions. This indicates that the PI controller was likely tuned for mid-to-high load conditions. Hunting results for the CHW and fan control show that implementing cascaded control reduced hunting in the third-floor unit and had minimal effect on the other floors.

Table 1. Building 0474 Hunting Analysis Results with green arrows showing improvements and red arrows showing decline when compared to the PI control baseline case.

Control	Type	AHU1	AHU2	AHU3	AHU4
CHW Valve	PI	2.29%	1.05%	11.4%	2.12%
	Cascaded	↓1.91%	↑3.79%	↓↓6.32%	↑3.35%
Fan Speed	PI	2.78%	0.17%	0.32%	0.52%
	Cascaded	↓1.24%	↓0.02%	↑0.42%	↓0.40%

The main benefits of cascaded control implementation at Building 0474 were improved tracking performance due to more aggressive performance afforded by the cascaded architecture. To fairly compare HVAC performance before and after implementation, weather disaggregation was applied to the data using the Degree Day (DD) method. A DD is related to how long and by how much outside ambient conditions stay above or below a baseline or balance temperature. Usually assumed to be 65 °F, this balance temperature is the ambient load condition under which a building requires no conditioning. Cooling and heating degree days, CDD and HDD, respectively, can be thought of as the area above or below the balance temperature for a given outside temperature profile. The DD is therefore a useful tool to compare HVAC data as it inherently normalizes for warmer or colder weather.

System performance is measured using the Root-Mean-Square (RMS) error given by Equation (10). For error to be calculated, the system must be ON and in cooling mode for more than 90 min. These criteria are important because, particularly on weekends, AHUs will cycle ON/OFF randomly for short periods of time to maintain building air quality. These bursts are not long enough for the AHUs to reach their setpoints and are not representative of the tracking ability of the valve controller. Detecting cooling mode is important as the chilled water valve can be saturated at 0% causing large error accumulation despite not being utilized. Criteria for detecting these conditions are given in Table 2 with cooling time found by the intersection of ON time and the negation of HEAT detection.

$$RMSE = \sqrt{\frac{\sum_{k=1}^{N}(T_{set}(k) - T(k))^2}{N}} \tag{10}$$

Table 2. Cooling Mode Detection Criteria for Building 0474.

Condition	Criteria	Comment
ON/OFF	$\omega_i = 0$ (Fan Speed)	Minimum ω_i when LOOP is active is 20%.
HEAT	$\delta_i = 0$ (Valve Opening)	Identified when true continuously for 90 min.

Improvements in system performance can be seen in Figure 7 that shows PI control data from 2017 and Cascaded Control (CC) data from 2018. For each floor, at least marginally, there is a reduction in dependence on load condition (i.e., flatter trend lines) and a much tighter dispersion of daily error with cascade control than PI control. This is seen visually and in the decrease in standard deviation from the trend line. Improved RMS error results show that the cascaded controller is better able to track setpoint changes and ensure occupant comfort.

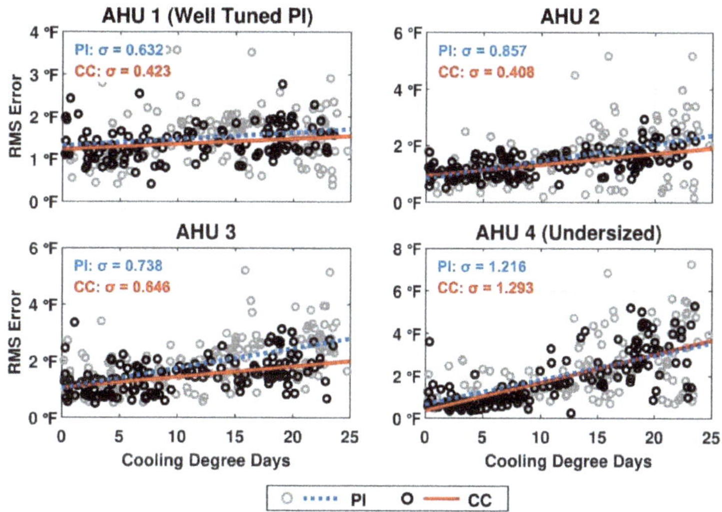

Figure 7. Performance comparison between PI and cascaded controllers at Building 0474. Cascaded control can be seen to provide tighter and more consistent performance.

The minimal improvements in AHU 1 and AHU 4 are the results of two main issues. For AHU 1, PI data from 2017 has less cold weather than CC in 2018. As these conditions tend to result in more error for this unit, the 2017 trend line is lower in this regime than expected. AHU 4 data is the result of the unit either being slightly undersized for observed loads or a system fault that restricts cooling capacity. In warm weather, AHU 4 was at maximum load with the valve and supply fan both operating continuously at 100%, but only slowly reaching command setpoints for static pressure and air temperature after several hours. This leads to large errors in warm weather that will be similar for both PI and CC control. However, there does appear to be an improvement in performance in cooler conditions. Overall, cascaded control was applied successfully to all AHUs at the YMCA building and showed performance benefits without introducing control hunting issues.

3.3. Building 1600 Results

Building 1600 is an approximately 85,000 ft^2 (7897 m^2) office and research facility completed in 1999 and consisting of three floors in a mostly L-shaped configuration with additional space on the ground floor. Each floor has a dedicated AHU and is numbered for the floor level it covers. AHU1, AHU2 and AHU3 have 63, 59, and 60 ton capacity with 22,050, 21,610, 20,160 CFM air flow capacity, respectively. There are 32 heating and cooling zones on the first floor, 40 on the second and 38 on the third floor roughly corresponding to the HVAC diagram given in Figure 8. Building 1600 has a 49 ton capacity Dedicated Outdoor Air System (DOAS) for its ventilation requirements that is functionally the same as a standard AHU except that 100% of its supply air is drawn from the outside (Figure 9). The DOAS supplies preconditioned ventilation air at maximum 7910 CFM to AHUs on each floor that have local cooling coils to make up for latent heat in the return air stream. Parallel fan powered VAV terminal boxes in each zone have reheat capabilities if necessary.

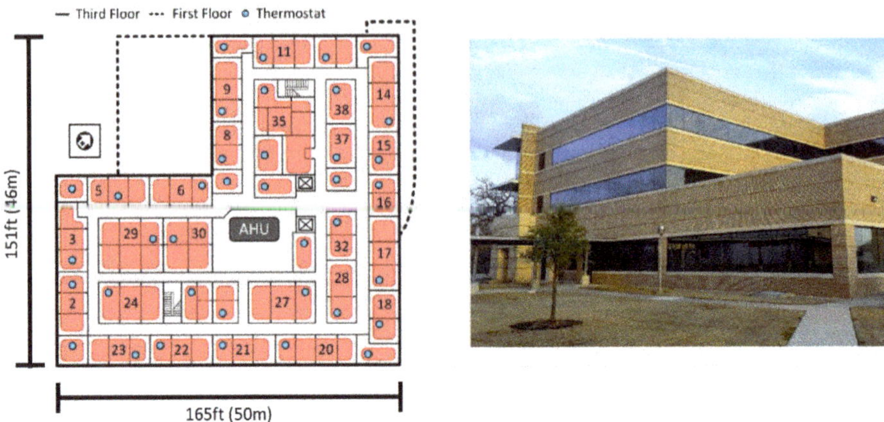

Figure 8. Layout of HVAC zones and exterior of Building 1600.

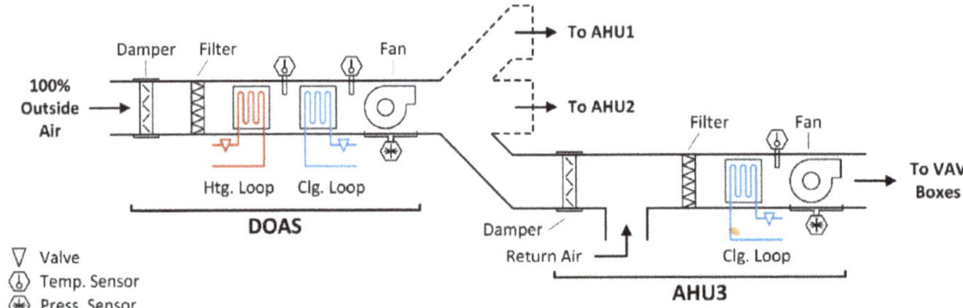

Figure 9. Building 1600 uses a dedicated outdoor air unit for ventilation supply to each floor's AHUs.

Historical data for this building was not available due to software limitations. However, dynamic trending of critical points for several months was facilitated by the university utilities office. This method of data collection records point values when signals vary above a threshold value with a maximum sampling time of 2 min. Data was initially collected from approximately 10:00 a.m. to 4:00 p.m. from November through December 2017 to capture original building operations. The nature of dynamic trending resulted in data sets with random sampling times. To utilize the hunting algorithm from [2], each dataset was resampled to enforce a 2 min sampling time.

Though Building 1600 is less than 20 years old and has an advanced HVAC system design, the AHU chilled water valve controls still have significant hunting issues. As seen in Figure 10, each floor's AHU valve control experiences some level of hunting behavior. AHU1 has a hunting period of approximately 60 min, AHU2 30 min, and AHU3 20 min. The level of hunting, in terms of amplitude and period, is again correlated with system load as it is significantly reduced/disappears when outdoor air temperature approaches 70 °F (21.1 °C). Apparent from the figure is the supply air fan for AHU2 also has a significant hunting issue. Fan speeds are allowed to vary within $\omega \in [20\%, 100\%]$ which accounts for the saturated appearance of the signal. Fan speeds for AHUs 1 and 2 vary only slightly or are constant during a normal day.

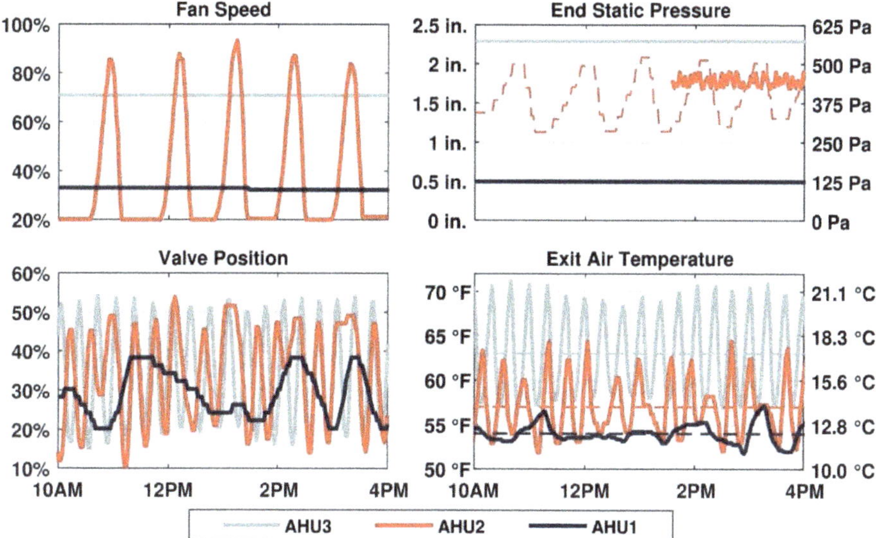

Figure 10. Performance data for Building 1600 on 5 December 2017 under original PI control shows significant levels of valve and fan hunting.

The tuning process at Building 1600 highlights several fundamental issues of practical building control. In particular, how hunting controllers can mask multiple system faults. The following sections detail issues discovered as they arose and how implementing cascaded control revealed the underlying problems.

3.3.1. Problem 1—Poorly Tuned Control Gains

Parsing building control code, the chilled water LOOP command settings for each AHU were found to vary widely as seen in Table 3. At issue are the vastly different sampling times seen in the upper floors. Due to the multiplication of the integral gain and sampling time (see Equation (4)), the effective integral gain for these systems is 30 times larger for upper floors than the first floor. Differences in gains help to explain the variation in loop performance between AHUs. Most likely, hunting behavior was observed in AHU3 and to compensate the magnitude of pg was reduced by an order of magnitude. Similarly, the integral gain for AHU1 was reduced to avoid oscillations.

Table 3. PI Control gains for Building 1600 AHU chilled water valves.

Unit	pg	ig	t_s (sec)	k_p	$k_i t_s$
AHU1	600	7.5	1	0.6	0.0075
AHU2	600	15	15	0.6	0.225
AHU3	60	15	15	0.06	0.225
DOAS	600	20	1	0.6	0.020

The main culprit of the nearly constant hunting in the initial dynamic data is therefore the large effective integral gains. From building data, however, there is still a clear dependence on operating conditions as warmer ambient temperatures reduce the prevalence of hunting. Implementing a properly tuned cascaded controller will therefore inherently eliminate oscillations due to poor tuning as well as reduce variations in performance due to changing operating conditions.

For initial cascaded tuning, the LOOP sampling time was $t_s = 1$ s with an initial inner loop gain of $k_L = 0.5$. The gains pg and ig for AHU1 were used as initial gains for the tuning process. The cascaded loop gains were therefore $k_{pc} = 0.2$ and $k_{ic} = 0.015$, which correspond to the initial LOOP gains $pg_c = 100$ and $ig_c = 7.5$ used with Algorithm 2. These calculations, including for the LOOP bias term, are given by Equations (11) and (12).

$$pg_c = pg - 1000k_L = 600 - 1000(0.5) = 100 \qquad \& \qquad ig_c = ig = 7.5, \tag{11}$$

$$B = B_2 - k_L B_1 = 50\% - \left(0.5\frac{\%}{°F}\right)\left(\frac{65\,°F + 55\,°F}{2}\right) = 20\%, \tag{12}$$

After some initial testing, the inner loop gain was increased to $k_L = 1$ to amplify the linearization effect of the cascaded controller. Due to the additional issues discussed below, the integral gain was slowly decreased to $ig_c = 2.5$. With these gains, the system showed a qualitative improvement in performance as seen in Figure 11. This improvement represents incremental progress with notable reductions in oscillation period and magnitude. After the remaining issues were fixed, the final integral gains for each unit were increased to 7.5, 10, 10, and 7.5, respectively.

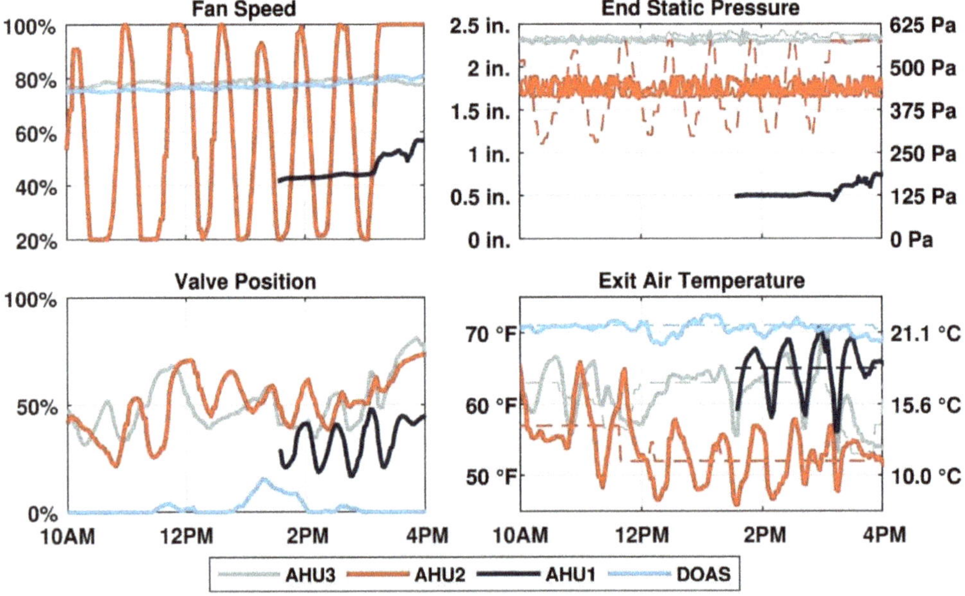

Figure 11. Performance data for Building 1600 on 20 March 2018 after initial cascaded loop tuning. Performance is improved but a fault with end static pressure sensors for AHU2 is exposed.

3.3.2. Problem 2—Failed End Static Pressure Sensors

As seen in Figure 11, fan speed for AHU2 hunts periodically throughout a normal day. The architecture of Figure 3 shows that the fan speed is used to maintain a certain static pressure at given points in the system ducting. Usually End Static Pressure (ESP) sensors are located at a point two-thirds along the longest path of the ducting. Given the L-shape of Building 1600, floors 2 and 3 have two ESP sensors.

In normal operation, the building code takes the minimum reading from the two ESP sensors as the input to the static pressure control loop. However, on floor 2 both sensors had failed, outputting a constant value that did not change with changes in supply fan speed. This had the effect of breaking the ESP feedback loop at the red mark shown in Figure 3, effectively introducing a constant disturbance between ESP setpoint and the fan speed control. While unmeasurable from the failed ESP sensors, the effect of the hunting fan speed was still observable through the damper command calculation. As dampers at each zones VAV box closed to accommodate rising ESP due to the increased fan speed, the ESP setpoint controller would lower the ESP setpoint. This process would reverse and eventually cause the observed sustained oscillation in the ESP setpoint. As soon as ESP sensors on floor 2 were replaced, the oscillations in AHU2 fan speed was eliminated giving the slightly improved results of Figure 12. Note that although AHU2 is parallel to AHU1 and AHU3, the hunting fan speed acted as a disturbance, affecting the distribution of fresh air being delivered to every AHU.

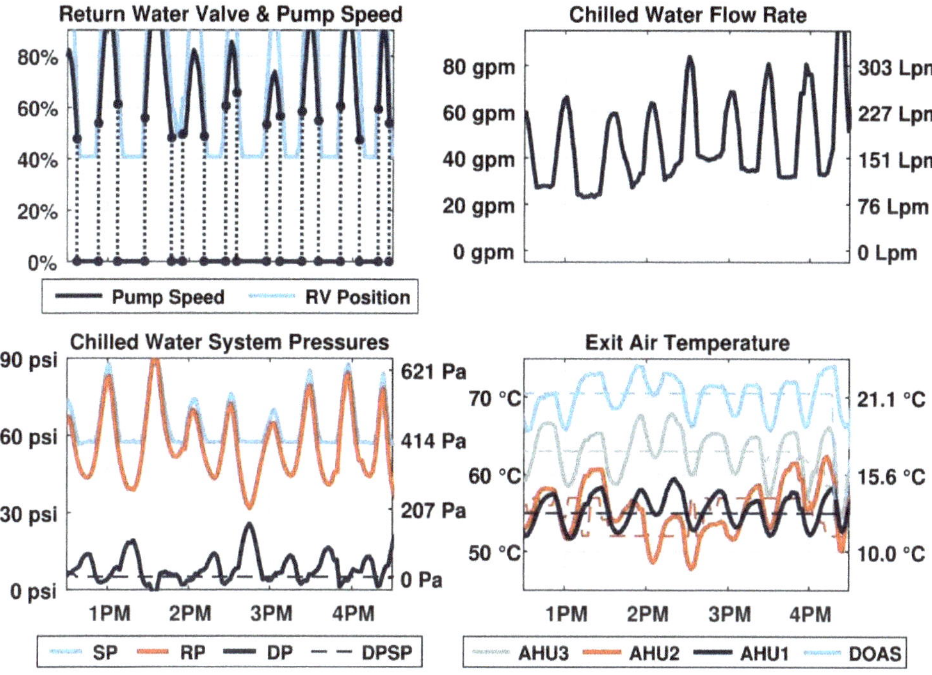

Figure 12. Data from Building 1600 on 16 April 2018 shows synchronized oscillations in AHU discharge air temperature due to short cycling of the building CHW pump, simultaneous actuation of the pump and the building return water valve, and a failed return water pressure sensor (RP).

3.3.3. Failed CHW System Pressure Sensor and Control Issue

After fixing the ESP sensor, a synchronized oscillation in all four AHUs in Building 1600 began to manifest (see Figure 12). Due to the configuration of the system, an issue with the DOAS was suspected as oscillations in discharge air temperature for that unit could propagate to the other three units. Trouble shooting proved inconclusive as simple valve stiction tests such as [19] failed to positively identify the issue.

In early April 2018, local weather conditions were cold enough that no conditioning of fresh air was needed from the DOAS. Despite the stable supply fresh air temperature being delivered to AHUs 1–3, discharge air temperatures still displayed the same synchronized oscillations. Their persistence strongly indicated that another upstream disturbance besides the DOAS was causing the oscillations.

Such a disturbance was determined to be coming from the building chilled water (CHW) supply system. As seen in Figure 13a, the system consists of two actuators (a pump and a valve) and sensors to measure Differential Pressure (DP). The CHW controller seeks to maintain a Differential Pressure Setpoint (DPSP) between the building supply and return water lines. DPSP is determined through a rule set that uses a time averaged Root-Mean-Square (RMS) valve position from the four AHUs. A PI controller operates on DP error to output DPLOOP $\in [20, 100]$, a demand variable that is interpolated to determine settings for the return water valve position and pump speed. A deadband block in the pump control is meant to prevent short cycling of the pump and to ensure that the pump and valve are actuating separately.

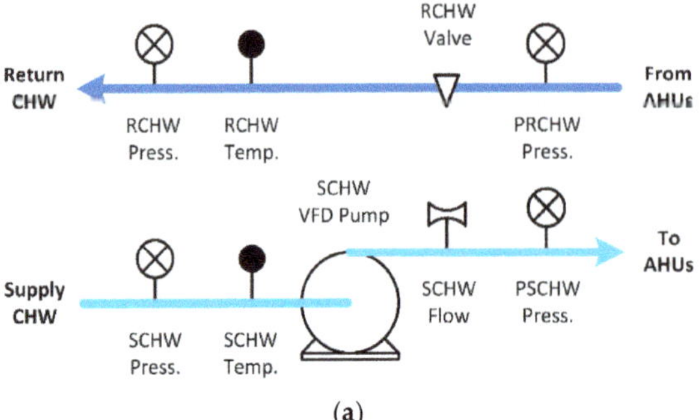

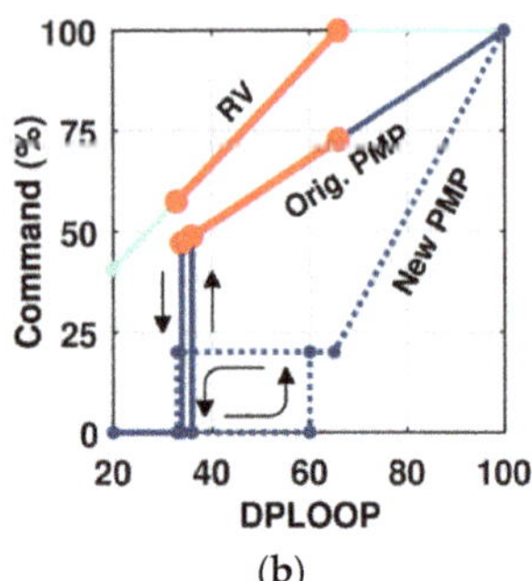

Figure 13. (**a**) Schematic of Building 1600 CHW supply system; (**b**) Comparison of original and new CHW system pump and valve control. Actuation overlap of return water valve and pump control (shown in red) resolved by adjusting deadband settings.

As seen in Figure 12, the building CHW pump short cycles ON/OFF several times throughout the day. These cycles correspond to the periodic oscillations seen in AHU discharge air temperature. The sudden changes in pump speed cause sharp changes in building CHW flow rate which affects flows to each individual AHUs simultaneously. The short cycling was due to several concurrent system issues. Firstly, the deadband region meant to prevent rapid pump cycles was extremely small turning ON the pump when DPLOOP rose above 36 and OFF when it dropped below 34. As DPLOOP would drop below 34 almost immediate after the pump switched ON, the pump would cycle OFF after the five-minute sampling time of the DPSP rules block. Additionally, because the linear interpolation for the return water valve was for $20 \leq$ DPLOOP ≤ 66, both the pump and the valve were actuating simultaneously for a significant range of operation shown graphically in Figure 13b. Secondly, the return CHW pressure (RP) sensor had a fault causing large swings in measurements. The resulting oscillation was propagated through the supply CHW PI controller causing the pump and valve to oscillate. Finally, the integral gain in the SCHW (Supply Chilled Water) PI loop was ig $= 125$ with a sampling time $t_s = 1$. The large integral gain caused DPLOOP to hunt even for small errors in DP. Each of these identified issues was fixed by working with campus utilities. The CHW program was changed to expand the deadband zone and alter interpolations to regions where the pump and valve actuate separately (see Figure 13b). The return pressure sensor was also replaced and calibrated and the DPLOOP PI controller was returned.

After fixing CHW supply issues, the system began to operate fault free. Initial results showed that hunting had been completely eliminated and that large disturbance oscillations due system faults had been removed. However, tracking performance was poor as cascaded controllers had been detuned to tolerate the many system faults. After retuning the controllers to improve tracking performance, system results are similar to those from Figure 14. Comparing with the original performance seen in Figure 10, implementing cascaded control and fixing the multiple faults revealed by the improved chilled water valve control has significantly improved building performance. Note that at the end of the tuning process, final cascaded LOOP gains were $k_L = 1$, $\text{pg}_c = 100$, and $\text{ig}_c = 10$ except for the DOAS whose integral gain was $\text{ig}_c = 7.5$.

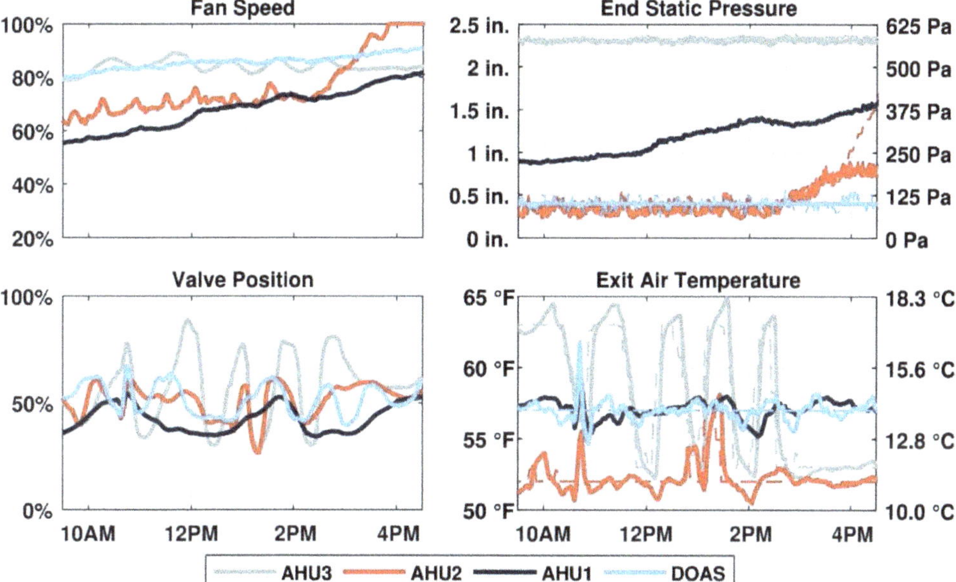

Figure 14. Data from Building 1600 on 14 May 2018 shows greatly improved building AHU discharge air temperature control due to cascaded control implementation and fixing revealed system faults.

4. Discussion

Having established that cascaded control can significantly improve the performance of AHU exit air temperature controllers, one final question is where the costs due to poor AHU valve control originate. The hidden and measurable costs of hunting behavior will be reviewed in this section, and the measured cost savings from the elimination of hunting results will be quantified.

4.1. Increased Replacement Costs

Most literature asserts that hunting will cause excessive component wear, eventually leading to increased replacement costs. While true, this cost is hard to estimate and is likely small because it only accounts for lost operation time as replacement actuators would be purchased regardless of hunting behavior.

4.2. Retuning Costs

Retuning costs due to occupant discomfort from hunting behavior are more easily estimated. Eliminating the time and expense of sending technicians to recommission each AHU on a seasonal basis has the potential for large savings in labor costs. However, such costs may vary across time and locations. In order to show the time-invariant effect of cascaded control, this paper will focus on quantifying the energy saving costs that require almost no investment costs.

4.3. Increased Energy Costs

There are additional energy costs associated with hunting behavior. Due to the nonlinear power consumption of most HVAC actuators (e.g., fan/pump power is cubically related to speed), more power is consumed above a nominal input than below. Thus, for oscillating signals, the average power consumed is greater than for the corresponding fixed signal.

This section quantifies the additional energy costs due to hunting behavior and the corresponding savings from the improved performance due to the implementation of cascaded controllers. As discussed in the previous section, equipment corrections at Building 1600 from the previous section have resulted in the HVAC system operating fault-free, and a comparison of daily energy usage and costs can be made by alternating between the original PI and new cascaded controllers. As the only difference will be the AHU temperature control architecture, assuming similar loads, any differences in energy usage are due to control type alone.

To estimate daily resource consumption, additional information about the building HVAC system was collected. The nominal power of the four AHU fans and CHW pump are known and given in Table 4. Each of these motors are variable speed, normally operating at some fraction of their maximum speed. The part load power can be found using standard fan/pump affinity laws leading to the instantaneous electrical power estimate of Equation (13) where $\omega_i \in [0, 100]$ are speeds and the subscripts '*oa*' and '*p*' are for the DOAS fan and SCHW pump, respectively. Each building on campus is billed at a rate of approximately 0.08 USD/kWh of electricity which represents the average cost of electricity production at the campus generation sites.

$$P_{elec} = 18.65\omega_1^3 + 18.65\omega_2^3 + 14.92\omega_3^3 + 5.595\omega_{oa}^3 + 14.92\omega_p^3 \quad [\text{kW}] \tag{13}$$

Table 4. Building 1600 HVAC Motor List.

Unit	AHU1	AHU2	AHU3	DOAS	SCHW
Type	Fan	Fan	Fan	Fan	Pump
Power	25 HP	25 HP	20 HP	7.5 HP	20 HP

The volume of chilled water used daily by the HVAC system in Building 1600 is monitored in real time. However, the associated costs must be estimated since campus utilities does not bill by volume, but by energy content. As all conditioning water is returned to the central processing plants, buildings that require more cooling will return warmer water. Solely billing on volume usage therefore does not capture the additional cost of re-cooling warmer return water. Calculating energy used by the HVAC system requires monitoring chilled water flow rate as well as the temperature differential between supply and return water. The instantaneous power delivered by the CHW is given by Equation (14).

$$P_{CHW} = c_p\, \rho\, \dot{V}\, \Delta T = 0.1463\, \dot{V}\, \Delta T \quad [\text{kW}] \tag{14}$$

This estimated power does not include costs associated with chilled water production. An estimate for production cost is found by assuming an efficiency from a comparable air-cooled chiller system. Coefficient of Performance (COP) curves for such a system are shown in Figure 15a, based on the model from [20]. The chiller was sized at 400 kW using the 98th percentile of instantaneous chilled water power observed for the period between May through December of 2018. This assures that the unit will meet almost all demand by the building chilled water system with nominal operation in a region of high COP. To calculate the chiller electric power, Equation (15) divides the instantaneous chilled water consumption by a cubic interpolation of the chiller COP based on part load (L_p) and outdoor air temperature (T_{oa}).

$$P_c = \frac{P_{CHW}}{\text{COP}(L_p, T_{oa})} \tag{15}$$

The COP curves shown in Figure 15a are used to estimate the additional costs associated with oscillations in chiller load. As discussed previously, hunting results in above average energy use for systems with nonlinear power profiles. Thus, for the chilled water, the additional cost of hunting is expected to be greatest in the regions where COP surface is the most nonlinear. However, hunting is most prevalent in times where the system part load is low (i.e., in cool weather with minimal demand). The COP curve in that region is essentially linear indicating that there will be minimal wasted energy due to hunting oscillations. To illustrate this effect, the cost penalties for a sinusoidal chilled water demand (5% variation around the nominal load) are given in Figure 15b, which shows wasted energy in the region of interest to be between 0.05 and 0.1%. This level of wasted energy might seem insignificant. However, a 5% variation around nominal load is a conservative estimation, and ±20% or more variation can often be observed (see Figures 10 and 11). Additionally, Figure 15b shows a sharp increase in wasted energy when the outside temperature is low, and the load is high. Systems tuned for high temperatures can suffer significantly with exacerbated level of wasted energy.

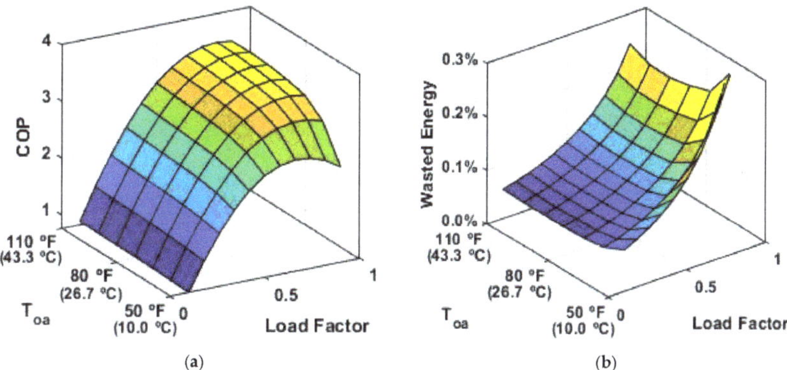

Figure 15. (**a**) COP surface for a rooftop air-cooled chiller system; (**b**) Estimated wasted energy due to ±5% sinusoidal hunting chiller load factor.

4.4. Estimated Cost Savings at Building 1600

The AHU discharge air temperature control (i.e., valve control) was switched between the original hunting PI control and the new cascaded control approximately every two weeks from May 2018 through May 2019. Leverage and standardized residual methods were used to filter outliers from daily data and to ensure a consistent comparison of the two approaches. More details on the statistical method used for the outliers can be found in Appendix A. Energy consumption, costs, and cooling degree days were calculated daily to generate Figure 16, comparing the two control architectures.

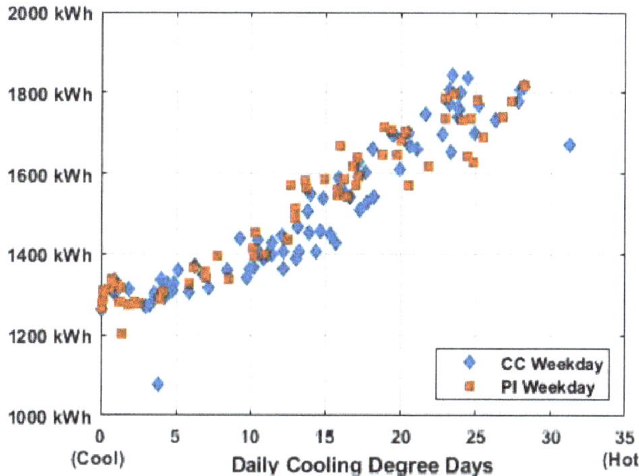

Figure 16. Daily HVAC energy production and usage costs for Building 1600 from May 2018 through May 2019.

Analysis of the two data sets (i.e., energy use with PI control and cascaded control) found statistically significant differences in bias values but not slopes. Equation (16) shows a difference of 54.6 kWh in daily AHU energy consumption for models fit with a constraint on equal slopes. The smaller intercept value for the cascaded controller indicates that it is better able to eliminate oscillatory behaviors that result in wasted energy and can better follow setpoints due to faster transient responses. The reduced bias value corresponds to a 2.2–4.4% savings in total energy consumed by the AHU system. This analysis, however, does not include other important cost factors of hunting. The true costs of hunting behavior would also include an increase in maintenance costs, resulting from the frequent actuation. When the maintenance and energy savings are combined with the economical and ease of implementation, cascade control in buildings is strongly recommended.

$$E_{PI} = 1488.3 + 35.8 \times CDD \quad [\text{kWh}]$$
$$E_C = 1393.7 + 35.8 \times CDD \quad [\text{kWh}]$$
(16)

5. Conclusions

Hunting behavior in buildings causes an increase in operating cost arising from: (1) increased replacement frequency of components due to excessive component wear, (2) retuning cost due to occupant discomfort, and (3) increased energy cost. Among these costs, this paper has focused on quantifying the energy savings from the detection and elimination of hunting behavior in several buildings on a university campus through the implementation of cascaded control loops. Shown in Figures 4 and 5, hunting in Building 1497 valve was significantly reduced from cycling 10 to 20 times per hour to no oscillations after the cascaded control implementation. As a result, exit air temperature that used to vary more than 2 °F was reduced within 0.5 °F. Shown in Figures 10 and 14, valve hunting in Building 1600 AHU2 and AHU3 decreased, with oscillations in valve position decreasing in magnitude from 30% to 20% after the implementation of cascaded control. Additionally, the actuation frequency decreased from approximately 3 to 4 cycles per hour to 1 cycle per hour for AHU3 and AHU2. Results at Building 0474 were mixed with a slight improvement in tracking performance but an overall improvement in the consistency of AHU discharge air temperature regulation. An estimation of the costs of poor AHU discharge air temperature control was presented for Building 1600. These results show 2.2–4.4% energy cost savings due to the elimination of chilled water valve hunting, with further potential savings associated with reduced maintenance costs. Further work and detailed analysis can be found in [4].

Results also show the mechanism for hunting behavior to cause a more measurable loss of efficiency in HVAC systems. While chilled water production may have minimal nonlinearity around a given operating point, fan and pump affinity laws have a consistent nonlinear relationship between speed and power. Should hunting be induced in those actuators due to their poor control or that of an upstream controller (i.e., chilled water control), energy savings will be more prevalent. This paper has shown that cascaded control improves tracking performance, reduces the need for seasonal retuning due to its inherently non-linearity limiting nature, and is easy to implement with a single LOOP command. While the scope of this paper has been on improving supply water control, many campus buildings examined by the authors have shown hunting behavior in their AHU supply fan loops. Implementation of cascaded control loops at these buildings can be used to more easily establish energy penalties related to poor PI control design.

Author Contributions: Conceptualization and research supervision, B.P.R.; methodology and investigation, C.P.; analysis and data curation, C.P. and D.P.; writing—original draft preparation, C.P.; writing—review and editing, C.P., B.P.R. and D.P. All authors have read and agreed to the published version of the manuscript.

Funding: This research was funded by National Science Foundation, grant number CMMI-1563361.

Data Availability Statement: Not applicable.

Acknowledgments: Special thank you to the Building Automation Supervisor, Christopher Dieckert, and the Texas A&M University Utilities and Energy Services for their support in conducting this research.

Conflicts of Interest: The authors declare no conflict of interest.

Nomenclature

B	Bias
C	Cooling demand
c_p	Heat capacity
CDD	Cooling Degree Day
COP	Coefficient of Performance
D	Damper demand
e	Error
E	Energy
F	Flow demand
G	System transfer function
k	Control gain
L	Inner loop transfer function
L_p	Part load
P	Power
PID	Proportional-Integral-Derivative controller
r	Reference input (set point)
$RMSE$	Root-Mean-Square error
s	Laplace variable
t	Time
u	Control signal
$\dot{V}$	Volumetric flowrate
y	Output
δ	Valve opening
θ	Damper position
ρ	Density
σ	Operating condition
ψ	Nonlinear gain
ω	Rotational speed
Subscripts	
C	Cascaded
CHW	Chilled water
D	Derivative
$elec$	Electric
i, L	Inner loop
I	Integral
o	Outer loop
oa	Outside air
p	Proportional
pc	Proportional-cascade
PI	Proportional-Integral
s	Sample
set	Setpoint

Appendix A

A set of linear regressions can be generalized into a matrix representation as in Equation (A1). Based on the regression fit, prediction of dependent variables can be accomplished using Equation (A2). With matrix manipulation, prediction can be expressed in terms of observation y as in Equation (A3). The matrix H can then be defined and maps the observation y to the prediction $\hat{y}$ as in Equation (A4).

$$Y = X\beta + e, \tag{A1}$$

$$\hat{y} = Xb, \tag{A2}$$

$$\hat{y} = X(X'X)^{-1}X'y, \tag{A3}$$

$$\hat{y} = Hy, \tag{A4}$$

The diagonal elements in the H matrix are called leverage points. These points represent the effect of observation y_i on prediction value $\hat{y}_i$. Points with high values of leverage points can be labeled as outliers and be filtered out. Another way to define leverage point is shown in Equation (A5). With expressions for the leverage points defined, a general rule-of-thumb of filtering criterion for leverage is presented in Equation (A6).

$$h_{ii} = \frac{1}{n} + \frac{(x_i - \bar{x})^2}{\sum_{j=1}^{n}(x_j - \bar{x})^2}, \tag{A5}$$

$$h_{ii} > \frac{6}{n}, \tag{A6}$$

As a second set of filters, the standardized residuals method was used to further process the building data. Standardized residual is defined as the ratio of the prediction error, e_i, over the standard deviation of the error (Equation (A7)). Points with standardized residual magnitude above 95% percentile confidence level of t distribution outlined in Equation (A8) were labeled as outliers and filtered out, where n is number of observations and k is the number of predictors.

$$e_i^* = \frac{e_i}{sd(e_i)}, \tag{A7}$$

$$t(n - k - 2), \tag{A8}$$

After the outliers had been removed, analysis of covariance was conducted to separate out the covariate effect of cooling degree days on the dependent variable, total energy consumption. In the analysis of covariance, cascade and PI control are classified by $\lambda = 1$ and $\lambda = 0$, respectively. In the analysis of data, the two different control algorithm distributions are fitted to one of the following cases:

Case I: Different intercepts and different slopes

$$Y = \beta_0 + \beta_1\lambda + \beta_2 X + \beta_3\lambda X + e, \tag{A9}$$

Case II: Different intercepts but same slopes

$$Y = \beta_0 + \beta_1\lambda + \beta_2 X + e, \tag{A10}$$

Case III: Same intercepts and same slopes

$$Y = \beta_0 + \beta_3 X + e, \tag{A11}$$

Data from Building 1600 was used to test which of these cases best fit the results. Using Case I, Table A1 is generated. For Case I, β_3 had high p-value and therefore, the two data sets have no significant difference in their slopes. Case II was checked for the two different controller data sets. p-values from Table A1 show significant differences in intercepts with same slopes. β_0 gives the intercept for the PI controlled dataset and $\beta_0 + \beta_1$ gives the intercept for the cascaded control dataset. Case III was not performed since Case II showed statical significance.

Table A1. Case I and II Fitted with Building Data.

	Coefficient		t	$P > |t|$
Case I	β_1	−92.8	−2.947	0.004
	β_3	2.9	1.441	0.152
Case II	β_0	1448.3	-	-
	β_1	−54.6	−3.202	0.002
	β_2	35.8	-	-

References

1. Price, C.; Liang, S.; Rasmussen, B. HVAC Nonlinearity Compensation Using Cascaded Control Architectures. *ASHRAE Trans.* **2015**, *121*, 217–231.
2. Chintala, R.; Liang, S.; Price, C.; Rasmussen, B. Identification and Elimination of Hunting Behavior in HVAC Systems. *ASHRAE Trans.* **2015**, *121*, 294–306.
3. Hägglund, T. A Control-Loop Performance Monitor. *Control Eng. Pract.* **1995**, *3*, 1543–1551. [CrossRef]
4. Price, C.R. Cascaded Control for Improved Building HVAC Performance. Ph.D. Thesis, Texas A&M University, College Station, TX, USA, 2018.
5. Hou, Z.-S.; Wang, Z. From Model-Based Control to Data-Driven Control: Survey, Classification and Perspective. *Inf. Sci.* **2013**, *235*, 3–35. [CrossRef]
6. Desborough, L.; Miller, R. Increasing Customer Value of Industrial Control Performance Monitoring -Honeywell's Experience. *AIChE Symp. Ser.* **2002**, *98*, 169–189.
7. Blevins, T.L. PID Advances in Industrial Control. *IFAC Proc. Vol.* **2012**, *45*, 23–28. [CrossRef]
8. O'Dwyer, A. *Handbook of PI and PID Controller Tuning Rules*, 3rd ed.; Imperial College Press: London, UK, 2009.
9. Li, Y.; O'Neill, Z.D.; Zhou, X. Development of Control Quality Factor for HVAC Control Loop Performance Assessment I—Methodology (ASHRAE RP-1587). *Sci. Technol. Built Environ.* **2019**, *25*, 656–673. [CrossRef]
10. Price, C.; Rasmussen, B. Optimal Tuning of Cascaded Control Architectures for Nonlinear HVAC Systems. *Sci. Technol. Built Environ.* **2017**, *23*, 1190–1202. [CrossRef]
11. Elliott, M.S.; Estrada, C.; Rasmussen, B.P. Cascaded Superheat Control with a Multiple Evaporator Refrigeration System. In Proceedings of the 2011 American Control Conference, San Francisco, CA, USA, 29 June–1 July 2011; pp. 2065–2070.
12. Price, C.R.; Rasmussen, B.P. Effective Tuning of Cascaded Control Loops for Nonlinear HVAC Systems. In *American Society of Mechanical Engineers Digital Collection, Proceedings of the ASME 2015 Dynamic Systems and Control Conference, Columbus, OH, USA, 28–30 October 2015*; American Society of Mechanical Engineers: New York, NY, USA, 2016.
13. Guo, C.; Song, Q.; Cai, W. A Neural Network Assisted Cascade Control System for Air Handling Unit. *Ind. Electron. IEEE Trans.* **2007**, *54*, 620–628. [CrossRef]
14. Elliott, M.; Walton, Z.; Bolding, B.; Rasmussen, B. Superheat Control: A Hybrid Approach. *HVAC&R Res.* **2009**, *15*, 1021–1043. [CrossRef]
15. *Applications Guide PID Control in Tracer Controllers*; American Standard Inc.: Piscataway, NJ, USA, 2001. Available online: https://www.trane.com/Commercial/Uploads/Pdf/1007/CNTAPG002EN.PDF (accessed on 31 August 2022).
16. Yin, X.; Wang, X.; Li, S.; Cai, W. Energy-Efficiency-Oriented Cascade Control for Vapor Compression Refrigeration Cycle Systems. *Energy* **2016**, *116*, 1006–1019. [CrossRef]
17. Bengea, S.C.; Kelman, A.D.; Borrelli, F.; Taylor, R.; Narayanan, S. Model-Predictive Control for Mid-Size Commercial Building HVAC: Implementation, Results and Energy Savings. In Proceedings of the Second International Conference on Building Energy and Environment, Boulder, CO, USA, 1 August 2012.
18. *APOGEE Powers Process Control Language (PPCL) User's Manual*; Siemens Building Technologies, Inc.: Florham Park, NJ, USA, 2000.
19. Choudhury, M.A.A.S.; Kariwala, V.; Shah, S.L.; Douke, H.; Takada, H.; Thornhill, N.F. A Simple Test to Confirm Control Valve Stiction. *IFAC Proc. Vol.* **2005**, *38*, 81–86. [CrossRef]
20. Yu, F.W.; Chan, K.T. Optimizing Condenser Fan Control for Air-Cooled Centrifugal Chillers. *Int. J. Therm. Sci.* **2008**, *47*, 942–953. [CrossRef]

buildings

Article

Model-Based Control Strategies to Enhance Energy Flexibility in Electrically Heated School Buildings

Navid Morovat [1,*], Andreas K. Athienitis [1], José Agustín Candanedo [1,2] and Benoit Delcroix [3]

[1] Center for Zero-Energy Building Studies, Concordia University, Montreal, QC H3G 1M8, Canada; andreask.athienitis@concordia.ca (A.K.A.); jose.candanedoibarra@canada.ca (J.A.C.)
[2] CanmetENERGY, Natural Resources Canada, Varennes, QC J3X 1P7, Canada
[3] Laboratoire des Technologies de l'Énergie, Hydro-Québec Research Institute, Shawinigan, QC G9N 0C5, Canada; delcroix.benoit@hydroquebec.com
* Correspondence: n_morova@encs.concordia.ca

Abstract: This paper presents a general methodology to model and activate the energy flexibility of electrically heated school buildings. The proposed methodology is based on the use of archetypes of resistance–capacitance thermal networks for representative thermal zones calibrated with measured data. Using these models, predictive control strategies are investigated with the aim of reducing peak demand in response to grid requirements and incentives. A key aim is to evaluate the potential of shifting electricity use in different archetype zones from on-peak hours to off-peak grid periods. Key performance indicators are applied to quantify the energy flexibility at the zone level and the school building level. The proposed methodology has been implemented in an electrically heated school building located in Québec, Canada. This school has several features (geothermal heat pumps, hydronic radiant floors, and energy storage) that make it ideal for the purpose of this study. The study shows that with proper control strategies through a rule-based approach with near-optimal setpoint profiles, the building's average power demand can be reduced by 40% to 65% during on-peak hours compared to a typical profile.

Keywords: energy flexibility; model-based control strategies; school buildings; measured data

Citation: Morovat, N.; Athienitis, A.K.; Candanedo, J.A.; Delcroix, B. Model-Based Control Strategies to Enhance Energy Flexibility in Electrically Heated School Buildings. *Buildings* **2022**, *12*, 581. https://doi.org/10.3390/buildings12050581

Academic Editor: Chi-Ming Lai

Received: 24 March 2022
Accepted: 27 April 2022
Published: 30 April 2022

Publisher's Note: MDPI stays neutral with regard to jurisdictional claims in published maps and institutional affiliations.

1. Introduction

Electric utilities consider demand-side management (DSM) a key solution to reduce peak power demand. In periods of peak power demand, using DSM is more cost-effective than operating peaking power plants or purchasing power from other jurisdictions [1]. DSM can have an even more significant effect on the grid when integrated with renewable energy sources (RESs). Buildings are important components of smart electricity grids; they can provide flexible services to reduce peak loads and shift demand in accordance with local RES production, such as energy storage in thermal mass and batteries [2,3], charging of electric vehicles [4], and HVAC system adjustments [5]. Ruilova et al. [6] defined energy flexibility in buildings as "the possibility to deviate the electricity consumption of a building from the reference scenario at a specific point in time and during a certain period". Annex 67 of the IEA Energy in Buildings and Communities Programme (IEA-EBC) defined energy flexible buildings as those with "the ability to manage [their] demand and generation according to local climate conditions, user needs, and grid requirements" [5]. Energy flexibility takes into consideration two-way communications between buildings and the power grid. In this way, buildings are regarded not as consumers but as prosumers [7]. Energy flexibility can be referred to in two ways: thermal energy storage and shifting equipment operation. According to the first approach, the energy consumption of a specific electrical device can be predicted based on the thermal properties of the device or building to minimize electricity consumption. In the second approach, some electrical devices can be controlled to shift the electricity demand to periods with lower electricity prices or greater renewable energy generation [8]. Based on the published literature and presented in IEA-EBC Annex 67, increasing energy flexibility for the design of smart energy systems and buildings is influenced by (1) physical features of the building [2,3], (2) heating, ventilation, and air conditioning (HVAC) systems [5], (3) appropriate control systems and strategies [5], and (4) IEQ requirements [9]. In this context, an effective application of control strategies within HVAC systems is essential for increasing buildings' efficiency [10–12] and energy flexibility [13].

Finck et al. [14] developed a method and tested it under real-life conditions, including the stochastic behavior of occupants and the dynamic behavior of the building and heating system. They used key performance indicators to quantify energy flexibility by considering: (1) energy and power, (2) energy efficiency, and (3) energy costs. They found that this categorization helps to make clear the benefits of using flexibility indicators in real-life applications. Junker et al. [15] presented a methodology for evaluating energy flexibility based on the flexibility function, to describe how a particular smart building or cluster of smart buildings reacts to a penalty signal. Coninck and Helsen [16] developed a methodology to quantify flexibility in buildings based on the cost curve. The methodology returns the amount of energy that can be shifted and the costs of this load shifting. Tauminia et al. [17] proposed a multidisciplinary approach to finding trade-offs between the need to limit environmental impacts and the trend toward higher building energy performance. They found that an oversized photovoltaic (PV) system is not the best solution for load-matching, grid interactions, and environmental impacts in the absence of storage systems. They noted that installing a storage system in conjunction with the appropriate size of a PV system would result in an improved load-matching of the building and reduce grid dependence at low generation times.

Montreal (Québec, Canada) is categorized in climate zone 6 [18], meaning it experiences extreme cold weather during winter. Québec generates most of its electricity (99.8%) from hydroelectric plants, and most commercial buildings rely on electricity as their primary or only energy source [19]. Thus, during cold weather in Québec, the morning peak load (6:00 a.m. to 9:00 a.m.) and evening peak load (4:00 p.m. to 8:00 p.m.) put a strain on the electrical grid [20]. Thus, it is imperative to analyze energy consumption and develop control strategies that effectively reduce and shift peak electricity demand due to heating in buildings. In this context, obtaining a model that provides reliable predictions and can be implemented in real controllers is crucial for optimizing building performance.

The Québec province has over 2600 schools, reaching over 1 million students and almost 100,000 teachers and other staff [21]. Therefore, quantification of energy flexibility in school buildings has a significant role in providing a safe and efficient operation of the future resilient grid. Additionally, indoor environmental quality (IEQ) has a considerable impact on the health and well-being of teachers and students. Thus, simultaneously meeting the need to improve energy flexibility as a grid requirement and the growing demands for environmental performance (especially during/after the COVID pandemic) is of utmost importance to be considered. To achieve these goals, we need to develop models for a school building that provide reliable predictions and can be generalized for widespread deployment in schools.

Although school buildings contribute considerably to the total energy needs, few studies have been focused on school buildings in Canada [22,23]. They examined the energy use intensity (EUI) of 129 elementary and junior high schools in Manitoba, Canada, using data collected from 30 school buildings over ten years. They found that the average EUI at a K-12 school is 127 kWh/m^2/year, 264 kWh/m^2/year at a junior high school, and 270 kWh/m^2/year at an elementary school. They stated that energy consumption might differ between K-12 and elementary and secondary schools due to differences in equipment and activities. Another study by Ouf et al. [23] examined the use of electricity and natural gas in Canadian schools. They divided the schools into three categories based on the year they were built: before 2004, between 2004 and 2013, and after 2013. They found that the electricity EUIs before 2004 was 58 kWh/m^2/year, between 2004 and 2013 was 116 kWh/m^2/year, and after 2013 was 125 kWh/m^2/year. Newly constructed schools are more energy efficient in heating and cooling, but school electricity usage has increased due to the electrification of heating systems and additional teaching equipment.

Building energy performance simulation (e.g., EnergyPlus^{TM} and TRNSYS) is a popular approach to studying school buildings. However, studies based on control-oriented models and measured sensor data of schools are relatively rare. This study investigates measured field data in an archetype electrically heated school building in cold regions. The main objectives of this paper are to: (1) develop a methodology to create data-driven control-oriented models that facilitate developing and assessing the impact of alternative control strategies in the schools; (2) propose different control strategies aimed to enhance the energy flexibility potential of school buildings; and (3) introduce key performance indicators (KPIs) as key parameters to quantify energy flexibility at the zone and building levels while considering IEQ.

2. Methodology

In recent years, international initiatives have recognized the need for a methodology to assess energy flexibility in buildings. This paper outlines the following steps in the method:

1. Identification of the system and load.
2. Characterization of flexibility.
3. Analyzing the impact of scenario modeling on the demand profiles.
4. Proposing key performance indicators to facilitate interaction between building operators, aggregators, and utility companies.

In addition, the method should be scalable and easy to implement [24]. Following these steps, this paper aims to model and enhance the energy flexibility of electrically heated school buildings through a rule-based approach with near-optimal setpoint profiles based on the archetype grey-box models. These archetype models can be adjusted depending on the specific features of the building. This study presents a practical methodology that facilitates the modeling and widespread implementation of appropriate control strategies in school buildings.

2.1. Data-Driven Grey-Box Model

Data-driven grey-box models ensure both physical insight and the reliability of measured data. Literature review indicates that grey-box models are also suitable for demand-side management in smart grids [25–28]. Gouda and Danaher [25] proposed a second-order model in which each construction element is modeled using three resistances and two capacitances. Candanedo and Dehkordi [26] presented a generalized approach for creating reduced-order control-oriented models. Their methodology can be implemented in building simulation tools to generate simplified models automatically. Bacher and Madsen [27] developed a statistical method for identifying models in building thermal studies. Reynders et al. [28] analyzed two detached single-family houses in Belgium. These two buildings represent two extreme cases of detached single-family houses in Belgium regarding insulation level (high and low insulation levels). They used data obtained from detailed building simulations with the IDEAS library in Modelica software. This study investigated five grey-box model types, ranging from first- to fifth-order models.

In grey-box models, choosing an appropriate level of resolution is essential, as it directly affects the parameter tuning and calculation time. A high-order model containing too many parameters requires information that is not often available with adequate accuracy. An oversimplified model may not be accurate enough to help make decisions. Therefore, obtaining a model that provides reliable predictions and can be implemented in real controllers is crucial for optimal building performance.

This paper investigates the accuracy of data-driven grey-box models for energy demand simulation and energy flexibility analysis. Providing a closer link between smart grids and smart buildings requires appropriate control strategies. Thus, this study presents an application of the developed model for control purposes based on smart grid requirements. The following steps are used to develop grey-box models and quantify energy flexibility:

1. Real building measurement data are collected from the smart meters installed in the archetype zones. Data included variables such as electricity consumption (kW), zone air temperature (°C), weather data, and specific data are related to each zone (e.g., floor heating temperature). All measurements are taken at intervals of 15 min.
2. Numerical models of thermal building control are developed. These models are based on RC model thermal networks.
3. The developed models are calibrated using the collected data. The important parameters are identified using the Sequential Least-Squares Programming (SLSQP) in Python.
4. Appropriate control strategies for zones with the convective system are presented to enhance the energy flexibility available from the building to the grid at specific times, depending on the grid requirement.
5. A building energy flexibility index (BEFI) is applied to quantify dynamic building energy flexibility at the zone and building levels.
6. Predictive control strategies for zones with the convective system and hydronic radiant floor system are presented to use the maximum thermal capacity of the concrete slab, reduce peak load during on-peak hours, and enhance energy flexibility when needed by the grid.

2.2. Governing Equation

A fully explicit finite difference approach is used to solve the energy balance equations at each node in the models. The fully explicit approach assumes that the current temperature of a given node depends only on its temperature and the temperature of the surrounding nodes at a previous time step. By using the heat balance for the control volume, a node's differential equation can be written as [29]:

$$T_i^{t+1} = \frac{\Delta t}{C_i}\left[\sum_j U_{ij}^t(T_j^t - T_i^t) + \sum_k U_{ik}^t(T_k^t - T_i^t) + \dot{Q}_i^t\right] + T_i^t \tag{1}$$

- U_{ij}: Thermal conductance between nodes i and j, W/K;
- U_{ik}: Thermal conductance between nodes i and k that node k has a known temperature, W/K.
- T: Temperature at node i, °C.
- C_i: Thermal capacitance at node i J/K.
- $\dot{Q}$: Heat source at node i, W.
- Δt: Time step, s.

The capacitance of the air node contains a factor C_T (air thermal capacitance multiplier) that accounts for phenomena such as (energy storage in furniture and objects, the time required for air mixing, delay due to ducting and other factors) that in a low-order model result in capacitance with an observed effective value significantly larger than the one calculated using only the physical properties of the air. The air thermal capacitance multiplier ranges between 6 and 10 [30]. In this paper, we use $C_T = 8$. Equation (1) in matrix form is:

$$\left\{\begin{array}{c} T_1 \\ \vdots \\ T_N \end{array}\right\}^{t+1} - \left\{\begin{array}{c} \frac{\Delta t}{C_1} \\ \vdots \\ \frac{\Delta t}{C_N} \end{array}\right\} \odot \left(\begin{bmatrix} -\sum_j^N U_{1j} - \sum_k^M U_{1k} + \frac{C_1}{\Delta t} & U_{12} & \cdots & U_{1N} \\ & \vdots & \vdots & \ddots & \vdots \\ U_{N1} & U_{N2} & \cdots & -\sum_j^N U_{Nj} - \sum_k^M U_{Nk} + \frac{C_N}{\Delta t} \end{bmatrix}\left\{\begin{array}{c} T_1 \\ \vdots \\ T_N \end{array}\right\}^t + \left\{\begin{array}{c} \dot{Q}_1 + \sum_k^M (U_{1kk}T_{kk}) \\ \vdots \\ \dot{Q}_N + \sum_k^M (U_{Nkk}T_{kk}) \end{array}\right\}\right) \tag{2}$$

where

- $\odot$: is an elementwise multiplication operator.
- N: is the number of nodes.
- M: is the number of nodes with known temperatures.

To assure numerical stability in the solution, the time step must be chosen according to the stability criterion defined in Equation (3):

$$\Delta t \leq min\left(\frac{C_i}{\sum U_i}\right) \tag{3}$$

Using Equation (4), the proportional–integral control (PI controller) calculates the heat produced by the heating system at each time step, and the integral part should be reset periodically.

$$\dot{Q}^{t+1} = k_{\text{p}} \cdot \left(T^t_{\text{setpoint}} - T^t_{\text{room.air}}\right) + k_{\text{i}} \int \left(T^t_{\text{setpoint}} - T^t_{\text{room.air}}\right) dt \tag{4}$$

where

- k_{p}: proportional gain of the controller, W/K.
- k_{i}: integral gain of the controller, W/K·s.

Equation (5) illustrates state-space representations of linear differential equation systems. In this equation, $(\mathbf{x})$ is the state vector with n elements, $(\mathbf{u})$ is the input vector with p elements, and $(\mathbf{y})$ represents output vectors with q elements. The vectors are linked by the following matrices: $\mathbf{A}$ (n × n), $\mathbf{B}$ (n × p), $\mathbf{C}$ (q × n), and $\mathbf{D}$ (q × p).

$$\dot{\mathbf{x}} = \mathbf{Ax} + \mathbf{Bu} \qquad \mathbf{y} = \mathbf{Cx} + \mathbf{Du} \tag{5}$$

Temperatures of thermal capacitances are generally considered the system's state in this approach since they have specific physical meaning and are relatively easy to measure [26]. Model identification refers to determining the physical properties of unknown systems based on some experimental or training data. In this paper, the Python function SLSQP is used to minimize the coefficient of variance of the root mean square error (CV-RMSE) as a fit metric. In accordance with ASHRAE Guideline 14, the model should not exceed a CV-RMSE of 30% relative to hourly measured data [31]. By minimizing CV-RMSE, the optimization algorithm determines the equivalent parameters for RC circuits. Equation (6) [31] is used to calculate CV-RMSE, where T_i represents the measurement data, $\hat{T}_i$ represents the simulation results, n corresponds to the total number of observations, and $\overline{T}$ represents the average of all measurements.

$$CV - RMSE(\%) = 100 \times \frac{\sqrt{\left[\sum_{i=1}^{n}\left(T_i - \hat{T}_i\right)^2 / n\right]}}{\overline{T}} \tag{6}$$

This methodology can be used to create archetype RC thermal networks for representative zones. For example, Figure 1 presents a third-order thermal network RC model (4R3C) for zones with a convective system, which is defined by three state variables.

The inputs to this RC thermal network model are outdoor temperature (T_{ext}), solar heat gain (Q_{SG}), internal heat gain from occupants and equipment (Q_{IG}), and heat delivered by local water–air heat pumps (Q_{aux}), and the output is the indoor air temperature. Montreal weather data are used to determine the outdoor temperature and solar irradiance [32]. Then, the performance of the RC thermal network model is validated with measured data. Table 1 presents an overview of the third-order thermal network parameters.

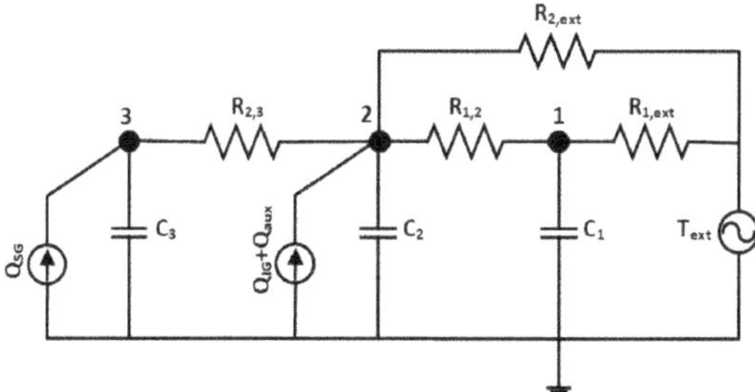

Figure 1. Thermal network model of the zones with convective system.

Table 1. Description of RC thermal network model parameters (4R3C).

Parameter	Description	Parameter	Description
$R_{1,\,ext}$	Resistance of Wall	1	Envelope node
$R_{1,2}$	Resistance between wall and air	2	Indoor air temperature node
$R_{2,3}$	Resistance between floor and air	3	Floor temperature node
$R_{2,\,ext}$	Resistance of infiltration	T_{ext}	Outdoor temperature
C_1	Capacitance of Envelope	Q_{SG}	Solar heat gain
C_2	Capacitance of effective air	Q_{IG}	Internal heat gain
C_3	Capacitance of floor	Q_{aux}	Heating power

2.3. Energy Flexibility Quantification

The term "building energy flexibility" refers to "the ability to deviate from the reference scenario at a specified point in time and for a specified period" [6]. Enhancing energy flexibility is essential for balancing supply and demand on the grid and incorporating renewable energy capacity to reduce peak demand at key periods for the grid. Flexibility is also essential for providing contingency reserves for emergencies (e.g., after a power outage) and enabling dynamic electricity trading. Flexible buildings can also meet immediate or short-term grid needs.

Thus, real-time energy flexibility should be predicted and calculated on short notice (e.g., kilowatts available over the next few hours). This paper presents dynamic building energy flexibility indexes (BEFI) to quantify energy flexibility in school buildings and their interaction with the smart grid. We have described the BEFI concept in two previous conference papers on the topic with a preliminary introduction and case studies [33]. By implementing the flexibility strategy and using the reference as-usual profile, Equation (7) calculates the average $\overline{BEFI}$ at time t for duration Dt, Equation (8) presents the BEFI as a percentage, and Equation (9) quantifies $BEFI\%$ at the building level.

$$\overline{BEFI}(t, Dt) = \frac{\int_t^{t+Dt} P_{ref}dt - \int_t^{t+Dt} P_{Flex}dt}{Dt} \tag{7}$$

$$BEFI\% = \frac{P_{ref} - P_{Flex}}{P_{ref}} \tag{8}$$

$$BEFI_{building} = \sum_1^n BEFI_{zone} \tag{9}$$

A model can be used to determine the difference in power demand (P, unit: Watt) between the reference case (P_{ref}) and the flexible case (P_{Flex}) to determine the available flexibility. This calculation gives the available flexibility at time t. Every hour, the calculation is repeated to give the available flexibility over the period $\overline{BEFI}(t, Dt)$.

3. Description of the Case Study: Electrically Heated School Building

The case study school (Figure 2) is an electrically heated building located in Sainte-Marthe-sur-le-Lac (near Montreal, QC, Canada). The total floor area of this two-story school building is 5192 m^2 (2596 m^2/story). The school includes the following features:

- In operation since 2017.
- Hydronic radiant floor systems in several zones (gym and offices).
- Convective systems in several zones (classrooms, library, kindergarten).
- A 28-loop geothermal system.
- An Electrically heated Thermal Energy storage device (ThermElect) with an 80 kW heating capacity. ThermElect converts electrical power into stored heat when the price of electricity is low (or when demand on the grid is low) and provides heat when demand is high.
- Water–air Heat pumps with a heating capacity of 40 kW at 36 terminals.
- Water–water heat pump with a capacity of 33 kW.

Figure 2. Electrically heated school building, Horizon-du-Lac (near Montreal, QC, Canada).

The building automation system (BAS) and dedicated electrical submeters at this school provide high-quality data with a sampling timestep of fifteen minutes. The thermocouples are T-types with a standard accuracy of 0.2 °C for the temperature range of 0 to 70 °C. The gym and offices have been designed with significant thermal mass, which helps to improve energy flexibility. Table 2 presents some of the key features of the school building:

Table 2. Key features of the school building.

General Information	
In operation since	2017
Site	Sainte-Marthe-sur-le-Lac, Québec, Canada
Latitude	45.5
ASHRAE climate zone	6
Heating degree days	4495
Net floor area (m^2)	2596 m^2/floor
Number of floors	2
Window type	Double-glazed argon low-e

Table 2. *Cont.*

Mechanical	
Space heating/cooling	Ground-source water–water HP and local water–air HPs
Ventilation system	Balanced mechanical ventilation with DCV, with centralized AHUs with rotary heat recovery. Centralized dedicated outdoor air system (DOAS) modulated based on CO_2
Main system, features	Ground-source heat pump (GSHP), energy recovery ventilator (ERV)
DHW source	Electricity boiler
Electrical	
Lighting, typical type, controls	LED-tube luminaires

System Description

Figure 3 illustrates the schematic of the building's heating system. The system consists of an integrated geothermal system, ThermElect, a water–water heat pump, local water–air HPs, and a hydronic radiant floor system. All heating systems are electrical devices and hence provide a link with the electrical grid. A predictive controller can exploit this link to help balance electricity production and demand, among other potential uses.

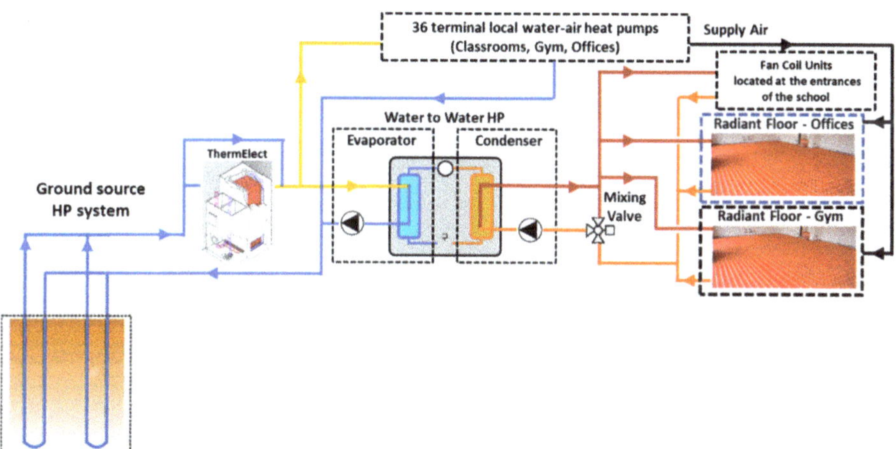

Figure 3. Process flow diagram (PFD) of the building heating systems.

The water–water HP has a capacity of 33 kW in two stages (16.5 kW per stage) and a maximum water supply temperature of 48.8 °C. Borehole Thermal Energy Storage (BTES) with 28 loops on the evaporator side of the HP generates low-temperature heat. A thermal energy storage device (ThermElect) pre-heats the water input to local water–air HPs and water–water HP. The supply water temperature of the HP to the zones is controlled through a thermostatic three-way valve with a maximum temperature setting of 48.8 °C. Thermostats regulate the indoor temperature in each zone separately. The hydronic radiant heating and convective systems supply space heating in the offices and the gym.

4. Modeling Results and Discussion

4.1. Archetype Zones with Convective System: Classrooms

Figure 4 shows one of the classrooms in this school. Classrooms are equipped with ground-source water–air heat pumps (1.5–2 tons each) with COP of 3.2 and proportional–integral control (PI) in the local-loop control of room temperature. The classrooms are typically 9.1 m long by 7.2 m wide, with ceiling–floor heights of 3.0 m. Figure 5 shows a schematic of a classroom equipped with a water-to-air HP:

Figure 4. Typical classroom with convective heating.

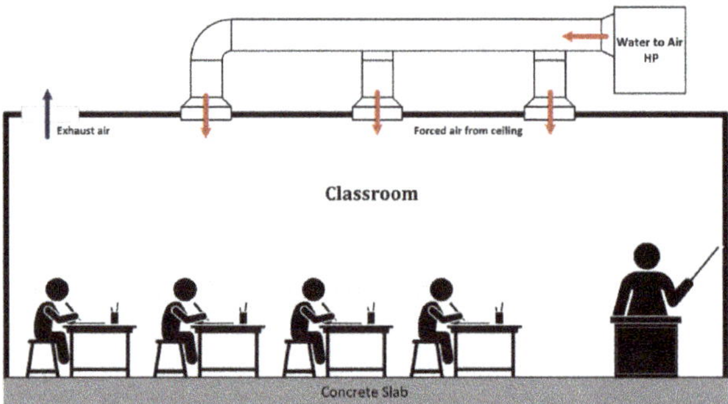

Figure 5. Schematic of a classroom with water–air heat pump.

Figure 6 presents the thermal network RC model structures for zones with convective systems. Figure 6a–c shows the first-order model (1R1C), second-order model (3R2C), and third-order model (4R3C), respectively. The inputs for the analyzed models include outdoor temperature (T_{ext}), solar heat gain (Q_{SG}), internal heat gain (Q_{IG}), and heat delivered by water–air heat pumps (Q_{aux}). Montreal weather data are used to determine outdoor temperature and solar radiation. In the measured data, the heat supplied by water–air HP is calculated by multiplying the measured electricity demand by COP of the HP and is used for comparison of models and measurements (shown in Figure 7). The solar heat gains, internal gains, and heating for the third-order model are distributed over the thermal capacitances (Figure 6c). The performance of the simplified RC models from the first-order to the third-order model is validated with measured data, as shown in Figure 7a–c.

As shown in Figure 7a, the first-order model cannot capture the system's dynamics well. The second-order model has better calibration results than the first-order model, but it still cannot capture details of the thermal dynamics of the system (Figure 7b). The calibration of the third-order model (Figure 7c) shows good accuracy and adequate statistical indices (CV-RMSE of 8% and a maximum difference of 0.4 °C). It should be noted that higher-order models require additional inputs, such as heat flux measurements, to guarantee observability. Since these measurements will not be available in most buildings, higher-order models' identity cannot be guaranteed [28].

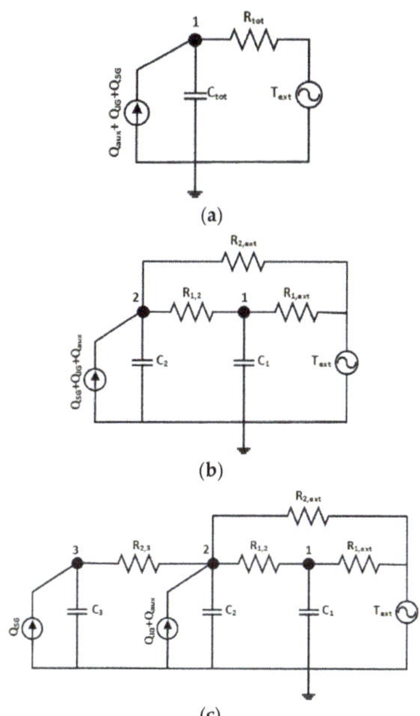

Figure 6. RC thermal network model: (**a**) first-order, (**b**) second-order, and (**c**) third-order models.

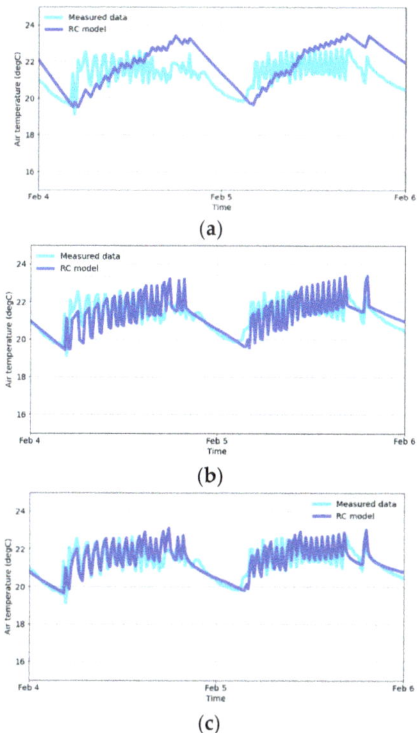

Figure 7. Calibration of RC thermal network models with measured data: (**a**) first-order, (**b**) second-order, and (**c**) third-order models.

4.1.1. Weather Conditions: Cold Winter Days

Weather data for Montreal's coldest days (5–8 February 2020) are selected because peak energy demand occurs under these conditions. Figure 8 presents the outdoor temperature and solar flux during these days. The weather data are measured data and were obtained from the hourly Montreal weather file [32].

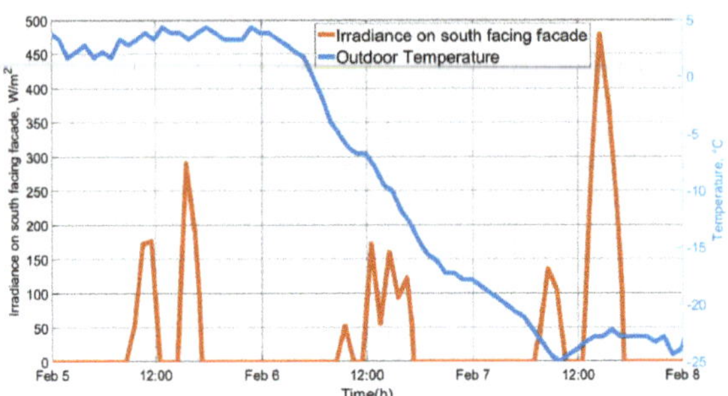

Figure 8. Outdoor temperature and solar flux-Montreal's coldest days.

4.1.2. Available Energy Flexibility in Contingency Event

Contingency reserves are amounts of power that a utility can use in the event of the loss of a generation unit or unexpected load imbalance. To address this need, real-time thermal load flexibility should be predicted ahead of time or calculated continuously and should be available at short notice (e.g., 10 min) over an hour or several hours. This section presents the contingency strategy to quantify the energy flexibility available from the zones with a convective system to the grid at specific times. In this case, a tolerance band setpoint profile is proposed. A flexible approach is proposed within the tolerance limits where the temperature is allowed to deviate from the reference setpoint. For example, during a flexibility event occurring at 2 p.m. for one hour, the temperature is allowed to drop by 2 degrees to provide a "temporary relief" to the heating system (Figure 9a). At this point, the setpoint is lowered two degrees (from 24 to 22 °C). Figure 9b shows the results of available energy flexibility during contingency events.

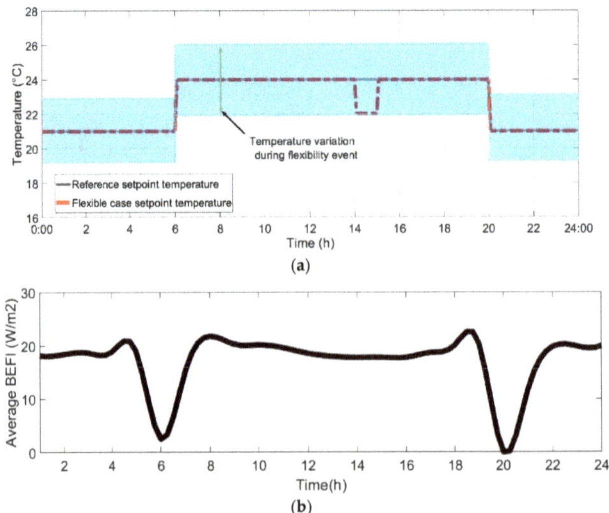

Figure 9. (**a**) Daily setpoint profile with acceptable temperature band and a flexibility event with a duration of 1 h at 2 pm; (**b**) energy flexibility curve, 1 h event.

According to Figure 9b, energy flexibility of around 20 W/m^2 can be provided to the grid in the event of loss of a generation unit or other unexpected power outages. The BEFI can be implemented in the BAS with a predictive model controller, which can optimize power flexibility for a known period of high demand. This makes BEFI appropriate for various grid requirements, including contingency reserves and load shifting.

4.2. Archetype Zones with Hydronic Radiant Floor and Convective Heating Systems

Figure 10 presents a schematic of the office zones equipped with local water–air HP and a hydronic radiant floor system on the school's first floor.

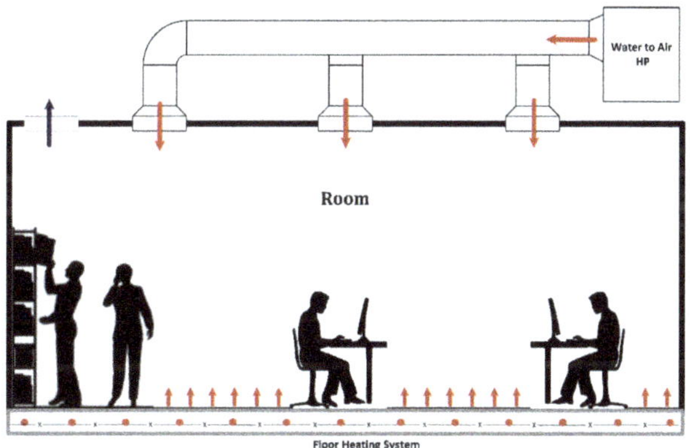

Figure 10. Schematic of the office zone equipped with hydronic radiant floor and convective systems.

The plan view of the offices and piping of the hydronic radiant floor system is shown in Figure 11. These offices are heated with hydronic radiant and local convective systems. Proportional–integral thermostats control heating systems. The thermocouples in the offices are T-types with a standard accuracy of 0.2 °C for the temperature range of 0 to 70 °C. In addition to air temperatures, floor temperatures are also measured at eight different locations, as shown in Figure 11.

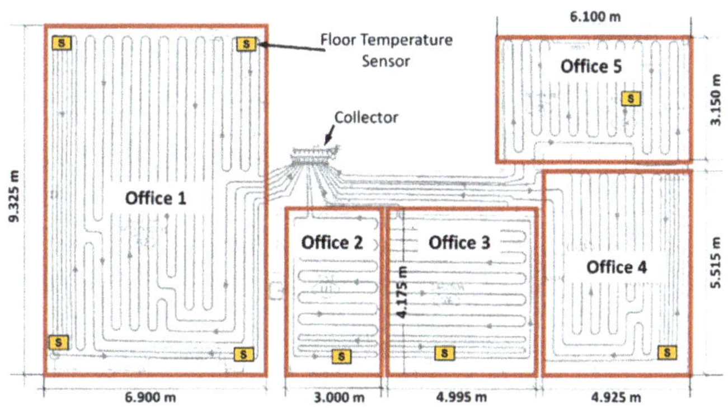

Figure 11. Plan view of the offices with hydronic radiant floor system.

Table 3 presents the radiant hydronic floor area in each zone and the piping length.

Table 3. Floor area and piping length of the offices with hydronic radiant system.

Thermal Zone	Area (m²)	Piping Length (m)
Office 1	64	244
Office 2	12	69
Office 3	21	91
Office 4	27	176
Office 5	19	87
Total	143	667

A front view of the slab with hydronic radiant piping can be seen in Figure 12. The floor is a concrete slab 15 cm thick insulated at the bottom with total thermal resistance of 5.64 m² K/W. The pipe is made of cross-linked polyethylene (PEX), has a diameter of 1.25 cm, and is located at the depth of 6 cm. The pipes were kept in place by a wire mesh before casting concrete, and the distance between the pipes was 30.4 cm.

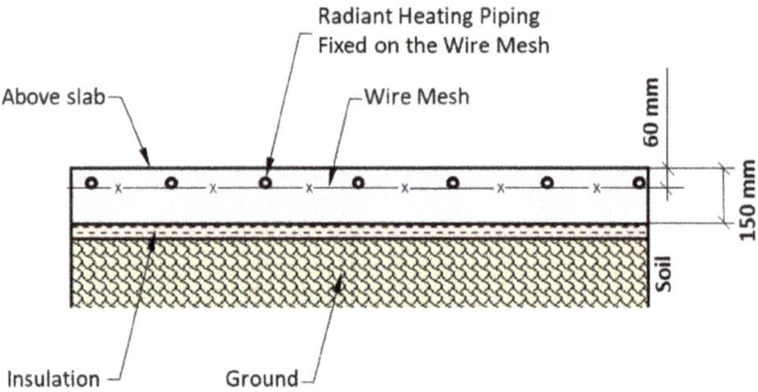

Figure 12. Slab cut with hydronic radiant heating piping.

Concrete's properties are affected by its age, temperature, humidity, and moisture content [34]. Following ASHRAE [35], a normal-density concrete has a conductivity of 1.7 W/(m·K), specific heat of 800 J/(kg·K), and a density of 2200 kg/m³. A water–water HP provides a controlled flow rate of 0.29 L/s with a maximum temperature of 48.8 °C. The HP has a nominal COP of 2.7 under full load conditions at 48.8 °C. Heating power to the hydronic radiant system is calculated by Equation (10):

$$Q = \dot{m} \times c_p \times \Delta T \tag{10}$$

According to the ASHRAE standard 55, the floor temperature must not exceed 29 °C [36]. Thus, a floor surface temperature of 26 °C is considered in this study. Several floor sensors and control valves protect the floor from overheating and enhance thermal comfort. The RC thermal network for the zones with hydronic radiant and convective heating/cooling systems is shown in Figure 13. The inputs are outdoor temperature (T_{ext}), solar gain (Q_{SG}), internal heat gain (Q_{IG}), heat delivered by the hydronic radiant system (q_{RF}), and heat delivered by the convective system (Q_{aux}). These inputs can be:

- *Controllable:* such as the heat delivered by the heating systems and the ventilation airflow rate.
- *Uncontrollable:* such as the outdoor temperature, solar gains, and internal gains.

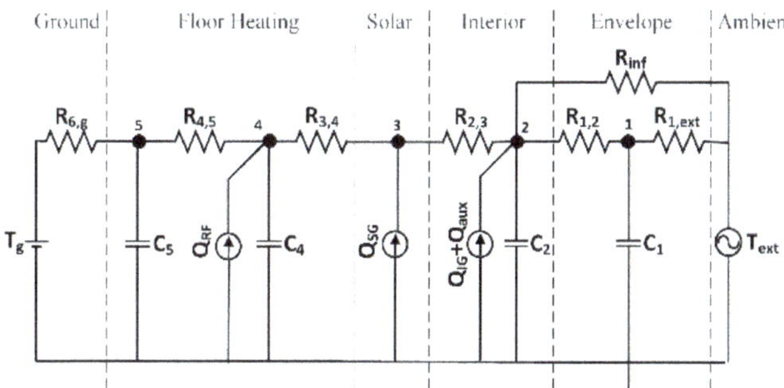

Figure 13. RC thermal network model of the zones with hydronic radiant floor system and convective system.

The performance of the simplified RC model is validated with measured data, as shown in Figure 14.

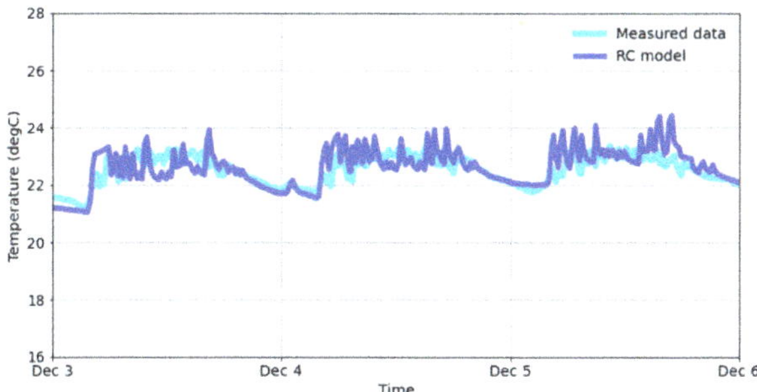

Figure 14. Calibration of RC thermal network model with measured data, zones with hydronic radiant floor system and convective system.

Table 4 provides an overview of the thermal network model parameters:

Table 4. Description of RC thermal network model parameters (7R4C).

Parameter	Description	Parameter	Description
1	Node of envelope	$R_{1,ext}$	Resistance of wall, (K/W)
2	Node of indoor air	$R_{1,2}$	Resistance between wall and air node, (K/W)
3	Node of floor surface	$R_{2,3}$	Resistance between floor and air node, (K/W)
4	Node of pipe	R_{inf}	Resistance of infiltration, (K/W)
5	Node of concrete (top)	$R_{3,4}$	Resistance between pipe and floor surface, (K/W)
T_{ext}	Temperature of outdoor, (°C)	$R_{4,5}$	Resistance between concrete and pipe, (K/W)
T_g	Temperature of ground, (°C)	$R_{6,g}$	Resistance between ground and concrete, (K/W)
Q_{SG}	Solar heat gain, (W)	C_1	Capacitance of envelope, (J/K)
Q_{IG}	Internal heat gain, (W)	C_2	Capacitance of effective Air, (J/K)
Q_{aux}	Heating power, (W)	C_4	Capacitance of floor (top), (J/K)
Q_{RF}	Heating of radiant floor, (W)	C_5	Capacitance of floor (below), (J/K)

4.2.1. Control Scenarios for Energy Flexibility Activation in Archetype Zones with Hydronic Radiant Floor and Convective Systems

This section investigates the heat supplied to the zones with hydronic radiant and convective systems. It will be possible to develop simple predictive control strategies that use thermal storage potential while also considering peak load and thermal comfort. In the reference case (a business-as-usual case), the hydronic radiant floor temperature setpoint (21.8 °C) is always lower than the air temperature setpoint (23 °C) during the daytime (Figure 15a). As a result, the convective system is the primary heating system, and the floor acts as a heat sink. In this study, alternative control scenarios for a cold winter day are examined and compared to current building operations as a reference case. Assumptions considered in designing control strategies include:

- To maintain the slab temperature within the comfort range, the slab surface is set to a maximum of 26 °C.
- The water–water HP can deliver up to 15 kW of heat to the radiant floor heating system, according to the observation from measured data.
- The operating temperature is considered to be the effective indoor temperature.
- During unoccupied hours (nighttime), the slab is charged and discharged during occupied hours (daytime).

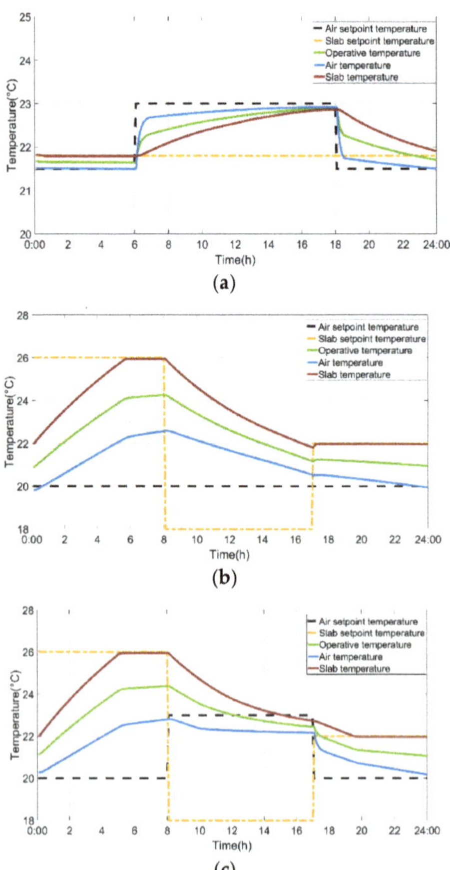

Figure 15. Temperature profile in different control strategies: (**a**) Control Strategy 1 (reference case); (**b**) Control Strategy 2; (**c**) Control Strategy 3.

1. Control Scenario 1 (Reference Case)

The reference case is presented in Figure 15a, which is the current operation of zones with hydronic radiant heating. In this scenario, the convective system is the primary heating system, and the hydronic radiant system is not commonly used. As seen in Figure 16a, in this case, the peak load is 10 kW and occurs during the on-peak hours (6 a.m. to 9 a.m.). Thus, in order to improve the energy flexibility of the building, the following two control scenarios are presented.

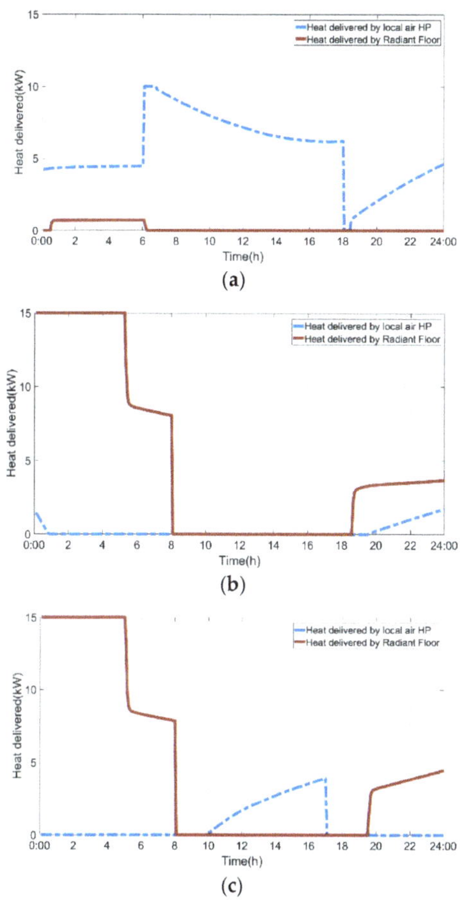

Figure 16. Heat delivered to the thermal zones in different control strategies: (**a**) Control Strategy 1 (reference case); (**b**) Control Strategy 2; (**c**) Control Strategy 3.

2. Control Scenario 2 (Constant Air Setpoint Temperature)

This control scenario involves preheating the slab from midnight to 8:00 a.m. with a setpoint temperature of 26 °C (Figure 15b). During occupied hours (from 8:00 a.m. to 5:00 p.m.), the slab's set point temperature is 18 °C, and then it is raised to 22 °C. It is considered that the air setpoint temperature is always constant and equal to 20 °C during occupied and non-occupied hours.

According to Figure 15b, the operative temperature varies between 21 and 24 °C, which is within the thermal comfort range for the occupants. In Figure 16b, the heat delivered to the thermal zones is calculated using Control Scenario 2. In this control scenario, the radiant floor system is the primary heating system, and the heating demand during occupied hours is reduced. It should be noted that in this control scenario, the ventilation system is off, resulting in poor air quality in the offices. Therefore, Control Strategy 3 is presented to address the air quality of the zones during occupied hours.

3. Control Scenario 3 (Variable Air Setpoint Temperature)

As part of this control scenario, the air setpoint temperature is increased to 23 °C during occupied hours (Figure 15c). As a result, the morning peak load can be reduced, and fresh air can be provided to the zones from 10:00 a.m. to 5:00 p.m. The energy consumption in this flexible scenario is 133.5 kWh, which is less than the reference case (136.6 kWh). The following section will address the slab's state of charge (SOC) (i.e., thermal storage), as well as the flexibility associated with reducing peak loads and energy consumption over peak periods.

4.2.2. State of Charge (*SOC*) of the Slab

The thermal inertia in the slab can provide the flexibility to reduce peak loads and shift the heat production of the radiant heating system in time. State of charge (*SOC*) is a concept that describes how much energy is stored at time t relative to the total capacity, as shown in Equation (11) [37]:

$$SOC = \frac{E_{th}(t) - E_{th,min}(t)}{E_{th,max}(t) - E_{th,min}(t)} \tag{11}$$

The *SOC* is the percentage of the stored thermal energy as a function of the minimum and maximum slab surface temperatures, as given by [38].

$$SOC = \frac{T_{slab}(t) - T_{th,min}(t)}{T_{th,max}(t) - T_{th,min}(t)} \tag{12}$$

where $T_{th,max}$ is the maximum slab surface temperature, set at 26 °C for indoor thermal comfort, and $T_{th,min}$ is the minimum slab surface temperature, considered equal to the average indoor air temperature.

Figures 17 and 18 illustrate heat storage and SOC of the slab in the reference case (Control Strategy 1) and flexible case (Control Strategy 3). It can be observed that in the reference case, the slab cannot be fully charged. Thus, the thermal energy storage capacity of the slab is not fully utilized; while using a flexible case (Control Strategy 3), the slab is fully charged during unoccupied hours and discharged during on-peak hours. This approach activates the thermal load flexibility of the school and allows the electricity grid to manage electricity demand when needed.

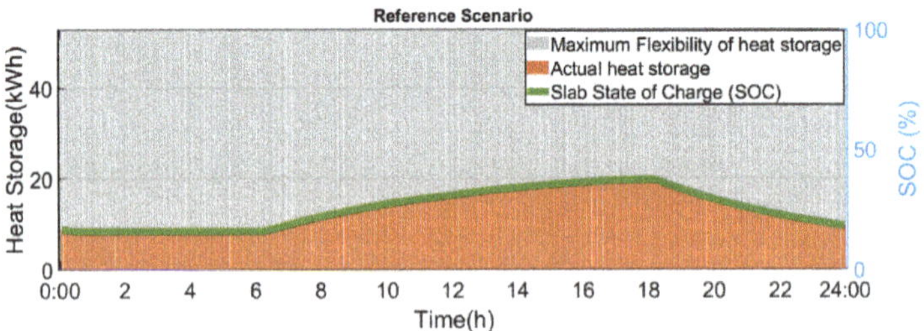

Figure 17. Heat storage and state of charge (SOC) of the slab (reference case).

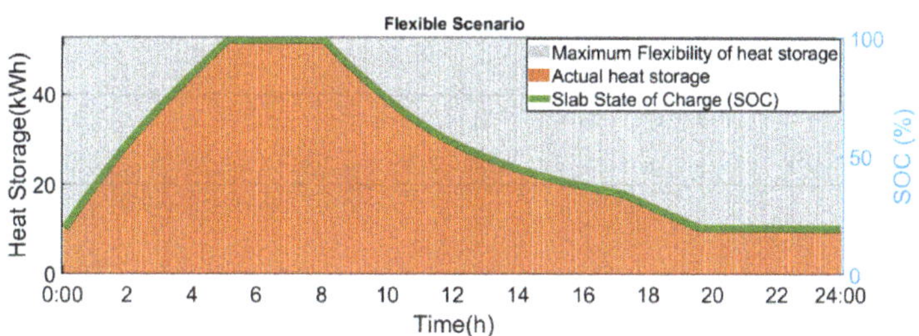

Figure 18. Heat storage and state of charge (SOC) of the slab (flexible case).

4.2.3. Thermal Load Flexibility in Archetype Zones with Hydronic Radiant System

Equations (7)–(9) calculate BEFI by implementing the flexibility strategy and comparing it with the reference as-usual profile. In Figure 19, a flexibility strategy is applied to zones with hydronic radiant and convective heating systems to calculate the hourly BEFI.

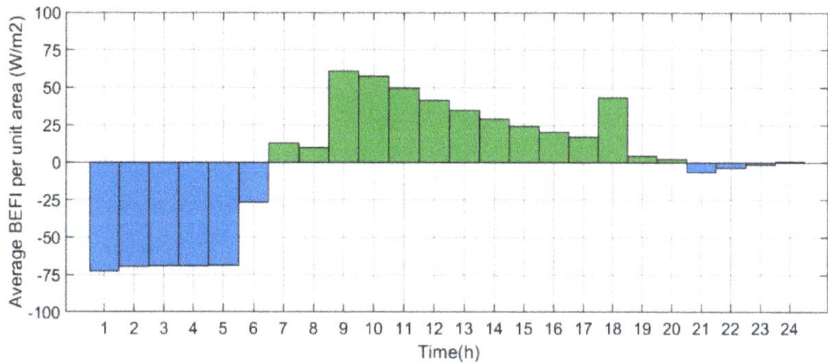

Figure 19. Average hourly BEFI (flexible control strategy).

By applying a flexible control strategy, available hourly BEFI provided to the grid during on-peak hours is positive, indicating the power reduction value available compared to the reference case. During nighttime (off-peak hours), the BEFI is negative, showing a higher power demand for charging the slab and preheating the zones. Based on Figure 19, around 60 W/m^2 energy flexibility can be provided to the grid in the morning and 45 W/m^2 in the evening (on-peak hours).

4.3. Building Level Energy Flexibility

Figure 20 presents energy flexibility at the building level. Zones with radiant floor and convective heating systems can provide around 60 W/m^2 energy flexibility. Additionally, classrooms and the library with convective heating systems can provide 20 W/m^2 during on-peak hours. In total, by implementing appropriate control strategies, the school building can provide energy flexibility from between 30 W/m^2 and 80 W/m^2 when needed by the grid.

In this school, the gym and offices' floor area is 586 m^2, and the classrooms', libraries', and kindergartens' floor is 2054 m^2. Therefore, the school at the building level has potential flexibility of between 50 and 80 kW, representing 40 to 65% building energy flexibility. This bottom-up approach opens the path towards labeling energy flexibility in school buildings as part of the future smart grid and smart cities.

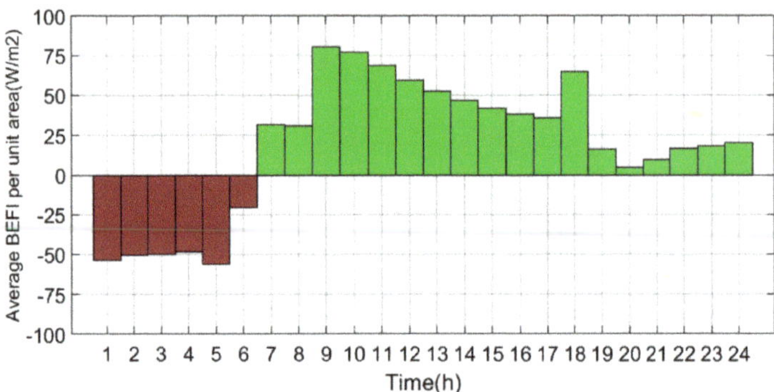

Figure 20. Hourly building energy flexibility in school.

5. Conclusions

School buildings are an important part of the building stock; they also represent a sizable portion of the total energy use in the building sector. Therefore, quantification of energy flexibility in school buildings has a significant role in providing a safe and efficient operation of the future resilient grid. This paper presented a practical methodology that facilitates the modeling and implementation of appropriate control strategies in school buildings. This paper also presented a methodology for defining and calculating a dynamic energy flexibility index for buildings. The dynamic building energy flexibility index (BEFI) is defined in terms of key performance indicators relative to a reference energy consumption profile at the zone level, building level, and as a percentage. The application of the BEFI was presented for an electrically heated school building in Canada. This study illustrated how low-order lumped parameter thermal network models could be utilized to calculate the BEFI. Furthermore, the activation of energy flexibility through a rule-based approach with near-optimal setpoint profiles is investigated. Results show that applying appropriate control strategies can enhance the school building energy flexibility by 40% to 65% during peak demand periods. In addition to improving energy flexibility in school buildings, these control strategies could reduce the size of HVAC units at the design stage, thereby lowering their operating and initial capital costs.

Author Contributions: Conceptualization, N.M., A.K.A. and J.A.C.; methodology, N.M.; software, N.M.; validation, N.M.; formal analysis, N.M.; investigation, N.M.; resources, A.K.A. and J.A.C.; data curation, N.M.; writing—original draft preparation, N.M.; writing—review and editing, N.M., A.K.A., J.A.C. and B.D.; visualization, N.M.; supervision, A.K.A. and J.A.C.; project administration, A.K.A. and J.A.C.; funding acquisition, A.K.A. and J.A.C. All authors have read and agreed to the published version of the manuscript.

Funding: This research was funded by NSERC/Hydro-Québec Industrial Research Chair at Concordia University, Grant Number IRCSA 452557.

Institutional Review Board Statement: Not applicable.

Informed Consent Statement: Not applicable.

Data Availability Statement: Not applicable.

Acknowledgments: Technical support from Hydro-Québec Research Institute (Laboratoire des Technologies de l'Énergie, Shawinigan, QC, Canada) under the NSERC/Hydro-Québec Industrial Research Chair held by Athienitis is greatly acknowledged. Technical support and data provided by Commission scolaire de la Seigneurie-des-Mille-Îles (CSSMI) are acknowledged with thanks. The authors would like to thank NSERC for funding this project through Candanedo's Discovery Grant and the NSERC/Hydro-Québec Industrial Research Chair ("Optimized operation and energy efficiency: towards high-performance buildings") held by Athienitis. The comments of colleagues at

CanmetENERGY are gratefully appreciated. The funding contribution of NRCan's Office of Energy Research and Development through the PERD programs is gratefully acknowledged.

Conflicts of Interest: The authors declare no conflict of interest.

Abbreviations

BAS	Building automation system
BEFI	Building energy flexibility index (W)
BTES	Borehole thermal energy storage
C	Thermal capacitance (J/K)
COP	Coefficient of performance
CV-RMSE	Coefficient of variance of the root mean square error
DSM	Demand-side management
Dt	Time (seconds/hours)
EUI	Energy use intensity ($kWh/m^2/year$)
HVAC	Heating, ventilation, and air conditioning
KPI	Key performance indicator
NZEB	Net zero energy building
P	Electric power (W)
PV	Photovoltaic
Q_{SG}	Solar gain (W)
Q_{IG}	Internal heat gain (W)
Q_{aux}	Heating power (W)
R	Thermal resistance (K/W)
RES	Renewable energy sources
RC	Resistance–capacitance
SOC	State of charge
T	Temperature (°C)
T_o	Outdoor temperature (°C)
T_{SP}	Setpoint temperature (°C)
Subscripts	
Flex	Flexible case
Ref	Reference case
SP	Setpoint

References

1. Davito, B.; Tai, H.; Uhlaner, R. The smart grid and the promise of demand-side management. *McKinsey Smart Grid* **2010**, *3*, 8–44.
2. Foteinaki, K.; Li, R.; Heller, A.; Rode, C. Heating system energy flexibility of low-energy residential buildings. *Energy Build.* **2018**, *180*, 95–108. [CrossRef]
3. Weiß, T.; Fulterer, A.M.; Knotzer, A. Energy flexibility of domestic thermal loads—A building typology approach of the residential building stock in Austria. *Adv. Build. Energy Res.* **2019**, *13*, 122–137. [CrossRef]
4. Afram, A.; Janabi-Sharifi, F. Theory and applications of HVAC control systems–A review of model predictive control (MPC). *Build. Environ.* **2014**, *72*, 343–355. [CrossRef]
5. Jensen, S.Ø.; Marszal-Pomianowska, A.; Lollini, R.; Pasut, W.; Knotzer, A.; Engelmann, P.; Stafford, A.; Reynders, G. IEA EBC annex 67 energy flexible buildings. *Energy Build.* **2017**, *155*, 25–34. [CrossRef]
6. Torres Ruilova, B. Evaluation of Energy Flexibility of Buildings Using Structural Thermal Mass. Master's Thesis, Universitat de Barcelona, Barcelona, Spain, 2017.
7. Ilic, D.; Da Silva, P.G.; Karnouskos, S.; Griesemer, M. An energy market for trading electricity in smart grid neighbourhoods. In Proceedings of the 2012 6th IEEE International Conference on Digital Ecosystems and Technologies (DEST), Campione d'Italia, Italy, 18–20 June 2012; pp. 1–6.
8. Lopes, R.A.; Chambel, A.; Neves, J.; Aelenei, D.; Martins, J. A literature review of methodologies used to assess the energy flexibility of buildings. *Energy Procedia* **2016**, *91*, 1053–1058. [CrossRef]
9. Reynders, G. Quantifying the Impact of Building Design on the Potential of Structural Storage for Active Demand Response in Residential Buildings. Ph.D. Thesis, KU Leuven, Heverlee, Belgium, 2015.
10. Afroz, Z.; Shafiullah, G.; Urmee, T.; Higgins, G. Modeling techniques used in building HVAC control systems: A review. *Renew. Sustain. Energy Rev.* **2018**, *83*, 64–84. [CrossRef]
11. Tabares-Velasco, P.; Christensen, C.; Bianchi, M. Simulated peak reduction and energy savings of residential building envelope with phase change materials. *ASHRAE Trans.* **2012**, *118*, 90–97.

12. Morovat, N.; Athienitis, A.K.; Candanedo, J.A.; Dermardiros, V. Simulation and performance analysis of an active PCM-heat exchanger intended for building operation optimization. *Energy Build.* **2019**, *199*, 47–61. [CrossRef]

13. Klein, K.; Herkel, S.; Henning, H.-M.; Felsmann, C. Load shifting using the heating and cooling system of an office building: Quantitative potential evaluation for different flexibility and storage options. *Appl. Energy* **2017**, *203*, 917–937. [CrossRef]

14. Finck, C.; Li, R.; Zeiler, W. Optimal control of demand flexibility under real-time pricing for heating systems in buildings: A real-life demonstration. *Appl. Energy* **2020**, *263*, 114671. [CrossRef]

15. Junker, R.G.; Azar, A.G.; Lopes, R.A.; Lindberg, K.B.; Reynders, G.; Relan, R.; Madsen, H. Characterizing the energy flexibility of buildings and districts. *Appl. Energy* **2018**, *225*, 175–182. [CrossRef]

16. De Coninck, R.; Helsen, L. Quantification of flexibility in buildings by cost curves–Methodology and application. *Appl. Energy* **2016**, *162*, 653–665. [CrossRef]

17. Tumminia, G.; Sergi, F.; Aloisio, D.; Longo, S.; Cusenza, M.A.; Guarino, F.; Cellura, S.; Ferraro, M. Towards an integrated design of renewable electricity generation and storage systems for NZEB use: A parametric analysis. *J. Build. Eng.* **2021**, *44*, 103288. [CrossRef]

18. Ashrae, A.S. *Standard 90.1-2016*; Energy Standard for Buildings Except Low Rise Residential Buildings. American Society of Heating, Refrigerating and Air-Conditioning Engineers, Inc.: Atlanta, GA, USA, 2016.

19. Distribution., H.-Q. État D'Avancement 2012 du Plan D'Approvisionnement 2011–2020. 2012. Available online: http://www.regie-energie.qc.ca/audiences/TermElecDistrPlansAppro_Suivis.html (accessed on 12 January 2022).

20. Hydro-Québec, A.r. Voir Grand Avec Notre Énergie Propre. 2019. Available online: http://www.hydroquebec.com/data/documents-donnees/pdf/rapport-annuel.pdf (accessed on 25 August 2021).

21. Statistic Canada. Elementary–Secondary Education Survey for Canada, the Provinces and Territories. 2017. Available online: https://www150.statcan.gc.ca/n1/dailyquotidien/171103/dq171103c-eng.htm (accessed on 6 November 2020).

22. Issa, M.H.; Attalla, M.; Rankin, J.H.; Christian, A.J. Energy consumption in conventional, energy-retrofitted and green LEED Toronto schools. *Constr. Manag. Econ.* **2011**, *29*, 383–395. [CrossRef]

23. Ouf, M.; Issa, M.; Merkel, P. Analysis of real-time electricity consumption in Canadian school buildings. *Energy Build.* **2016**, *128*, 530–539. [CrossRef]

24. O'Connell, S.; Reynders, G.; Seri, F.; Keane, M. Validation of a Flexibility Assessment Methodology for Demand Response in Buildings. In *IOP Conference Series: Earth and Environmental Science*; IOP Publishing: Bristol, UK, 2019; p. 012014.

25. Gouda, M.; Danaher, S.; Underwood, C. Building thermal model reduction using nonlinear constrained optimization. *Build. Environ.* **2002**, *37*, 1255–1265. [CrossRef]

26. Candanedo, J.A.; Dehkordi, V.R.; Lopez, P. A control-oriented simplified building modelling strategy. In Proceedings of the 13th Conference of International Building Performance Simulation Association, Chambéry, France, 26–28 August 2013.

27. Bacher, P.; Madsen, H. Identifying suitable models for the heat dynamics of buildings. *Energy Build.* **2011**, *43*, 1511–1522. [CrossRef]

28. Reynders, G.; Diriken, J.; Saelens, D. Quality of grey-box models and identified parameters as function of the accuracy of input and observation signals. *Energy Build.* **2014**, *82*, 263–274. [CrossRef]

29. Athienitis, A.; Santamouris, M. *Thermal Analysis and Design of Passive Solar Buildings*; Routledge: London, UK, 2002.

30. Hong, T.; Lee, S.H. Integrating physics-based models with sensor data: An inverse modeling approach. *Build. Environ.* **2019**, *154*, 23–31. [CrossRef]

31. ASHRAE Guideline. *Guideline 14-2002, Measurement of Energy and Demand Saving*; American Society of Heating, Ventilating, and Air Conditioning Engineers: Atlanta, GA, USA, 2002.

32. Hydro-Québec. Simulation Énergétique des Bâtiments. Available online: https://www.simeb.ca:8443/index_fr.jsp (accessed on 12 October 2021).

33. Athienitis, A.K.; Dumont, E.; Morovat, N.; Lavigne, K.; Date, J. Development of a dynamic energy flexibility index for buildings and their interaction with smart grids. In Proceedings of the 2020 Summer Study on Energy Efficiency in Buildings, Virtual, 17–21 August 2020.

34. Marshall, A. The thermal properties of concrete. *Build. Sci.* **1972**, *7*, 167–174. [CrossRef]

35. Ashrae, I. *2009 ASHRAE Handbook: Fundamentals*; American Society of Heating, Refrigeration and Air-Conditioning Engineers: Atlanta, GA, USA, 2009.

36. *ASHRAE 55*; Thermal Environmental Conditions for Human Occupancy. American Society of Heating, Refrigeration and Air-Conditioning Engineers: Atlanta, GA, USA, 2017.

37. Reynders, G.; Lopes, R.A.; Marszal-Pomianowska, A.; Aelenei, D.; Martins, J.; Saelens, D. Energy flexible buildings: An evaluation of definitions and quantification methodologies applied to thermal storage. *Energy Build.* **2018**, *166*, 372–390. [CrossRef]

38. Gouvernement du Québec. *Plan Directeur en Transition, Innovation et Efficacit Energetiques du Quebec 2018–2023*; Gouvernement du Québec: Québec City, QC, Canada, 2018.

Article

Comparison of Model Complexities in Optimal Control Tested in a Real Thermally Activated Building System

Javier Arroyo [1,2,3,*], Fred Spiessens [2,3] and Lieve Helsen [1,2]

[1] Department of Mechanical Engineering, KU Leuven, 3001 Leuven, Belgium; lieve.helsen@kuleuven.be
[2] EnergyVille, Thor Park, Waterschei, 3600 Genk, Belgium; fred.spiessens@vito.be
[3] Flemish Institute for Technological Research (VITO), 2400 Mol, Belgium
* Correspondence: javier.arroyo@kuleuven.be

Abstract: Building predictive control has proven to achieve energy savings and higher comfort levels than classical rule-based controllers. The choice of the model complexity needed to be used in model-based optimal control is not trivial, and a wide variety of model types is implemented in the scientific literature. This paper shares practical aspects of implementing different control-oriented models for model predictive control in a building. A real thermally activated test building is used to compare the white-, grey-, and black-box modeling paradigms in prediction and control performance. The experimental results obtained in our particular case reveal that there is not a significant correlation between prediction and control performance and highlight the importance of modeling the heat emission system based on physics. It is also observed that most of the complexity of the physics-based model arises from the building envelope while this part of the building is the most sensitive to weather forecast uncertainty.

Dataset: https://doi.org/10.17632/xzdy23nzvj.1

Keywords: model predictive control; advanced controls; control-oriented models; energy efficiency; optimization; thermal comfort

Citation: Arroyo, J.; Spiessens, F.; Helsen, L. Comparison of Model Complexities in Optimal Control Tested in a Real Thermally Activated Building System. *Buildings* **2022**, *12*, 539. https://doi.org/10.3390/buildings12050539

Academic Editor: Etienne Saloux

Received: 24 March 2022
Accepted: 19 April 2022
Published: 23 April 2022

Publisher's Note: MDPI stays neutral with regard to jurisdictional claims in published maps and institutional affiliations.

1. Introduction

During the last decade, there has been a clear interest growth in using advanced control techniques for heating, ventilation, and air conditioning (HVAC) systems in buildings. This increased interest is motivated by the large amount of energy needed for building acclimatization, accounting for 15% of the world's final energy use [1], and by the recent introduction of demand response techniques in the building sector [2]. Buildings with a large thermal mass like those equipped with thermally activated building structures (TABS) can notably benefit from advanced control thanks to the large flexibility potential offered by their high thermal mass. Moreover, typical control strategies can difficultly handle the wide variety of time constants involved in these building systems.

Advanced controllers are often predictive, as in model predictive control (MPC), and require a building system model to anticipate future behavior and optimize the controls at supervisory or local-loop levels. The model is a core element of the controller, which relies on the model predictions to decide actions. Therefore, the choice of the modeling approach is fundamental for setting up a predictive controller and heavily influences the implementation [3]. Many practical factors influence the development of models for MPC [4], and the choice of the model technique and complexity directly determines the optimization or the state estimation methods that can be used.

Three paradigms are used when modeling the building envelope, emission, and production systems, namely white-, grey-, and black-box modeling. These paradigms range from a purely physics-based approach to an entirely data-driven approach, respectively.

Although there are fuzzy boundaries among these paradigms, this terminology has always helped the community in defining the approaches followed by each case study. Authors usually master only one of these approaches and advocate their paradigm by highlighting its advantages and stressing the weaknesses of the other paradigms. Nevertheless, it remains unclear if there is a best approach or if particular contexts ask for a certain modeling paradigm.

The models can be evaluated in both prediction and control performance. A model is evaluated in prediction performance when historical data is used to study how the model maps inputs to outputs. The historical inputs are introduced into the model to investigate how accurately the model outputs fit the measured outputs. The fitness is assessed based on metrics like the Root Mean Square Error (RMSE) or the n-step-ahead prediction error. A model is evaluated in control performance when implemented into a predictive controller to decide actions that are applied to the actual building. In this case, the performance is evaluated based on the metrics optimized by the controller, like thermal discomfort or operational cost.

Previous studies have evaluated and compared different modeling approaches for optimal control, although usually the models are assessed in prediction performance only, and a correlation is assumed between this metric and the control performance. However, factors other than the model prediction accuracy influence the control performance, namely the suitability of a controller model for optimization, the extrapolation capabilities of the model, and the robustness of the controller to forecasting errors. Hence, the acid test for these models is their implementation in a predictive controller of an actual building to evaluate how the controller can accomplish its predefined objective when using a particular model type.

The main goal of this paper is to share practical aspects of implementing the three main modeling paradigms for optimal predictive control in a real case. To this end, a direct comparison of a representative model of each approach is performed in an actual building test case. First, one model representative of each modeling paradigm is selected, configured, and calibrated. Second, the prediction and control performances of the three models are assessed when implemented into the same MPC framework and test case building. Finally, guidelines for building modeling in optimal control are provided. While simple and unoccupied, the envisaged test case is prone to real weather disturbances that may not be captured by simulations, such as actual forecast uncertainty, measurement errors, or hidden infiltration losses.

Therefore, the main contribution of this paper is the comparison of the three main modeling paradigms for predictive control in a real test building. A thorough analysis is performed for different aspects of the models, and special care is taken to ensure a fair assessment. Moreover, the authors of this work assemble a mix of physics- and data-driven backgrounds, which is imperative to avoid any bias or prejudice concerning the evaluated modeling techniques.

The outline of the paper is as follows: Section 2 summarizes other studies that have addressed a similar research topic; Section 3 elaborates on the building test case and the experimental setup used to gather data and perform the experiments; Section 4 explains the modeling approach followed for each modeling paradigm and makes the first comparison in prediction performance; Section 5 describes the control task and how each of the obtained models has been implemented into the same MPC framework to evaluate their control performance. The correlation between simulation and control performance is also investigated in this section; Section 6 discusses the main insights obtained when developing and implementing each of the modeling paradigms into the same MPC in a real building. Finally, Section 7 draws the main conclusions.

2. Related Work

The benefits of optimal predictive control based on weather forecast data have already been proven several times [3,5–9]. These studies highlight that the main bottleneck for the widespread adoption of MPC in the building sector is the need for methods to obtain building models that are reliable and suitable for optimization. An extensive review on control-oriented thermal modeling, in general, can be found in [10]. The advantages and disadvantages of each modeling approach are listed, and a decision tree is provided to aid in deciding a modeling method for control-oriented modeling of multi-zone buildings. This study is mainly theoretical and leans on previous scientific literature, but it does not directly compare modeling paradigms. Contrarily, Blum et al. [4] directly compared the controller model complexity, among other elements, in optimal predictive control. They identified seven practical factors that influence the development of MPC in buildings and systematically analyzed each factor's influence in simulations. The controller model complexity was studied by implementing three grey-box models, gradually increasing their order from one thermal capacitance up to four. They showed that higher-order models can improve the performance of a given MPC up to 20%, although they acknowledged that better initial guesses are required in the parameter estimation process as the number of parameters increase.

Similarly, Picard et al. [11] compared controller model complexities using simulations of six-zones residential buildings with different insulation levels. It was indicated that a low prediction accuracy of the controller model can heavily influence the MPC performance and that low order models may not suffice to capture all the building system dynamics. They systematically reduced the order of controller models derived from a linearized plant model used as an emulator to assess performance. Although they did not compare different modeling paradigms, they clearly indicated the relevance of utilizing a suitable model complexity. Arroyo et al. [12] evaluated different grey-box modeling complexities in control performance. Their focus was the comparison of single-zone to multi-zone RC model structures. Centralized and decentralized multi-zone grey-box model architectures were compared to each other and to a single-zone grey-box model architecture. It was found that modeling inter-zonal effects was beneficial when using multi-zone grey-box models for optimal control. However, their comparison did not involve different modeling paradigms either.

Picard et al. [13] compared a grey-box model to a purely white-box model for an existing 12-zones office building. The models were validated with real data, and their control performance was evaluated in simulation. They showed that both approaches can lead to an efficient MPC as long as accurate identification data sets are available. In the considered simulation case, the white-box MPC led to better thermal comfort while using 50% less energy than the best grey-box model. However, it should be noted that the white-box controller model was derived from the plant model used for the simulations, something that would not be possible in practice. Moreover, the authors stated that further research was still needed to confirm the strength of the white-box approach in the presence of all uncertainties.

A white-box model was compared to a black-box model in [14]. In this case, the white-box model was a lumped RC representation of the plant, and the black-box model used an artificial neural network architecture. Again, the physics-based model was better than the data-driven approach, although the RC parameter values were obtained assuming perfect knowledge of the thermal and geometrical building characteristics. A direct comparison across the three main modeling paradigms was performed by Arendt et al. [15]. They implemented the white-, grey-, and black-box paradigms by selecting representative models for each approach, similar to what is done in this paper. However, they only analyzed the fitting accuracy to indoor temperature on historical monitoring data of a building. The differences in control performance were not explicitly investigated.

The studies above indicate a common understanding of the importance of polishing the modeling techniques to adopt advanced control. However, despite using simulations, the test cases were always different, and each work followed a different approach for the analysis. To solve this issue, the recently developed building optimization testing (BOPTEST) framework [16] enables the direct comparison of different controllers for the same building and boundary conditions by using simulations. The aim is to derive guidelines for optimal control by benchmarking controllers in the same test cases.

While simulations offer a clear, practical advantage for assessment purposes, they do not allow coping with real model mismatch and forecasting uncertainty, which is only possible in actual implementations. Nevertheless, access to real buildings for testing and research purposes is more difficult and expensive. Hence, there are fewer efforts that implement MPC in operational buildings. In this context, the work of Široký et al. [5] excels in achieving energy savings between 15% and 28% in a two months experiment performed on an actual building in Prague, Czech Republic. The setup of this study was unique because two identical building blocks enabled the cross-comparison of two different controllers by switching the control strategies between the blocks to compensate for the changing boundary conditions. However, their focus was on showing the energy savings potential of implementing MPC compared to a heating curve based control strategy. A novel modeling approach for the same building based on partial least squares was introduced in [17]. Other demonstrations of MPC in actual buildings are described in [6,18–20]. White- [18], grey- [6], and black-box [19] models have been used in practical implementations of MPC in real buildings separately. However, none of them performed a comparison of any kind between different models for the same predictive controller.

3. Experimental Setup

The test building of this study is called the Vliet building and is located on the Arenberg campus of the KU Leuven University in Heverlee, Belgium. A detailed overview of the building materials, properties, and geometry can be found in [21], and a brief description of the building is provided here for completeness. The building was constructed in 1996, and is equipped with a local weather station. It comprises separate testing modules to conduct research about different aspects of building physics and HVAC systems. The module under consideration in this paper was constructed in 2011 and is a small unoccupied room with an elementary rectangular geometry of 3.45 m long, 1.80 m wide, and 2.4 m high. Two walls connect to an adjacent room that is always conditioned at a constant temperature of 21 °C. The other two walls connect to the outer space, and one of them has a window of 1.25 m wide and 1.60 m high. The window is integrated into the southwest oriented façade. The building exterior, the control setup, the heat production system, the heat distribution system, and the interior of the test room are shown in Figure 1.

The test room is conditioned using hydronic heat production, distribution, and emission systems. The heat production system is an electric boiler that always maintains water at 60 °C ready for delivery. The water is distributed to the TABS emission system with embedded pipes in the concrete floor and ceiling (two circuits), although only the floor TABS are used in the experiments. Each circuit uses a pump for distributing the hot water and has a three-way valve to mix the water coming from the heat production system with the water returning from the TABS, which enables control of the water supply temperature to the concrete.

A LabView software interface has been developed as described in [22], which allows automatic data acquisition and control at a higher configuration level. The field and automation layers of the infrastructure are interfaced through modules of National Instruments. The interface controls the primary actuators of the system, namely the circulation pumps and the circuit valves. Most of these actuators are relays with a simple on/off control. For the three-way valves of each distribution circuit (floor and ceiling), a continuous signal is used between 0 and 10 V to prescribe a state between a fully closed or opened valve position. When the valve is fully closed, there is no water coming from the heat production system, and the water through the TABS is recirculated. When the valve is fully opened, there is no recirculation, meaning that all water going to the TABS is delivered directly by the heat production system at 60 °C, without mixing with the return water. Any signal in between specifies partial recirculation as an intermediate case between the two scenarios described above.

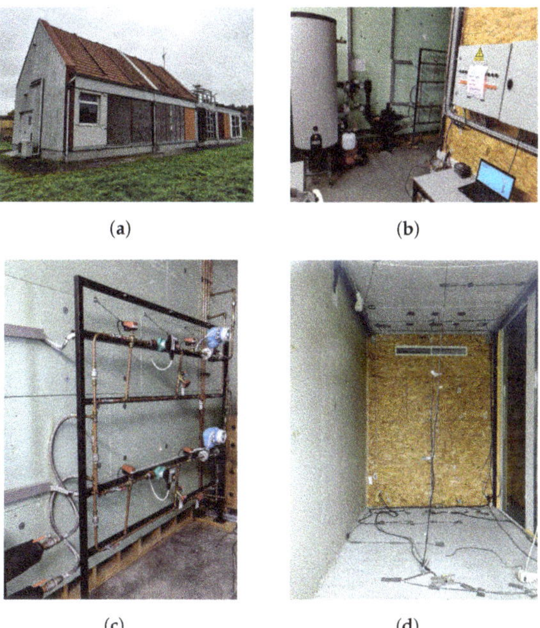

Figure 1. Overview of the Vliet building. (**a**) Building exterior; (**b**) Production system and control setup; (**c**) Distribution system; (**d**) Test room interior.

The measurements available are the supply and return temperatures to and from the TABS, the water flow rate through each circuit, and four air temperatures at different heights in the test room. The water temperatures are measured with PT-1000 sensors inside the pipes, and flow-meters provide the water flow rate through each circuit. The electric power use profile is needed to calculate the operational cost when subject to a dynamic electricity price profile, but there is no electric power measurement available in the setup. Therefore, it is assumed that there is a direct electric water heater that instantaneously converts electricity to heat with a constant electric-to-thermal power efficiency of $\eta = 0.9$. The instantaneous heat can be directly calculated from the supply and return water temperatures and the mass-flow rate of the water through the TABS circuit as defined in Equation (1). All these variables are measured in the setup.

$$P(t) = \frac{\dot{Q}(t)}{\eta} = \frac{\dot{m}(t)c_p(T_s(t) - T_r(t))}{\eta} \tag{1}$$

In Equation (1) $\dot{m}$ is the mass flow rate of the water through the floor TABS, and c_p is the specific heat capacity of the water. The electric power of the circulation pump is neglected.

Figure 2 shows the main components with P the electric power from the grid, u the valve opening signal, $\dot{m}$ the mass flow rate through the floor TABS, T_s and T_r the water supply and return temperatures to and from the TABS, respectively, T_z the average zone air temperature, referred to as the zone air temperature hereafter. The latter temperature is obtained as the simple average of the four temperature sensors located at different heights. Finally, $\dot{Q}$ is the thermal heat delivered by the floor TABS. The green-coloured element indicates the controllable signal, which is the valve opening in this case since the circulation pump is automatically switched on in a post processing step when $u > 0$. The grey-coloured elements indicate measured signals. Finally, the non-coloured circles indicate variables that are derived from measurements.

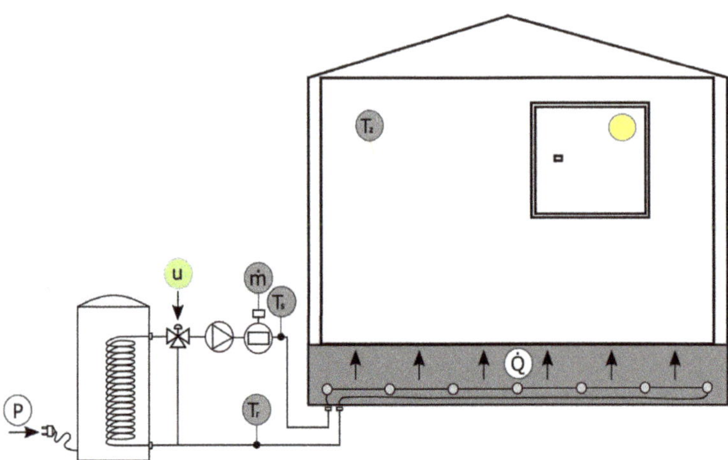

Figure 2. Scheme of the Vliet test building.

4. Modeling

This section explains the steps taken to identify and calibrate the models that are later implemented as system equations in the MPC in Section 5. First, the modeling task is described. Second, the data gathering process is explained. Third, the selection of a representative model of each modeling paradigm is justified. The identification and calibration processes of each model are also described. Finally, the models are compared in prediction performance using auto- and cross-validation data sets from the obtained operational data.

4.1. Modeling Task

The controlled variable of the optimal control problem is the zone air temperature, and the controllable variable is the opening of the three-way valve that determines the water supply temperature to the floor TABS and as a consequence the heat delivered to the room. The two most important system disturbances are the ambient temperature T_a and the solar irradiation $\dot{Q}_{rad}$ (Note that the room is unoccupied, thus disturbances associated to occupancy are not considered.). The predictive controller thus needs to know the effect of the valve opening, ambient temperature, and solar irradiation on the zone air temperature. The controller also needs to estimate the electric power used to compute the operational cost. Hence, a multi-input multi-output model is required to represent the system, and the modeling task is to find a function F that estimates T_z and P from u, T_a, and $\dot{Q}_{rad}$, as indicated by Equation (2). Notice that the estimated variables may depend not only on the inputs and disturbances, but also on previous outputs and system states. The models may use a vector of hidden internal states $\mathbf{X}$ to capture those regressive effects. Finally, a set of parameters $\boldsymbol{\theta}$ characterize the model. These parameters are considered constant in this study. The control toolbox used in this work does not require the models to be linear nor convex for optimization.

$$\dot{T}_z(t), P(t) = F(u(t), \dot{Q}_{rad}(t), T_a(t), T_z(t), \mathbf{X}(t), \boldsymbol{\theta}) \tag{2}$$

4.2. Data Gathering

Data are needed to identify, train and calibrate the models and test the models' performance. Historical data obtained from 24 January until 21 February 2018 is used for these tasks. That is, a total of four weeks of historical data are gathered for system identification and model calibration. During this period, the heating of the test room is alternated between heating and free-floating periods, i.e., periods with hot water circulating through the

TABS and periods with no water circulation, respectively. The first heating period lasts for three days and tests and tunes the software interface and the whole experimental setup. During this period, large excitations are introduced, leading to a zone air temperature up to 33.35 °C. After these excitations, the building is left to free-float. By studying the evolution of the zone air temperature during this period, a settling time of six days is observed, from where it is deduced that the system's time constant is approximately two days. Notice that this time constant is only an approximation because an actual building system never fully settles because of the effect of the disturbances.

A PID controller is implemented for the second heating period to deliver heat to the floor TABS following a pseudo-random binary sequence (PRBS) of 1 kW amplitude. This signal type is suggested by [23] and has been used by other authors to gather data for building modeling, e.g., in [24]. This input sequence is beneficial for system identification because it ensures that there is no correlation between the controllable inputs and the disturbances. The PRBS is designed to put the heating power input signal in the range relevant for control. Finally, the building is left to free-float for the remaining days. The evolution of the variables of interest during these periods is plotted in Figure 3.

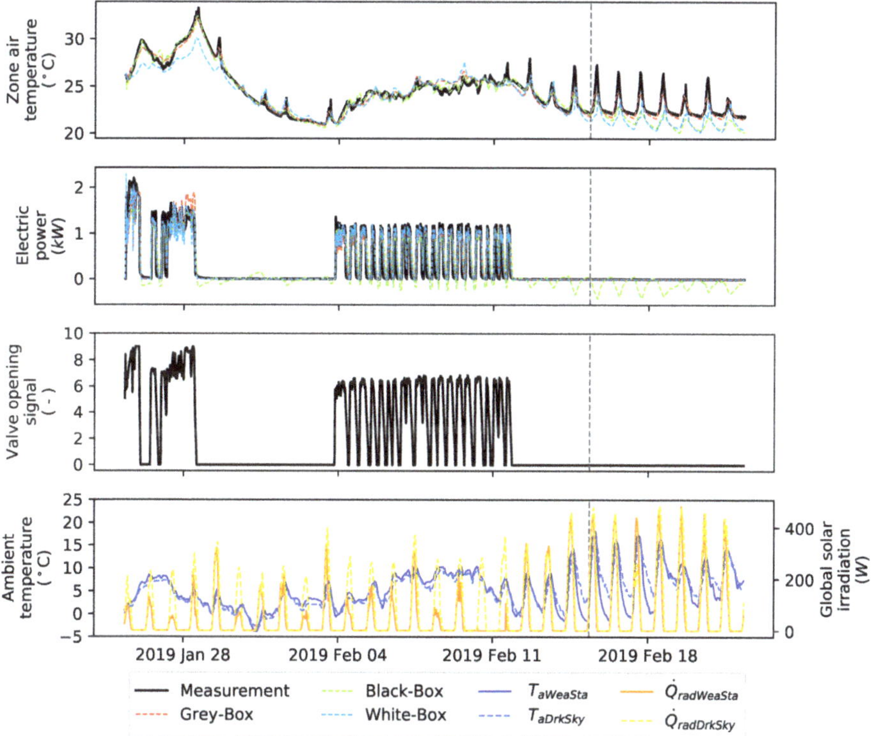

Figure 3. Comparison of the white-, grey-, and black-box models in prediction performance, together with the inputs.

Measurement data from the weather station of the Vliet building was also obtained during this period to compare with the data provided by the DarkSky web-service, the weather forecast provider used by the MPC. The objective is to assess the reliability of the weather forecast service. DarkSky is not limited to climate forecast but also provides current weather condition information for any location which is what is shown by the dashed lines in Figure 3. Even though DarkSky can mostly follow the disturbances' evolution, it is observed that it occasionally shows significant deviations. These deviations go up to 6.72 °C for ambient temperature and up to 366.67 W for solar irradiation. Forecasting is an even more challenging task than estimating current weather for a given location. Hence, errors are expected to be more frequent and larger when forecasting these variables. Forecasting errors are unavoidable, so the MPC will have to deal with these inaccuracies.

4.3. General Considerations

A general-purpose modeling, simulation, and optimization framework is required to model all three paradigms. The toolchain should be powerful to support simulation, optimization, and data handling, and at the same time, it should be flexible to prototype the different modeling approaches. This work uses the JModelica toolbox [25] since it meets all requirements listed above. This toolbox uses Modelica [26], a general-purpose modeling language for configuring the models. Additionally, it provides solvers for Optimica [27], an extension of Modelica for the definition of optimization problems. This choice is also motivated by the development of several open-source Modelica libraries for building energy simulations like *Buildings* [28], *IDEAS* [29], *AixLib* [30], *BuildingSystems* [31], and *FastBuildings* [32]. These libraries facilitate the implementation of white-, grey-, and black-box modeling techniques and continue their development within IBPSA Project 1 [33].

4.4. White-Box Modeling

White-box modeling uses *only* physical insights obtained from the meta-data of the building to represent its thermal behavior. Information like the building orientation, construction components, dimensions, and thermal properties of the materials must be known. The correctness of these data is essential to obtain an accurate model since all parameters are derived from this meta-data. The *IDEAS* [29] library is used for representing the main components of the Vliet building. The main components used from the *IDEAS* library to model the building are listed in Table 1. For the sake of conciseness, we refer to [29] for a detailed description of the main equations and assumptions used in these component models.

Table 1. Modelica components used for the white-box model.

Building zone	`IDEAS.Buildings.Components.Zone`
	`IDEAS.Buildings.Components.SlabOnGround`
	`IDEAS.Buildings.Components.OuterWall`
	`IDEAS.Buildings.Components.Window`
	`IDEAS.Buildings.Components.BoundaryWall`
Heating system	`IDEAS.Fluid.Interfaces.PrescribedOutlet`
	`IDEAS.Fluid.Actuators.Valves.ThreeWayEqualPercentageLinear`
	`IDEAS.Fluid.HeatExchangers.RadiantSlab.EmbeddedPipe`
Boundary conditions	`IDEAS.BoundaryConditions.SimInfoManager`

An overview of the heating system and the building envelope model is provided in Appendix A (Figure A1a,b). The building envelope and the heating system models are connected through the heating port that interfaces both models. Since the model relies on physical principles, it explicitly represents the water mass-flow rate and the supply and return water temperatures to and from the TABS. Hence, it is possible to directly estimate the heat to the TABS and the electric power from these variables.

It is important to note that the white-box model uses some additional weather inputs for the building envelope, namely the dew point temperature, the atmospheric pressure, the relative humidity, and the wind direction and speed. All these variables are obtained from the DarkSky web service. The solar irradiation is also split into direct normal irradiation and diffuse horizontal irradiation based on a preprocessing step [34]. Additionally, some simplifications are required from the original *IDEAS* library components to allow the introduction of this model into the MPC optimization problem. The introduced changes are the following: (1) The window glazing absorption and transmission properties are set constant, (2) the ReaderTMY3 class from *IDEAS* is transformed to bypass the external solar irradiation data from DarkSky, and (3) the air properties are represented using the air model obtained from [35]. This air model reduces the number of algebraic loops when compared to the air model used in the *IDEAS* library. The resulting system of differential-algebraic equations (DAE) has a total of 4420 unknowns and equations. After translation, the system has 94 continuous-time states and 636 time-varying variables, that is variables calculated from the states.

The parameters are derived from information about the building's geometry and materials, and the historical data-set described in Section 4.2 is only used for testing. The calibration of the parameters is performed by an iterative process of changing individual parameters manually and testing to evaluate the simulation accuracy in the historical data. This manual calibration of the white-box model to achieve a proper fit has been the most cumbersome task of implementing the three modeling paradigms. The reason is that a simulation needs to be performed to evaluate every new combination of parameter values, many of them having counteracting effects on the outputs. Automatic parameter estimation is avoided here to respect the white-box model nature strictly. However, a systematic process to identify the parameters that primarily influence the outputs of a complex physics-based model and to train these parameters from monitoring data would be highly beneficial.

The blue dashed lines in Figure 3 show the evolution of the zone air temperature and electric power over the period for which data was gathered for model calibration. Note that these are the outputs with the implemented simplifications and after the calibration process. The solid black lines indicate the measured variables for the same period.

4.5. Grey-Box Modeling

A grey-box model uses some physical insights with lumped parameters trained by monitoring data to learn the system's thermal behavior. In this case, the accuracy of the meta-data of the building is not critical since the model also utilizes monitoring data to train its parameters. Contrarily, the richness of the gathered monitoring data is crucial because it is used to train the model parameters through prediction error methods. In this paper, the parameters are estimated through standard least squares. The Grey-Box Toolbox [32] is used for this task and has been chosen for compatibility with the Modelica language through the *FastBuildings* library. A forward selection procedure is employed to decide on the model structure. This process increases the model complexity step-wise by adding more thermal resistances and capacitances. Additionally, the optimized values of the previous more simple models are introduced as initial guesses for the new, more complex models. A 5R3C model architecture is selected for the building envelope. An overview of this model is provided in Appendix B. Notice that there is an additional thermal resistor in an outer layer of the model not shown in Figure A2. This thermal resistor connects the test room with a constant temperature source that represents the adjacent room.

The RC model does not explicitly represent the water mass-flow rate nor the supply and return water temperatures to and from the TABS. Because of this, the heat to the TABS needs to be estimated in another way, by using a quadratic polynomial from the three-way-valve opening u as represented in Equation (3).

$$\dot{Q} = a_0 + a_1 u + a_2 u^2 \tag{3}$$

The polynomial parameters (a_0, a_1, and a_2) are trained together with the thermal resistances and capacitances of the model, although they lack any physical interpretation. Only the thermal capacitance representing the zone air temperature ($cZon$) is given an initial guess value computed from system knowledge. All other parameters are provided with reasonable initial guesses according to what they represent and are allowed a wider search space because of the larger uncertainty on their value. These are the wall capacitance $cWal$, the TABS embedded capacitance $cEmb$, the wall internal and external thermal resistors *rWalInt, rWalExt*, the embedded internal and external thermal resistors *rEmbInt, rEmbExt*, and the window transmittance *gA*. For every model structure considered, a latin hypercube sample is generated on the parameter space to aid in the search for a good local minimum. Additionally, the target variables, i.e., T_z and P, are normalized with their maximum and minimum values along the training data-set to facilitate the parameter estimation process. This process has proven extremely helpful when training a model with outputs of different orders of magnitude. The main components used from the *FastBuildings* library to prototype the grey-box model are listed in Table 2. We refer to [32] for a complete description of the main equations and assumptions used in these type of models.

Table 2. Modelica components used for the grey-box model.

Building zone	`FastBuildings.Zones.BaseClasses.Capacitor`
	`FastBuildings.Zones.BaseClasses.Resistance`
	`FastBuildings.Zones.Windows.Window_gA`
Heating system	Quadratic polynomial from the three-way-valve opening (Equation (3))
Boundary conditions	`FastBuildings.Input.SIM_Inputs`

Three weeks of measured data are used for training and the last week for cross-validation. The parameter estimation process successfully finds a satisfactory set of model parameters for a model of three states. A substantial reduction of DAE complexity is observed when compared to the white-box model. In this case, the DAE has a total of 88 unknowns and equations. After translation, the system has 3 continuous-time states and 18 time-varying variables. The red dashed lines in Figure 3 show the outputs of the selected grey-box model in auto- (left of the vertical dashed line) and cross-validation (right of the vertical dashed line).

4.6. Black-Box Modeling

Finally, a purely data-driven model is identified as representative of the black-box paradigm. From the many existing black-box modeling approaches it is decided to use a state-space representation to follow the same mathematical architecture as the one used for the grey-box models. In contrast to grey-box models, the parameters of a black-box state-space model directly coincide with the elements of the state-space matrices of the model. Because it lacks any physical interpretability, a schematic presentation has no added value, one block relates the outputs with the inputs.

Analogously to grey-box modeling, the first three weeks of monitoring data are used for training, and the last week is used for cross-validation. Again, the least-squares method is used to maximize the fitting on training data, and a forward selection procedure is implemented that increases the order of the state-space model one by one. A model of order three is found to provide a good fitting on training data. Coincidentally, this order matches the number of thermal capacitances of the grey-box model. This is convenient since both present the same structure and dimension, which elucidates further their comparison. The DAE complexity is further reduced to 17 unknowns and equations. After translation, the model has three continuous-time states and 13 time-varying variables. The green dashed lines in Figure 3 show the outputs of the selected black-box model in auto- and cross-validation.

4.7. Prediction Performance Assessment

Three models have been identified, trained and calibrated in the previous sections. Each of them represents one modeling paradigm. A summary of the model structures and their fitting to historical data can be found in Table 3. In this table, n_e is the number of equations of the original model; n_x and n_v are the number of states and variables after symbolic manipulation, respectively. The RMSE is used for the evaluation of goodness of fit because that is the target metric to be minimized during the parameter estimation process. The RMSE is calculated as defined in Equation (4).

$$\text{RMSE} = \sqrt{\frac{\sum_{k=1}^{\mathcal{M}} (e_{z,k})^2}{\mathcal{M}}} \tag{4}$$

where:

$$e_{z,k} = y_{z,k} - m_{z,k} \qquad \forall k \in 1, \ldots, \mathcal{M} \tag{5}$$

In Equation (4), $e_{z,k}$ are the residuals and $\mathcal{M}$ is the number of measurements. In Equation (5), $y_{z,k}$ indicates the model output, $m_{z,k}$ the measurement at time index k.

Table 3. Summary of model structures and their fitting to historical data. n_e is the number of equations of the original model; n_x and n_v are the number of states and variables after symbolic manipulation, respectively. The RMSE is evaluated in auto- (RMSE_a) and cross- (RMSE_c) validation.

Model	Translated Model			T_z [K]		P [kW]	
	n_e	n_x	n_v	RMSE_a	RMSE_c	RMSE_a	RMSE_c
White-box	4420	94	636	1.00	1.13	0.18	0.003
Grey-box	88	3	18	0.52	0.64	0.20	0.002
Black-box	17	3	13	0.92	2.26	0.22	0.15

Particularly, RMSE_a is the RMSE in the auto-validation period, and RMSE_c is the RMSE in the cross-validation period. Notice that this separation is made for the training of the data-driven models only, but it is maintained for the evaluation of the white-box model for clarity. The number of equations refers to the original model size. The translated model is the model after symbolic manipulation i.e., after eliminating aliases and performing index reduction.

What stands out in Table 3 is the strong contrast between the white-box and the data-driven model sizes. The white-box model is 50 times larger than the grey-box model and 260 times larger than the black-box model when comparing the number of equations in the original models to represent the same building system. After translation, the physics-based model has 31 times more states than both data-driven models. The white-box model is thus the most computationally demanding for both simulation and optimization. While the computational complexity may not be critical for simulation, it plays an important role in optimization, as shown in the following section.

The notable model size difference stems from the amount of detail that the white-box model needs in order to describe the main building physics, e.g., the heat-flow through each of the layers of the walls, roof, and window; the thermal properties of the fluid media like the zone air or the water through the pipes of the heating system; or the air infiltration and the pressure differences. Interestingly, most of the model details come from the building envelope, which is the part of the system mostly exposed to the forecast uncertainties. The model of the building envelope has 91 states and 566 variables after translation, whereas the model of the heating system has only 3 states and 217 variables after translation.

The data-driven models present a simpler structure and rely on the parameters trained by historical data to capture the building dynamics. The selected data-driven models are very similar in their mathematical architecture. Basically, the two main differences between the grey- and the black-box models are that: (1) the black-box model lacks any physical insight, and that (2) its parameters are allowed a much wider search space in the parameter estimation process. Their simpler model structure does not hamper their prediction performance. Contrarily, these models show better accuracy in auto-validation when compared to the very complex white-box model. This indicates that an increased model complexity does not necessarily lead to increased accuracy. Moreover, calibrating a complex physics-based model is found to be a cumbersome and very time-consuming task. Most of the parameters of the white-box model are located in the building envelope, with most of them having a weak influence on the model outputs compared to the heating system parameters. The higher accuracy of the data-driven models may be related with their more flexible structure which facilitates to learn the correlations between the inputs and outputs of the data-set. However, the white-box model is expected to better explain the dynamic coupling of the building system. Figure 3 and Table 3 reveal that the black-box model only poorly extrapolates beyond the training conditions.

5. Implementation and Control

The acid test of a controller model is the assessment of its performance when implemented in a predictive controller. This section implements all models described in the previous section one by one in the same MPC formulation for the Vliet building. Three different experiments are designed to challenge the controllers in different ways. In this section first, the control task is introduced. Second, the model predictive control and the solution method of the optimization are explained. Third, the experiments are described. Finally, the control performance of the controllers using each paradigm is analyzed.

5.1. Control Task

The objective of the MPC is to maintain indoor thermal comfort at the lowest possible operational cost. The comfort assessment is based on zone air temperature since it is the main factor influencing the thermal comfort perceived by the occupants in a building [36]. Note that zone air temperature is used in this work instead of zone operative temperature because it is the only measured temperature in the zone. Most standards are based on the zone operative temperature, but the air temperature can be a good proxy for buildings with TABS systems where the radiant temperature is frequently not too different [37]. The comfort range is inspired by ISO 7730 Class A, which establishes a comfort band of 22 ± 1 °C. However, this range is shifted two degrees upwards to increase the heat demand intentionally. The motivation is that the space adjacent to the test room is maintained at 21 °C which stabilizes the indoor temperature. A higher heat demand increases the degrees of freedom to test control. Hence, thermal discomfort is defined as the cumulative deviation of the zone air temperature out of the temperature range 24 ± 1 °C, and it has the units Kelvin-hour (Kh). A highly dynamic pricing tariff is contemplated by fictitiously exposing the controller to the 2019 Belpex prices, i.e., the day-ahead prices of the year before the experiments start. No additional taxes or transportation fees are considered in the price signal, to exploit the highly dynamic profile.

Therefore, the control task consists of deciding every step a valve opening value that minimizes the overall discomfort and operational cost. To this end, the controller is provided at the beginning of every control step with only one measurement: the zone air temperature. Also the price signal and the weather forecast along the prediction horizon are provided. It is worth mentioning that the experimental setup does not have any auxiliary fast-reacting system, making control complicated because of the lack of controllability. Since no cooling system is available, the only way for the controller to deal with overheating that can be induced by solar irradiation is by heating less, which might lead to too low temperatures, something that the controller should carefully balance. Still, an MPC with a proper system model should be able to minimize discomfort while aiming for the lowest operational cost.

5.2. Model Predictive Control Formulation

All controller models explained in Section 4 are implemented in the same MPC. The objective function l of the MPC is defined according to the control task explained in Section 5.1 as a multi-objective function with two counteracting terms: discomfort and operational cost. The formulation of this objective function is specified in Section 5.3

because it varies with each experiment carried out. The set of Equation (6a–e) defines the optimal control problem that is general to all experiments.

$$\min_{u} \int_{t=t_i}^{t_i+T_h} l(t)dt \tag{6a}$$

$$\dot{T}_z(t), P(t) = F(u(t), \dot{Q}_{rad}(t), T_a(t), T_z(t), \mathbf{X}(t), \boldsymbol{\theta}) \tag{6b}$$

$$\underline{T}_z(t) - \delta^{T_z}(t) \leq T_z(t) \leq \overline{T}_z(t) + \delta^{T_z}(t) \tag{6c}$$

$$\underline{u}(t) \leq u(t) \leq \overline{u}(t) \tag{6d}$$

$$\delta^{T_z}(t) \geq 0 \tag{6e}$$

where t_i is the initial time of the current control step, δ^{T_z} are the deviations of the indoor temperature out of the comfort range, $\underline{u}$, and $\overline{u}$ are the technical constraints (lower and upper values) of the input signal, respectively, and $\underline{T}_z$ and $\overline{T}_z$ are the lower and upper bounds of the comfort range, respectively. The prediction horizon T_h is set of two days based on the estimated time constant of the test room. A control step of 15 min is chosen as a reasonable trade-off between controllability and granularity of the prediction horizon.

One of the main advantages of using the JModelica toolchain is that the MPC optimization problem can be easily constructed by extending the models with the Optimica language to formulate the objective and constraints of the optimization problem. JModelica utilizes the direct collocation discretization scheme with CasADi [38] for defining the nonlinear program and for algorithmic differentiation. More efficient discretization methods exist for the mostly linear mathematical structures of building envelopes [39]. However, direct collocation is known for its versatility and robustness. Moreover, symbolic elimination as implemented in [40] is used to eliminate many of the algebraic variables in a preprocessing step and to improve the numerical efficiency.

The MPC module described in [41] is used to effectively reuse the fixed discretization scheme of the receding horizon controller every time step. The module has been extended to allow mutable external data, which was not implemented but is required to expose the optimization to the continuously changing boundary condition data. Additionally, the MPC uses the result from the solution obtained in the previous step to warm-start every new optimization problem. An unscented Kalman filter is implemented as described in [42] to estimate the initial value of the hidden states of the models every control step after measuring the zone air temperature. The sigma points are chosen according to [43], and the covariance matrices are constructed according to the fitting information obtained during training and calibration as summarized in Table 3.

5.3. Description of the Experiments

Three experiments are designed with slightly different formulations of the MPC objective function and constraints. The goal is to challenge the controllers, and thus the controller models, in different ways. All three modeling approaches are tested by implementing their representative model into the MPC formulation of each experiment. The three models are alternated such that the MPC runs for at least two weeks with each model in every experiment. The two-week period is decided as a reasonable period length to trade-off between shuffling the models and enabling a representative behavior of each. Note that shuffling is beneficial because it minimizes the influence of seasonal boundary condition changes.

The experiments always use the same comfort bounds: $\underline{T}_z = 23\,^{\circ}\text{C}$ and $\overline{T}_z = 25\,^{\circ}\text{C}$, and the same lower limit on the valve opening: $\underline{u} = 0$. Moreover, only the zone air temperature measurement is available to the controllers in all experiments and the valve opening is the control variable. The DarkSky web-service is used to retrieve and update the expected weather forecast every control step. More specifically, all weather variables required by each model as described in Section 4 are obtained from this forecast provider. That is, the ambient temperature and solar irradiation for the data-driven models, and the additional weather variables required by the white-box model. The variations introduced for each experiment on the objective function and the valve opening upper limit are explained below.

Experiment 1 uses the objective function defined in Equation (7a), where $p^e(t)$ is the dynamic Belpex electricity price, and w is a weight that has been carefully tuned to account for the different orders of magnitude between the operational cost and the thermal discomfort. This experiment also sets the upper constraint of the valve opening to its maximum technical allowed value: 10 V. This experiment is interesting from a control point of view because it exploits the full technical range allowed by the only controllable variable: the three-way valve opening. However, it is not representative of an actual TABS building where the typical water supply temperature is never higher than 30 °C. A fully opened valve leads to water supply temperatures that can be high as 50 °C and beyond. This motivates the definition of the second experiment.

$$l(t) = p^e(t)P(t) + w\delta^{T_z}(t) \tag{7a}$$

$$\bar{u} = 10 \tag{7b}$$

Experiment 2 uses the same objective as *Experiment 1*, which has been reintroduced in Equation (8a) for clarity. However, the opening of the three-way valve is limited to 3 V to avoid very high water supply temperatures. This limit leads to water supply temperatures that never surpass 30 °C, which reflect real TABS conditions much better.

$$l(t) = p^e(t)P(t) + w\delta^{T_z}(t) \tag{8a}$$

$$\bar{u} = 3 \tag{8b}$$

Experiment 3 maintains the same limit of the valve opening as *Experiment 2* but squares the operational cost in the objective function aiming to steer more flexibility from the test room during operation. Squaring the operational cost reduces the peaks of energy use when prices are higher. The experiment formulation is summarized in the set of Equation (9a,b).

$$l(t) = (p^e(t)P(t))^2 + w\delta^{T_z}(t) \tag{9a}$$

$$\bar{u} = 3 \tag{9b}$$

Two key performance indicators (KPIs) are defined to benchmark and compare the performance of the MPC in all experiments while using each of the controller models. Since the duration of the implementation of each model into the MPC does not last for exactly the same time period, all KPIs are normalized per day. Additionally, three days of initialization are used every time a new modeling approach is implemented into the MPC. These three days are not included in the KPI calculation. The aim is to avoid that the outcome of one experiment influences the results of the next one.

The first performance indicator is the cumulative discomfort D as defined in Equation (10).

$$D(t_0, t_f) = \int_{t_0}^{t_f} \delta^{T_z}(t)dt \tag{10}$$

where t_0 and t_f indicate the initial and final time of the evaluation period, respectively.

The second performance indicator relates to the cumulative operational cost. In order to fairly compare the operational cost of different controllers in a real testbed, it is required to normalize the incurred cost by a proxy of the energy needs of the building. The reason is that each controller is implemented in different periods with different energy demands, which might bias the cross-comparison in performance among the controllers. A typical approach (e.g., Široký et al. [5]) is to normalize the cost with the heating degree days as a measure for the building energy demand. The heating degree days are computed as the difference between the indoor temperature setpoint and the ambient temperature. Inspired by this, the normalized cumulative operational cost is defined by Equation (11).

$$C(t_0, t_f) = \int_{t_0}^{t_f} \left(\frac{p^e(t)P(t)}{T_z(t) - T_a(t)} \right) dt \tag{11}$$

Both the thermal discomfort and the normalized operational cost are direct indicators of the MPC performance because they represent what is being minimized in the objective function.

5.4. Control Performance Assessment

5.4.1. General Evaluation

This section analyzes the MPC results in a real test building when using the different controller models. Figure 4 provides a complete overview of the period when the MPC was actively controlling the test room heating system. The first graph shows the evolution of the zone air temperature and the price signal; the second graph shows the evolution of the three-way valve opening as decided by the predictive controller; the last graph shows the evolution of the main weather disturbances, i.e., ambient temperature and solar irradiation. The duration of each experiment is shown at the top of the figure, and the background colors indicate the periods when each model was implemented into the MPC in the following order: grey-, black-, and white-box model.

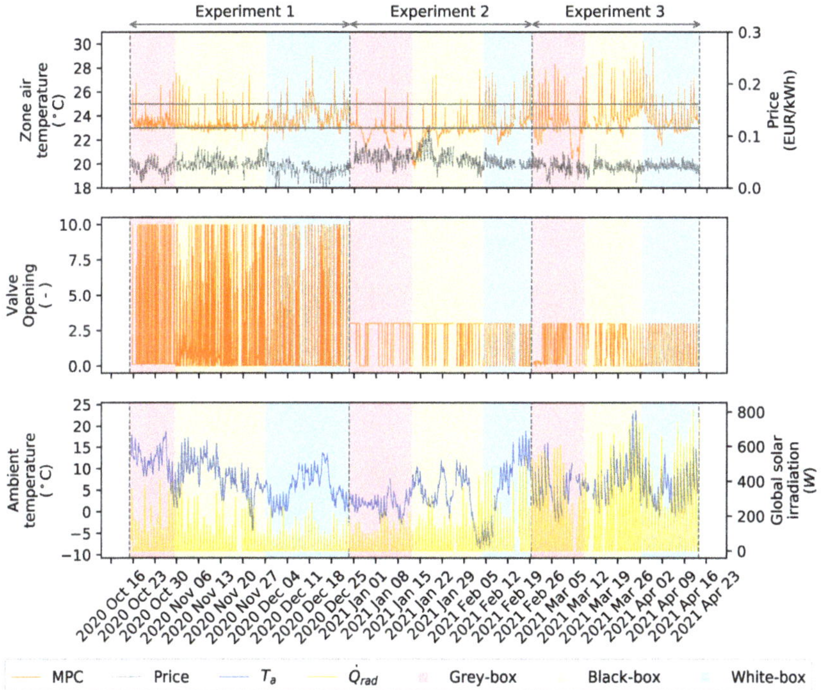

Figure 4. Experimental results with MPC applied in the real test building.

An overview of the obtained thermal discomfort and normalized operational cost for each experiment and controller model is shown in the left column of Figure 5. The results of this figure should be interpreted cautiously because the comparison of control performance over changing boundary conditions is always subject to uncertainty (in this case, mainly related to the weather forecast). It is observed that no controller model always outperforms the others for all three experiments. In fact, it is the MPC formulation that primarily influences the results. This is observed the significant cost reductions obtained when limiting the heat input in *Experiment 2* and when squaring the cost term of the objective function in *Experiment 3*, an effect common to all controller models.

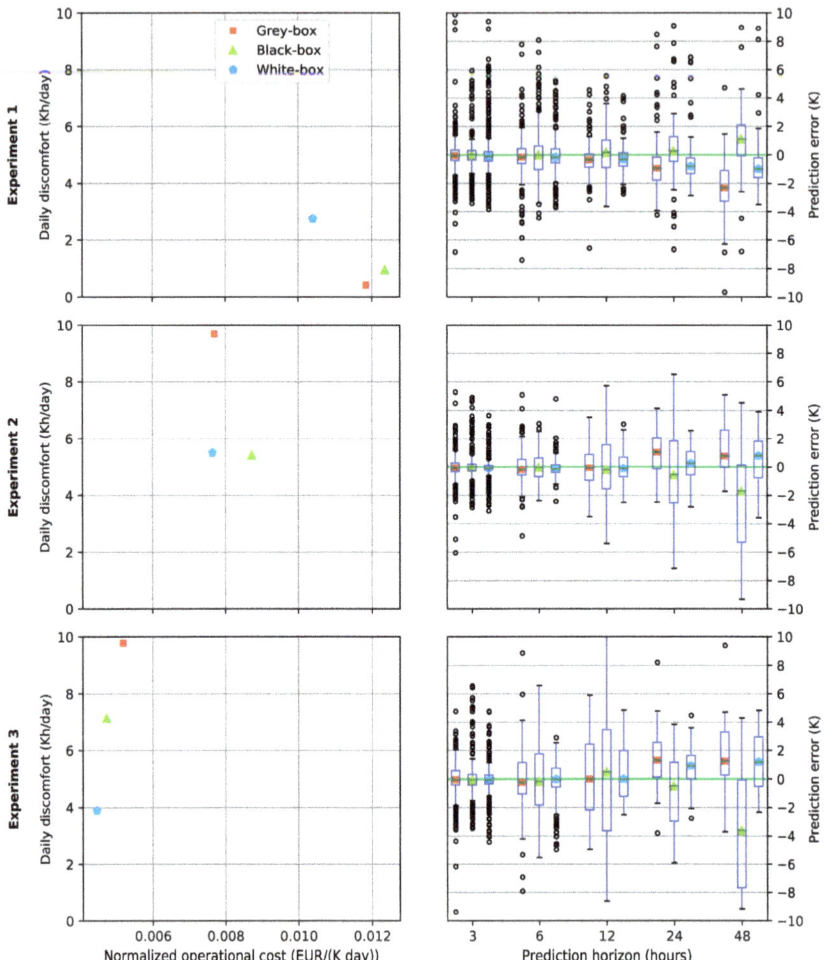

Figure 5. Summary of control KPIs (**left**) and the n-step-ahead prediction errors for each model and experiment (**right**).

The data-driven models show a substantial increase in discomfort during Experiments 2 and 3. This is explained by the different operational conditions of these experiments compared to the training data that do not include information of opening the valve within the range $0 < u \leq 3$. The detrimental effect of changing operating conditions might have been prevented by exciting the system using the constraint input range when generating the training data-set. This stresses the importance of having high-quality, i.e., rich in information, training data for data-driven approaches. Contrarily, the white-box model does not substantially increase its thermal discomfort when tested out of the conditions used for its calibration since it relies on first-principle physical equations and does not lean on the limited information of the calibration data. Note that, although the grey-box model uses simplified physics to represent the building envelope, its heating system is modeled using a polynomial without physical interpretability. This likely damages the performance of the grey-box model in Experiments 2 and 3 and highlights the importance of modeling the heating system based on physical insights. A way to amend this effect can be to base the polynomial on a full performance map. On the other hand, only the white-box model presents algebraic loops after symbolic elimination. These are sets of variables that depend on each other, increasing the computational complexity of the optimization problem. The three largest blocks have sizes of 12, 9, and 8 variables.

5.4.2. Analyzing the Correlation between Prediction and Control Performance

A preliminary inspection of the correlation between prediction and control performance can be made based on comparing the results from Table 3 and Figure 5. It is observed that only for *Experiment 1*, where the operational conditions are similar to the training conditions, there is a correlation between the RMSE in auto-validation and control performance. However, the analysis based on the results of Section 4 is limited because there is no heating during the available cross-validation period. The RMSE in auto-validation represents only the one-step-ahead prediction accuracy. In order to investigate the correlation between prediction and control performance, it is useful to analyze the prediction accuracy for longer horizons and using test data obtained in conditions that differ from the training data. Such analysis can throw light on the robustness of the models to make predictions beyond the training conditions. Hence, the data from the control experiments are used offline to evaluate the prediction error of the zone air temperature as a function of the prediction horizon.

The same state estimator of the MPC is utilized to implement the *a priori* estimates of the zone air temperature for different prediction horizons. Every step, the estimate is cached, and the model state is updated with the historical measurement to perform a new prediction estimate. The process is repeated for every model and experiment, and the results are shown in the boxplots in the right column of Figure 5. The centered line gives the mean, the box gives the first and third quartiles, and the whiskers mark the range of the non-outlier data defined as 1st/3rd quartile ± 1.5 times the interquartile range (IQR). Using the data obtained from the experiments allows to conveniently display the prediction errors together with the control performance obtained during each experiment.

Overall, the predictions are centered and not skewed. Only from horizons longer than 24 h, predictions start showing some bias, most notably for the black-box model. The black-box model also shows the most scattered error distribution, which is more noticeable as the prediction horizon increases. The white-box model generally shows the most precise predictions closely followed by its grey-box counterpart. Despite using the same data for prediction evaluation as the data obtained from the control experiments, it is still impossible to make a strong conclusion on any significant correlation between prediction and control performance.

5.4.3. Analyzing the *n-Step ahead Control Deviation*

To further assess the control strategies, the variability of the optimization results obtained with each model is investigated. The goal is to assess how effectively the MPC implements its devised strategy. At every control step, the complete optimization result trajectories are stored to compare each prediction with the eventual value observed at the prediction time. We call this metric the *n-step ahead control deviation*. Notice that it differs from the classical *n-step ahead prediction error* in the sense that the predictions are not obtained as the outcome of simulations carried out offline, but as the full trajectory results of the MPC optimizations performed every control step. Therefore, this metric does not only evaluate the controller model, but the combination of all elements that interact within the MPC, namely the model, the weather forecast and the state estimator. Hence, the complete MPC machinery is being evaluated when using a particular model. A small control deviation might suggest a good control behavior because it indicates that the strategy devised by the MPC is effectively implemented. However, it is important to note that small control deviations do not guarantee improved performance. Figure 6 illustrates the process to compute the control deviations for a variable V along a horizon with h optimization steps. In this figure, the subscript $(\cdot)_{p|q}$ indicates a value that is estimated for time p based on the prediction of time q. The red lines illustrate the control deviation for each step of the prediction horizon and an example for the one-step ahead control deviation calculation. Further steps are not shown for clarity.

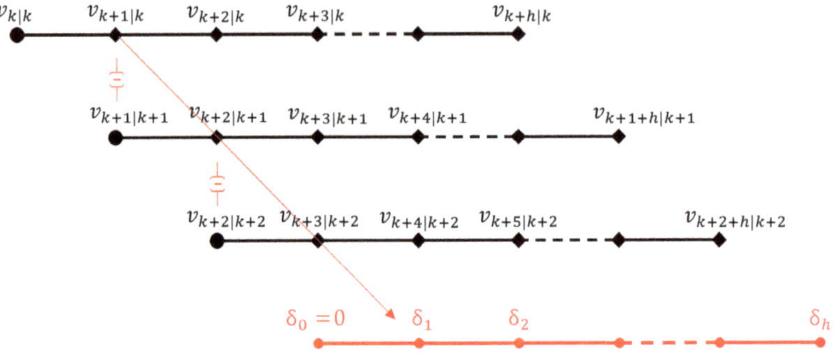

Figure 6. Illustration of the control deviation calculation process.

The distributions of the obtained control deviations are determined for all prediction horizons and shown in Figure 7. Specifically, Figure 7 shows the mean and standard deviation of the control deviations as a function of the prediction horizon for each controller model and experiment. In general, the control deviations are centered and remain at reasonable bounds. It can be seen that the deviations of the MPC using the white-box model are the largest, especially for the zone air temperature. These deviations are enhanced during Experiments 2 and 3 where solar irradiation is substantially increased during the periods that the white-box model is implemented into the MPC. The black-box model is also exposed to increased solar irradiation in *Experiment 3*, but it does not lead to enhanced deviations. Solar irradiation is also large during the period that the black-box model is implemented in Experiment 3. However, this model does not lead to enhanced deviations The daily pattern of the control deviations of the zone air temperature evidences the stronger influence of the weather forecast uncertainty in the white-box approach. This effect is not observed for the valve opening control deviations that are not significantly biased and do not show a daily pattern for any model.

The stronger influence of the weather conditions uncertainty on the white-box model predictions can be explained by the increased number of disturbances that this model needs to handle. Interestingly, the white-box model leads to the most accurate predictions of the zone air temperature when using historical data while showing the largest control deviations for the same variable. This lack of correlation reveals the larger sensitivity of this model to the weather forecast uncertainty, which can be considerable, as indicated by Figure 3. In *Experiment 3* the white-box model shows the most significant control deviations in the zone-air temperature. However, it also leads to the least thermal discomfort of all models. The detrimental effect of a higher uncertainty influence is effectively amended by the MPC machinery that undergoes forecast and state updates every control step. Also, the more reliable representation of the heating system aids in balancing this effect.

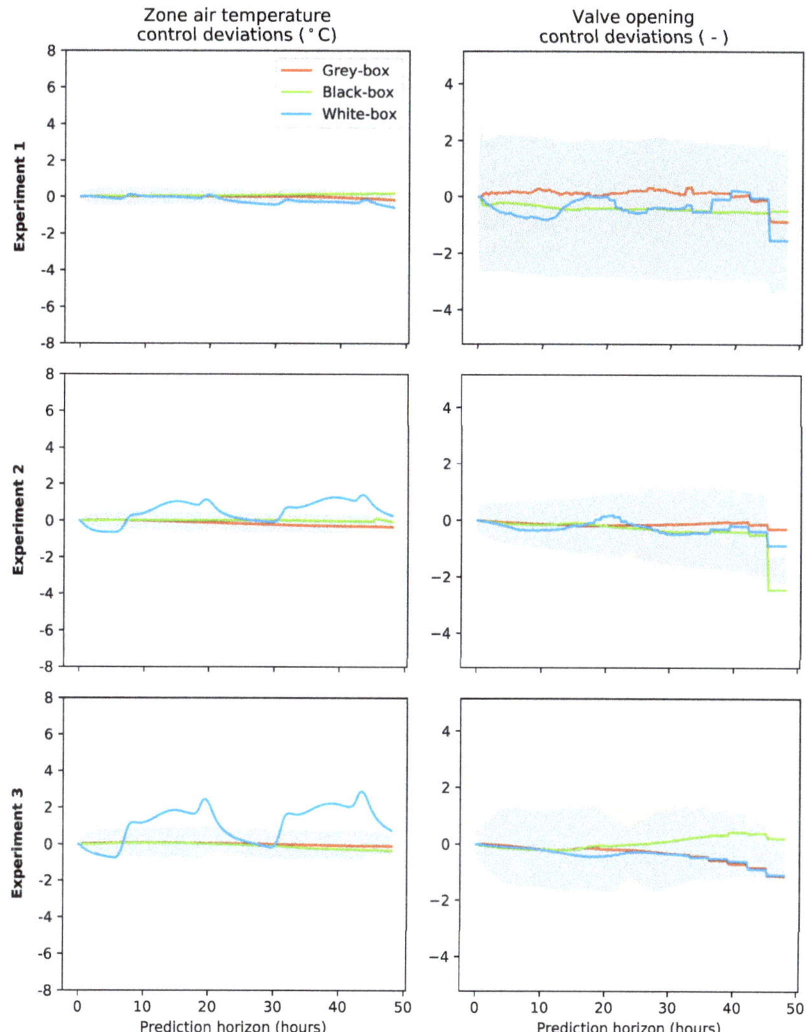

Figure 7. Mean and standard deviation of the control deviation as function of the prediction horizon. The graphs in the left column show the control deviations for the zone air temperature in each experiment and have units of (°C). The graphs in the right column show the control deviations for the valve opening in each experiment and have no units (-). Note that negative values are possible since both positive and negative control deviations can occur between predictions and observed values.

6. Discussion

There are strong reasons to believe that Modelica will constitute the future of building modeling and optimization, namely the support of IBPSA and several tools, the active development of multiple building component libraries, the language flexibility, and the possibility of extending the models for gradient-based optimization algorithms. An essential choice in the study has been the selection of the representative models. The white-box model is prototyped with the IDEAS library. The grey-box model architecture is decided from a forward-selection procedure. Finally, the black-box model is chosen to have the same representation as the grey-box model but not being constrained by any physical insight. Despite the scope of this study being limited to one model per paradigm, the selected models constitute a good representation of the main characteristics of each modeling paradigm.

This paper has highlighted practical aspects that are relevant for the implementation of each modeling approach into MPC. The most notable aspect is the significant size difference between the physics-based and the data-driven models. Even though there has been considerable progress with the development of numerous Modelica libraries that ease the development of white-box models, there is still a substantial engineering effort required to prototype and calibrate them. One of the main advantages of the white-box modeling approach is that it theoretically does not require monitoring data to be assembled. However, we experienced here that historical data is still needed to validate and calibrate the model. In fact, despite having complete access to all geometry and material properties, we observe that the calibration process is *essential* to achieve a good prediction performance. There exist efforts to alleviate this issue by generating the models directly from building information models [44], from geographic information systems [45], or from building floor plans and meta data [46].

Even if historical data is still required, the physics-based approach brings other relevant advantages. First, the richness of the data used for calibration and validation is not critical because the model is reinforced with physical insights. Second, constraining the model with physical principles increases its robustness and reliability so that the model may be well suited to make predictions for the full range of operating conditions. Although this notion may not be new, we experience that not modeling the heating system can seriously hamper the control performance of the data-driven models when working out of the operating conditions captured by the training data-set. Finally, the physics-based approach brings other implicit advantages like the possibility to interpret its results or its increased suitability for fault detection and diagnosis.

Another relevant aspect that has been investigated is the correlation between prediction and control performance. It is found that a better prediction performance does not necessarily indicate an improved control performance. Contrarily, increasing the amount of physical detail in the model increases the robustness in prediction and control performance. This was confirmed by Blum et al. [4] who evaluated the prediction performance using the RMSE in cross-validation. We come to the same conclusion by also analyzing the n-step-ahead prediction errors for horizons up to two days. In fact, the white-box model shows the worst accuracy of all models in training data, while generalizing better and outperforming in control. Increasing the amount of physical detail is, however, a double-edged sword: the model size is substantially increased, which is particularly critical for the solution method of the optimization problem. Two approaches can be followed: either the model complexity is reduced or the optimization solver is improved. The transcription method used is particularly critical in this regard since it can easily magnify the optimization complexity. Another disturbing matter when increasing the model size of the building envelope is its increased sensitivity to weather forecast uncertainty. The increased sensitivity to weather forecast uncertainty is revealed by the daily pattern of the control deviations evaluated for the zone air temperature, common to the three experiments and magnified by the high solar irradiation.

Overall, we observe that including physical insights in a model is beneficial, but the increased complexity should be carefully handled. On the one hand, the analysis of the models' architecture in Section 4 shows that most of the complexity of a white-box model comes from the building envelope: only 3 out of the 94 model states belong to the heating system. On the other hand, the control performance assessment of Section 5 highlights the importance of modeling the heating system in detail. Moreover, it is suggested that a more detailed building envelope model can be more sensitive to weather forecast uncertainty. Consequently, we advocate a modeling approach that synergizes the physics- and data-driven paradigms. Primarily the heating system should be modeled in detail and based on first principles, which is motivated by three main aspects that are treated in this paper:

(1) the relevance of the heating system in control performance, (2) its low complexity overhead, and (3) its low sensitivity to uncertainty during operation. On the other hand, the building envelope can substantially benefit from using operational data for training and making some simplifications, motivated by: (1) the benefits of utilizing a systematic approach to calibrate the parameters, (2) the high complexity overhead of the building envelope, and (3) its high sensitivity to uncertainty during operation. Still, physical insights may be used as much as possible to derive good initial guesses of the parameters and obtain an intelligible model.

7. Conclusions

This paper investigates the strengths and weaknesses of the three main modeling paradigms for optimal building predictive control. A representative model of each paradigm is selected, and the three models are configured using the same optimal control framework for a thermally activated building located in Leuven, Belgium. The models are analyzed and evaluated in prediction and control performance. The main difference between the physics- and data-driven models is the disparity in model size, which stems from the number of equations and states that the white-box model needs to describe the main building physics completely.

The physics-based model has 94 states for the envisaged case, while the data-driven models use only 3 states. It is shown that most of the states of the white-box model belong to the building envelope, with only three states belonging to the heating system. Despite its increased level of detail, the white-box model has the highest RMSE in historical monitoring data, and its calibration process is a very cumbersome task. However, the results suggest that increasing the amount of physical detail is beneficial for prediction and control. Specifically, it is shown that not modeling the heating system based on first principles can seriously hamper the control performance. Prediction and control performance are also compared, but a correlation between both metrics cannot be found, even when using the same data for prediction assessment as those obtained from the control experiments. Finally, the *n-step ahead control deviation* is proposed to assess if the strategies devised by the MPC are effectively implemented. The analysis of this metric and the prediction performance indicates that the very detailed building envelope model may be more sensitive to forecast uncertainty.

To conclude, we suggest a modeling approach that synergizes the physics- and data-driven approaches. On the one hand, the heating system may be mainly modeled based on physical principles because of its lower complexity overhead and less exposure to uncertainty. On the other hand, the building envelope may introduce some simplifications while using physics to derive a good initial guess and monitoring data to train its parameters.

Author Contributions: J.A.: Conceptualization, Methodology, Software, Formal Analysis, Data Curation, Writing—Original Draft, Writing—Review, Editing, Visualization, Funding acquisition. F.S. and L.H.: Conceptualization, Methodology, Writing—Review, Editing, Supervision, Funding acquisition. All authors have read and agreed to the published version of the manuscript.

Funding: This research is financed by VITO through a PhD Fellowship under the Grant No. 1710754 and by KU Leuven through the C2 project Energy Positive Districts (C24M/21/021). This work emerged from the IBPSA Project 1, an international project conducted under the umbrella of the International Building Performance Simulation Association (IBPSA). Project 1 will develop and demonstrate a BIM/GIS and Modelica Framework for building and community energy system design and operation.

Data Availability Statement: The data obtained during the experiments and the Python scripts to generate the figures of this paper can be found at: https://data.mendeley.com/datasets/xzdy23nzvj/1 (accessed on 24 March 2022).

Conflicts of Interest: The authors declare no conflict of interest.

Appendix A. White-Box Modelica Model

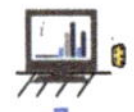

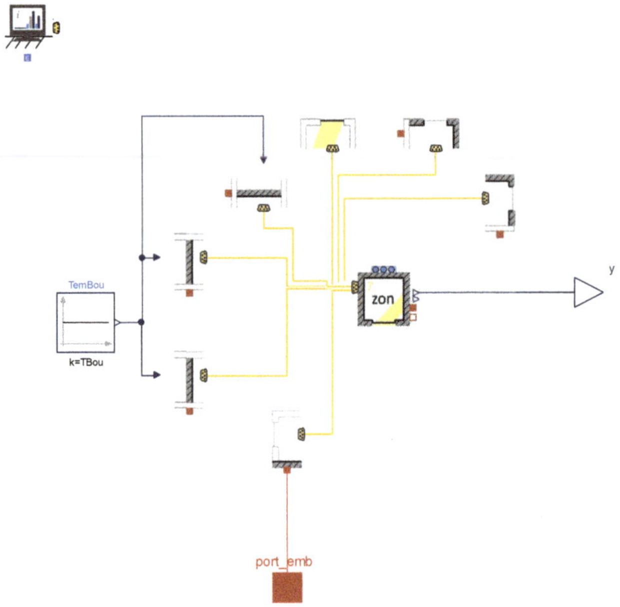

Figure A1. White-box model of the Vliet building test room and heating system. (**a**) White-box model of the building envelope of the test room; (**b**) White-box model of the heating system.

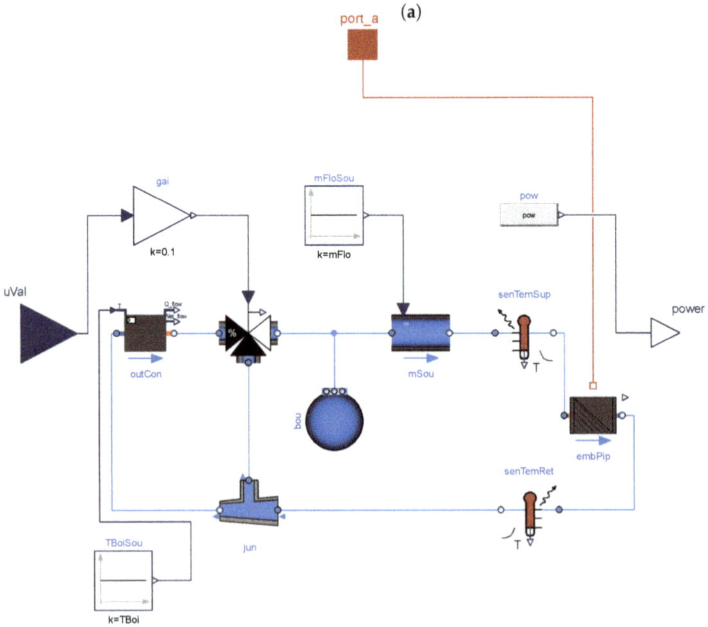

Appendix B. Grey-Box Modelica Model

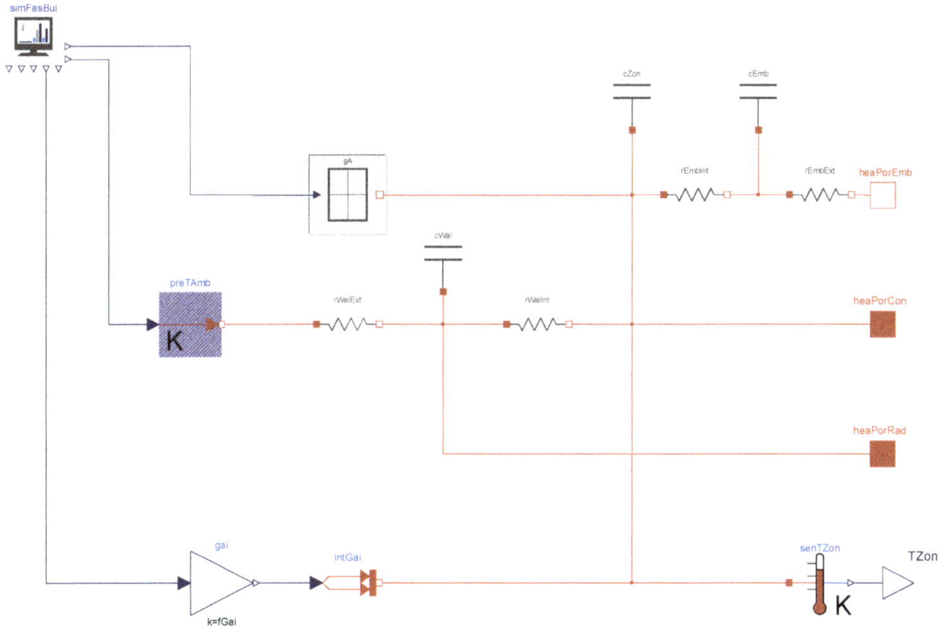

Figure A2. Grey-box model of the building envelope of the Vliet building test room.

References

1. IEA. *Global Status Report for Buildings and Construction: Towards a Zero-Emissions, Efficient and Resilient Buildings and Construction Sector*; Technical Report; International Energy Agency: Paris, France, 2019.
2. Dupont, B.; Dietrich, K.; De Jonghe, C.; Ramos, A.; Belmans, R. Impact of residential demand response on power system operation: A Belgian case study. *Appl. Energy* **2014**, *122*, 1–10. [CrossRef]
3. Sturzenegger, D.; Gyalistras, D.; Morari, M.; Smith, R.S. Model Predictive Climate Control of a Swiss Office Building: Implementation, Results, and Cost-Benefit Analysis. *IEEE Trans. Control Syst. Technol.* **2016**, *24*, 1–12. [CrossRef]
4. Blum, D.; Arendt, K.; Rivalin, L.; Piette, M.; Wetter, M.; Veje, C. Practical factors of envelope model setup and their effects on the performance of model predictive control for building heating, ventilating, and air conditioning systems. *Appl. Energy* **2019**, *236*, 410–425. [CrossRef]
5. Široký, J.; Oldewurtel, F.; Cigler, J.; Prívara, S. Experimental analysis of model predictive control for an energy efficient building heating system. *Appl. Energy* **2011**, *88*, 3079–3087. [CrossRef]
6. Coninck, R.D.; Helsen, L. Practical implementation and evaluation of model predictive control for an office building in Brussels. *Energy Build.* **2016**, *111*, 290–298. [CrossRef]
7. Drgoňa, J.; Arroyo, J.; Cupeiro Figueroa, I.; Blum, D.; Arendt, K.; Kim, D.; Ollé, E.P.; Oravec, J.; Wetter, M.; Vrabie, D.L.; et al. All you need to know about model predictive control for buildings. *Annu. Rev. Control* **2020**, *50*, 190–232. [CrossRef]
8. Oldewurtel, F.; Parisio, A.; Jones, C.N.; Gyalistras, D.; Gwerder, M.; Stauch, V.; Lehmann, B.; Morari, M. Use of model predictive control and weather forecasts for energy efficient building climate control. *Energy Build.* **2012**, *45*, 15–27. [CrossRef]
9. Cieśliński, K.; Tabor, S.; Szul, T. Evaluation of Energy Efficiency in Thermally Improved Residential Buildings, with a Weather Controlled Central Heating System. A Case Study in Poland. *Appl. Sci.* **2020**, *10*, 8430. [CrossRef]
10. Atam, E.; Helsen, L. Control-Oriented Thermal Modeling of Multizone Buildings: Methods and Issues: Intelligent Control of a Building System. *IEEE Control Syst. Mag.* **2016**, *36*, 86–111. [CrossRef]
11. Picard, D.; Drgoňa, J.; Kvasnica, M.; Helsen, L. Impact of the controller model complexity on model predictive control performance for buildings. *Energy Build.* **2017**, *152*, 739–751. [CrossRef]
12. Arroyo, J.; Spiessens, F.; Helsen, L. Identification of multi-zone grey-box building models for use in model predictive control. *J. Build. Perform. Simul.* **2020**, *13*, 472–486. [CrossRef]
13. Picard, D.; Sourbron, M.; Jorissen, F.; Váňa, Z.; Cigler, J.; Ferkl, L.; Helsen, L. Comparison of Model Predictive Control Performance Using Grey-Box and White-Box Controller Models of a Multi-zone Office Building. In Proceedings of the 4th International High Performance Buildings Conference, West Lafayette, IN, USA, 11–14 July 2016.

14. Mugnini, A.; Coccia, G.; Polonara, F.; Arteconi, A. Performance Assessment of Data-Driven and Physical-Based Models to Predict Building Energy Demand in Model Predictive Controls. *Energies* **2020**, *13*, 3125. [CrossRef]
15. Arendt, K.; Jradi, M.; Shaker, H.; Veje, C. Comparative Analysis of White-, Gray- and Black-box Models for Thermal Simulation of Indoor Environment: Teaching Building Case Study. In Proceedings of the Building Performance Modeling Conference and SimBuild, Chicago, IL, USA, 26–28 September 2018.
16. Blum, D.; Arroyo, J.; Huang, S.; Drgoňa, J.; Jorissen, F.; Taxt Walnum, H.; Yan, C.; Benne, K.; Vrabie, D.; Wetter, M.; et al. Building Optimization Testing Framework (BOPTEST) for Simulation-Based Benchmarking of Control Strategies in Buildings. *J. Build. Perform. Simul.* **2021**, *14*, 586–610. [CrossRef]
17. Prívara, S.; Cigler, J.; Váňa, Z.; Oldewurtel, F.; Žáčeková, E. Use of partial least squares within the control relevant identification for buildings. *Control Eng. Pract.* **2013**, *21*, 113–121. [CrossRef]
18. Drgoňa, J.; Picard, D.; Helsen, L. Cloud-based implementation of white-box model predictive control for a GEOTABS office building: A field test demonstration. *J. Process Control* **2020**, *88*, 63–77. [CrossRef]
19. Jain, A.; Smarra, F.; Reticcioli, E.; D'Innocenzo, A.; Morari, M. NeurOpt: Neural network based optimization for building energy management and climate control. *arXiv* **2020**, arXiv:2001.07831.
20. Castilla, M.; Álvarez, J.; Berenguel, M.; Rodríguez, F.; Guzmán, J.; Pérez, M. A comparison of thermal comfort predictive control strategies. *Energy Build.* **2011**, *43*, 2737–2746. [CrossRef]
21. Leenknegt, S. Numerical and Experimental Analysis of Passive Cooling through Night Ventilation. Ph.D. Thesis, KU Leuven, Leuven, Belgium, 2013.
22. Perarnau Olle, E. Design and Experimental Implementation of a Data Interface for the Optimal Control of Thermal Systems. Master's Thesis, KU Leuven, Leuven, Belgium, 2018.
23. Madsen, H.; Schultz, J. *Short Time Determination of the Heat Dynamics of Buildings*; BYG-Rapport; Report No. 243; Department of Civil Engineering, Technical University of Denmark: Kgs. Lyngby, Denmark, 1993.
24. Bacher, P.; Madsen, H. Identifying suitable models for the heat dynamics of buildings. *Energy Build.* **2011**, *43*, 1511–1522. [CrossRef]
25. Åkesson, J.; Årzén, K.E.; Gäfvert, M.; Bergdahl, T.; Tummescheit, H. Modeling and optimization with Optimica and JModelica.org-Languages and tools for solving large-scale dynamic optimization problems. *Comput. Chem. Eng.* **2010**, *34*, 1737–1749. [CrossRef]
26. Mattsson, S.E.; Elmqvist, H. Modelica—An international effort to design the next generation modeling language. *IFAC Proc. Vol.* **1997**, *30*, 151–155. [CrossRef]
27. Åkesson, J. Optimica—An Extension of Modelica Supporting Dynamic Optimization. In Proceedings of the 6th International Modelica Conference, Bielefeld, Germany, 3–4 March 2008.
28. Wetter, M.; Zuo, W.; Nouidui, T.; Pang, X. Modelica Buildings library. *J. Build. Perform. Simul.* **2014**, *7*, 253–270. [CrossRef]
29. Jorissen, F.; Reynders, G.; Baetens, R.; Picard, D.; Saelens, D.; Helsen, L. Implementation and verification of the IDEAS building energy simulation library. *J. Build. Perform. Simul.* **2018**, *11*, 669–688. [CrossRef]
30. Müller, D.; Lauster, M.; Constantin, A.; Fuchs, M.; Remmen, P. AixLib—An Open-Source Modelica Library within The IEA-EBC Annex 60 Framework. In Proceedings of the BauSIM IBPSA Conference, Dresden, Germany, 12–14 September 2016; pp. 3–9.
31. Nytsch-Geusen, C.; Banhardt, C.; Inderfurth, A.; Mucha, K.; Möckel, J.; Rädler, J.; Thorade, M.; Tugores, C. Buildingsystems—Eine modular hierarchische Modell-Bibliothek zur energetischen Gebäude und Anlagensimulation. In Proceedings of the BauSIM IBPSA Conference, Dresden, Germany, 12–14 September 2016; pp. 473–480.
32. Coninck, R.D.; Magnusson, F.; Åkesson, J.; Helsen, L. Toolbox for development and validation of grey-box building models for forecasting and control. *J. Build. Perform. Simul.* **2016**, *9*, 288–303. [CrossRef]
33. Wetter, M.; van Treeck, C.; Helsen, L.; Maccarini, A.; Saelens, D.; Robinson, D.; Schweiger, G. IBPSA Project 1: BIM/GIS and Modelica framework for building and community energy system design and operation—Ongoing developments, lessons learned and challenges. In Proceedings of the Sustainable Built Environment Conference, Seoul, Korea, 12–13 December 2019; p. 012114.
34. Brinsfield, R.; Yaramanoglu, M.; Wheaton, F. Ground level solar radiation prediction model including cloud cover effects. *Sol. Energy* **1984**, *33*, 493–499. [CrossRef]
35. Jorissen, F. Modelica_MPC. 2020. Available online: https://github.com/Mathadon/modelica_MPC (accessed on 15 November 2020).
36. Barbato, A.; Bolchini, C.; Geronazzo, A.; Quintarelli, E.; Palamarciuc, A.; Piti, A.; Rottondi, C.; Verticale, G. Energy Optimization and Management of Demand Response Interactions in a Smart Campus. *Energies* **2016**, *9*, 398. [CrossRef]
37. Kontes, G.D.; Giannakis, G.I.; Horn, P.; Steiger, S.; Rovas, D.V. Using Thermostats for Indoor Climate Control in Office Buildings: The Effect on Thermal Comfort. *Energies* **2017**, *10*, 1368. [CrossRef]
38. Andersson, J.; Gillis, J.; Horn, G.; Rawlings, J.; Diehl, M. CasADi—A software framework for nonlinear optimization and optimal control. *Math. Program. Comput.* **2019**, *11*, 1–36. [CrossRef]
39. Jorissen, F.; Boydens, W.; Helsen, L. TACO, an automated toolchain for model predictive control of building systems: Implementation and verification. *J. Build. Perform. Simul.* **2018**, *12*, 180–192. [CrossRef]
40. Magnusson, F. Numerical and Symbolic Methods for Dynamic Optimization. Ph.D. Thesis, Department of Automatic Control, Lund University, Lund, Sweeden, 2016.
41. Axelsson, M.; Magnusson, F.; Henningsson, T. A Framework for Nonlinear Model Predictive Control in JModelica.org. In Proceedings of the 11th International Modelica Conference, Versailles, France, 21–23 September 2015; pp. 301–310. [CrossRef]

42. Sun, F.; Li, G.; Wang, J. Unscented Kalman Filter using augmented state in the presence of additive noise. In Proceedings of the IITA International Conference on Control, Automation and Systems Engineering, Zhangjiajie, China, 11–12 July 2009; pp. 379–382.
43. Wan, E.A.; Merwe, R.V.D. The Unscented Kalman Filter for nonlinear estimation. In Proceedings of the IEEE 2000 Adaptive Systems for Signal Processing, Communications, and Control Symposium, Lake Louise, AB, Canada, 4 October 2000; pp. 153–158. [CrossRef]
44. Reynders, G.; Andriamamonjy, R.A.L.; Klein, R.; Saelens, D. Towards an IFC-Modelica Tool Facilitating Model Complexity Selection for Building Energy Simulation. In Proceedings of the 15th International Conference of IBPSA, San Francisco, CA, USA, 7–9 August 2017.
45. De Jaeger, I.; Reynders, G.; Saelens, D. Impact of spatial accuracy on district energy simulations. In Proceedings of the 11th Nordic Symposium on Building Physics, Trondheim, Norway, 11–14 June 2017; Volume 132, pp. 561–566.
46. Jorissen, F.; Picard, D.; Six, K.; Helsen, L. Detailed white-box non-linear model predictive control for scalable building HVAC control. In Proceedings of the 14th International Modelica Conference, Linköping, Sweden, 20–24 September 2021.

Article

Data-Driven Model-Based Control Strategies to Improve the Cooling Performance of Commercial and Institutional Buildings

Etienne Saloux * and Kun Zhang

CanmetENERGY, Natural Resources Canada, Varennes, QC J3X 1P7, Canada
* Correspondence: etienne.saloux@nrcan-rncan.gc.ca

Abstract: The increasing amount of operational data in buildings opens up new methods for improving building performance through advanced controls. Although predictive control has been widely investigated in the literature, field demonstrations still remain rare. Alternatively, model-based controls can provide similar improvement while being easier to implement in real buildings. This paper investigates three data-driven model-based control strategies to improve the cooling performance of commercial and institutional buildings: (a) chiller sequencing, (b) free cooling, and (c) supply air temperature reset. These energy efficiency measures are applied to an existing commercial building in Canada with data from summer 2020 and 2021. The impact of each measure is individually assessed, as well as their combined effects. The results show that all three of the measures together reduce building cooling energy by 12% and cooling system electric energy by 33%.

Keywords: advanced controls; energy efficiency; model-based controls; free cooling; chiller sequencing; temperature reset strategies

Citation: Saloux, E.; Zhang, K. Data-Driven Model-Based Control Strategies to Improve the Cooling Performance of Commercial and Institutional Buildings. *Buildings* **2023**, *13*, 474. https://doi.org/10.3390/buildings13020474

Academic Editors: Christopher Yu-Hang Chao and Kian Jon Chua

Received: 5 January 2023
Revised: 26 January 2023
Accepted: 7 February 2023
Published: 9 February 2023

1. Introduction

1.1. Motivation

Despite the efforts made toward control optimization, most existing buildings are still not operated at a high level of energy efficiency. It is estimated that the annual energy use of existing buildings could be reduced by up to 30% [1] with the improvement of control and operation, as well as the detection and correction of equipment problems and inefficiencies [2]. Besides energy savings, more efficient building operation could reduce maintenance costs by 20% [3]. Furthermore, with the advent of increasingly efficient technologies, buildings are becoming more and more complex. Thus, the adequate optimization of building operation has become a key enabler for harvesting the full potential of these energy systems. Kramer et al. [4] conducted a vast campaign to prove the business case for building analytics. They categorized commercial products that are available on the market for optimizing building controls into three categories: energy information system, fault detection and diagnostics, and automated system optimization. They tested 85 different software on 6500 buildings from 104 organizations and showed that the median annual energy savings could reach up to 9% with a two-year simple payback.

Moreover, sub-hourly data have become increasingly available in buildings and provide an untapped opportunity for improving existing building controls. This vast amount of data could be leveraged to support the development of advanced control strategies, which could eventually be further integrated into existing building automation systems (BAS) or building optimization commercial platforms. Model-based predictive controls (MPC) are a compelling example and have been intensively investigated in the past decades [5]. This method consists of the use of a control-oriented model along with forecasts (e.g., weather, occupancy) to predict the future behavior of a building hours ahead and optimize its heating and cooling system operation accordingly. Although it shows good promise in theory, its implementation in existing buildings remains relatively scarce and somewhat challenging [6,7]. On the other hand, operational data could still be leveraged through simpler approaches such as data-driven rule- or model-based controls, which are much easier to implement in existing buildings.

This paper aims to investigate three data-driven model-based control strategies, which can be readily implemented in actual building controls to improve the cooling performance of Commercial and Institutional (CI) buildings equipped with BAS.

1.2. Literature Review

Various control strategies for optimizing building operation have been investigated in the past and inventoried in review papers. For instance, Wang and Ma [8] investigated supervisory and optimal control for building Heating, Ventilation and Air Conditioning (HVAC) systems while Afram and Janabi-Sharifi [9], and more recently Taheri et al. [10], conducted reviews of MPC strategies for HVAC systems. In addition, some authors focused on specific aspects of building operation. Thieblemont et al. [11] and Yu et al. [12] investigated MPC for buildings equipped with energy storage devices. Péan et al. [13] reviewed control strategies for heat pump systems for enhancing flexibility. Darwazeh et al. concentrated on peak load management strategies [14]. Finally, Mirakhorli and Dong [15] focused on occupancy behavior-based MPC while Park et al. [16] conducted a review of field implementations of occupant-centric building controls. Although a good proportion of these publications focus on applications related to building heating performance, advanced controls have also been widely used for improving the cooling performance. Some of these applications include, among others:

- Building precooling in a time-of-use tariff structure, targeting the reduction of building energy use during mid-peak and high-peak hours [17].
- The optimization of HVAC operating conditions, targeting the reduction of building energy use [18] or building electric power [19].
- The optimization of fresh air intake based on occupancy, aiming to provide a suitable amount of fresh air depending on the actual number of occupants [20].
- Natural and hybrid ventilation, targeting the incorporation of more fresh air into the building at critical times of the day to reduce mechanical cooling [21,22].
- The optimization of cooling system performance, with the goal of improving the chiller performance at the part-load ratio [23–25].
- The management of ice banks, targeting the reduction of building energy use during peak hours [26–28].
- The management of radiant slab systems to reduce building energy use in the morning while improving thermal comfort [29,30].

Three applications are further discussed since they are the foundations of the proposed measures: local HVAC controls, chilled water system performance at the part-load ratio, and free cooling.

Recently, ASHRAE Guideline 36 (G36) has been released and provides high-performance control sequences to improve the operation of HVAC systems [31]. Such standardized control sequences will help reduce engineering time and reduce programming and commissioning time while improving energy efficiency and indoor air quality. This guideline has focused on airside equipment, mainly variable air volume (VAV) systems and terminal units, as well as more recently, on waterside equipment related to heating and cooling plants. Zhang et al. [18] estimated the potential savings obtained by retrofitting existing controls to new control sequences based on this guideline and applied it to multi-zone VAV systems. They tested various scenarios under different climates, building operating hours, and internal load magnitudes, and found that G36 control sequences could reduce energy use by 2–75% with an average of 31%. Three control strategies played a significant role in this reduction: supply air temperature reset, duct static pressure reset, and zone airflow control.

For the cooling system operation at the part-load ratio, Thangavelu et al. [23] investigated a multi-chiller plant and evaluated its electric power as the contribution of that of the chiller, the circulating pumps, and the cooling tower fans. They proposed a novel methodology, which optimizes the flowrates of chilled and condenser water, chilled and condenser water temperatures, and cooling tower air flowrates, to maximize cooling system performance. They validated the approach using three case studies and found that the energy savings could reach 20%, 40%, and 42%, respectively, compared to conventional controls. Fan et al. [32] studied optimal control strategies for a multi-chiller system to determine the optimal sequencing based on probability density distribution of the cooling load ratio. They obtained savings of up to 4% compared to the original control strategy. Liao and Huang [24] developed a hybrid predictive sequencing control for a multi-chiller plant, which optimizes the chiller sequencing based on the forecasted cooling load. For this purpose, chiller performance curves were derived from three months of historical data from a real chiller and the objective function was intended to enhance system stability (i.e., number of chiller switches) and reduce operational costs while maintaining thermal comfort and energy efficiency. It is worth mentioning that operational costs were calculated as the sum of the chiller plant energy consumption, the start-up cost of chiller plants (chillers and pumps), and the depreciation cost of all devices. The results showed that the switch number of chillers was reduced by 20% and the operational cost decreased by 4%. A similar approach was developed by Gunay et al. [25] for equipment sequencing in a central heating and cooling plant. The hourly cooling peak load was forecasted for the next day and the chiller sequencing was adjusted accordingly. Operational data were used to derive data-driven performance curves for the five chillers of the cooling plant. The authors obtained cooling energy savings of 25% compared to the current operation. The importance of chiller scheduling has also been reported by Chen et al. [33]. The authors studied the impact of multi-chiller plants on design and operation optimization and considered 13 centrifugal chillers. They showed that it is a good practice to select chillers with different capacities and that energy consumption could be reduced by 20% with the best chiller design option compared with the worst option.

Free cooling is an effective strategy to improve HVAC system performance when outdoor air conditions are favorable. Broadly speaking, "free cooling" strategies can be implemented in various ways such as by using airside and waterside economizers, or natural and hybrid ventilation. An airside economizer increases fresh air intake and reduces the return air by adjusting the dampers in the air handling unit (AHU) system. ASHRAE Standard 90.1 specifies various operating conditions for economizer high limits and devices (e.g., fixed and/or differential dry bulb or enthalpy) [34]. Hybrid ventilation offers a great opportunity for taking advantage of favorable outdoor air conditions to precool buildings at night in order to discharge building thermal mass, which decreases operative temperature, and to reduce mechanical cooling during the day [22]. A hybrid ventilation system consists of a possibly fan-assisted system (e.g., windows and motorized dampers), which allows the outdoor air to enter the building and that is used along with a mechanical cooling system (e.g., chillers) to provide cooling to the building at all times [21]. Stasi et al. [35] investigated nearly zero-energy buildings three hybrid ventilation strategies: earth-to-air heat exchanger, night hybrid ventilation, and free cooling mode, and showed that cooling energy was reduced by 14–21%, and electricity consumption by 8%, respectively.

1.3. Paper Objectives and Contributions

This paper intends to develop and evaluate energy efficiency measures to improve the cooling system performance of CI buildings. Specifically, these measures are:

- Chiller sequencing based on data-driven performance curves;
- Free cooling strategy based on electric power estimation of the whole cooling system, including both the waterside (chilled water system) and airside system (AHUs);
- Supply air temperature reset strategies.

The contributions of the paper include the following:

- The proposed measures build upon operational data, virtual energy meters, and control-oriented models, which do not require prediction and forecasting, and aim to be readily implementable in existing buildings while being more robust and less prone to errors compared to more complex methods such as model predictive control.

- The proposed measures tackle the existing systems and do not require the installation of additional equipment (e.g., sensors, meters, motorized operable windows for natural ventilation, etc.). The implementation only requires minimum code modification in the BAS, making it applicable to a wide range of CI buildings with different HVAC configurations.

- This work evaluates both the individual and combined impacts of energy efficiency measures in the same building, which has rarely been investigated in the past to the best of our knowledge.

The paper is organized as follows: Section 2 describes the case study, an existing large commercial building. The three proposed measures are described in Section 3, where the methodology to assess the performance is also explained. Section 4 reports the performance of each measure, as well as their combined performance, and discusses the generalizability of the measures. Section 5 concludes the paper.

2. Case Study: A Large Commercial Building

2.1. Building Description

The case study building is a 36,000-m^2, 11-floor commercial building located in Montreal, Quebec. Figure 1 shows a schematic of the chilled water network. The cooling plant is composed of three chillers: one centrifugal chiller connected to two wet cooling towers and two screw chillers connected to two air-cooled condensers or dry cooling towers. A heat exchanger is connected to the condenser water network of the air-cooled condensers as a waterside economizer. It is activated when the outdoor air temperature is low. The chilled water is distributed to cooling coils in the AHUs which are located on different floors of the building.

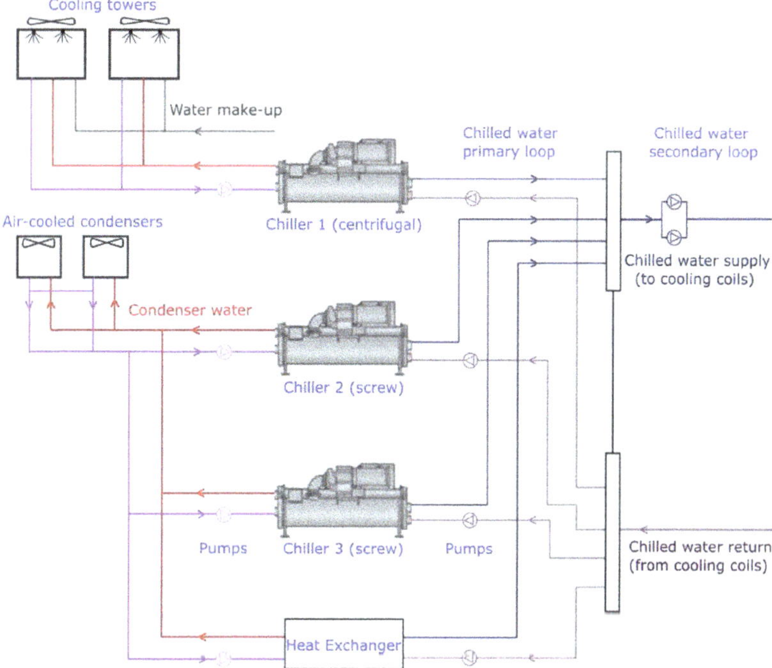

Figure 1. Schematic of the chilled water system in the large commercial case study building.

The airside system is a primary-secondary AHU system. The primary AHU system is composed of two fresh air pre-treatment units equipped with gas burners, precooling coils, humidifiers, and heat recovery wheels to pre-treat the outdoor air before being sent to the secondary AHU system. This secondary system consists of 22 cooling coils in the AHU located from the 2nd floor to the 11th floor. Once conditioned in the primary AHU system, the outdoor air is mixed with the return air in each AHU in the different floors, further cooled down if required, and discharged to building zones through single- or dual-duct systems. For dual-duct systems, a fraction of the cooled air flowrate is bypassed via the hot deck before being mixed again with the cold deck. Figure 2 shows a schematic of the air handling unit system.

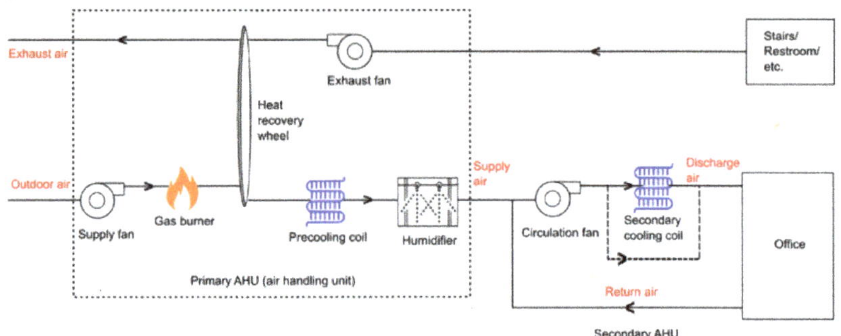

Figure 2. Schematic of the air handling unit system in the large commercial case study building [36].

2.2. Operational Data

Building operational data such as temperature, relative humidity, pressure, flow rates, valve openings, chiller modulation, and electric currents (amps) were recorded at five-min intervals. More specifically, the following variables were used for the modelling and calculations:

- *Primary loop of chilled water network*: amps of circulation pumps;
- *Secondary loop of chilled water network*: flowrate along with supply and return water temperature to estimate building cooling load; amps of circulation pumps;
- *Chillers*: % Rated Load Amps (RLA) for each chiller, used to derive electric power;
- *Condenser water loop*: amps of fans in cooling towers and air-cooled condensers; amps of circulation pumps;
- *Air handling units*: supply and discharged air temperatures; outdoor air temperature and relative humidity; return air temperature and relative humidity; air flowrates; amps from the fans.

The data collected from 1 June 2020 to 1 September 2020 and from 1 June 2021 to 1 September 2021 were used to develop and validate the proposed data-driven measures. The outliers and missing data were removed from the dataset based on statistical means and expert knowledge. The available data allowed for calculation of the cooling thermal power at the building level, although not at the chiller level due to the lack of measurement devices. Therefore, chiller performance curves were generated for combinations of chillers in operation and the building cooling power was used as a proxy. More information on the uncertainty analysis of the building cooling power can be found in [36]. Spot measurements were used to derive electric power from % RLA for each chiller. When spot measurements could not be collected, manufacturer specifications were used as an alternative.

2.3. Energy Use of the Building

This subsection provides an overview of the electric and thermal energy usage in the case study building, especially the contribution of chillers, pumps, and cooling towers to the building electric power and the impact of ventilation on the building cooling load. This analysis aims to better put in perspective the potential energy savings of the proposed measures.

Figure 3 shows the electricity use of the chillers, chiller auxiliary equipment (i.e., circulation pumps, cooling tower fans, air-cooled condenser fans), the AHU fans (for both primary and secondary systems), and the remaining load, deduced from the total building electricity consumption. Note that chiller auxiliary equipment concentrates on the major energy loads (pumps and fans) and excludes the consumption of smaller loads such as valves and damper actuators. In total, the chilled water system and the AHU fans contribute 60–70% of the total building electricity use.

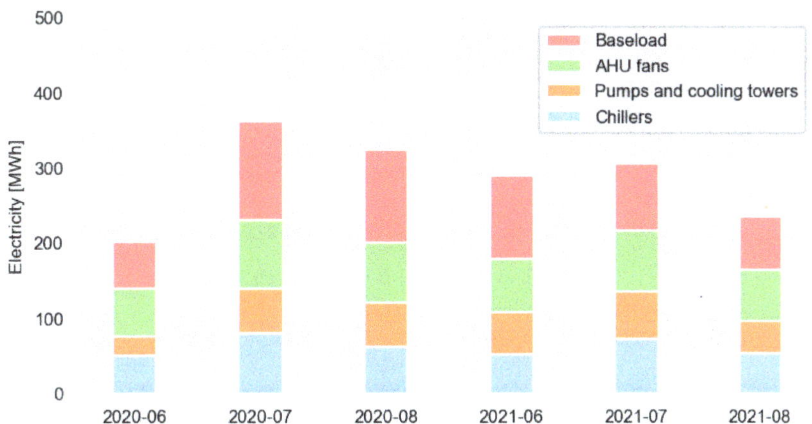

Figure 3. Electricity use of the case study building: "Baseload" refers to the non-HVAC load; "AHU fans" includes all of the fans in the air distribution network; "Pumps and cooling towers" refers to all the pumps in the water network and all the fans in the cooling towers and air-cooled condensers; "Chillers" include all three chillers.

A similar calculation was done for the building cooling load. Figure 4 shows the mechanical cooling load provided by the chilled water system as well as the contribution from fresh-air pretreatment, which accounts for ventilation requirements for occupancy and air quality, as well as space cooling (i.e., room conditioning), deduced from the difference between the chilled water system and fresh-air pretreatment. The fresh air pretreatment thermal power ($\dot{Q}_{\text{fresh air pretreat}}$) is calculated as follows:

$$\dot{Q}_{\text{fresh air pretreat}} = \dot{m}_{\text{fre,occ}}(h_{\text{OA}} - h_{\text{EA}}), \tag{1}$$

where $\dot{m}_{\text{fre,occ}}$ is the fresh air flowrate related to occupancy, and h_{OA} and h_{EA} are the enthalpy of outdoor air and exhaust air. A negative value means that fresh air pretreatment cools down the building by introducing cool fresh air; in contrast, a positive value means that it requires additional cooling to pretreat the air to indoor conditions. Figure 4 shows that fresh air pretreatment significantly contributes to the total cooling load, especially when it was hot and humid in July and August 2020, and August 2021. In June 2020 and 2021, and July 2021, its average was close to zero, meaning that the free cooling and cooling load introduced by the fresh air intake is approximately equal in those months.

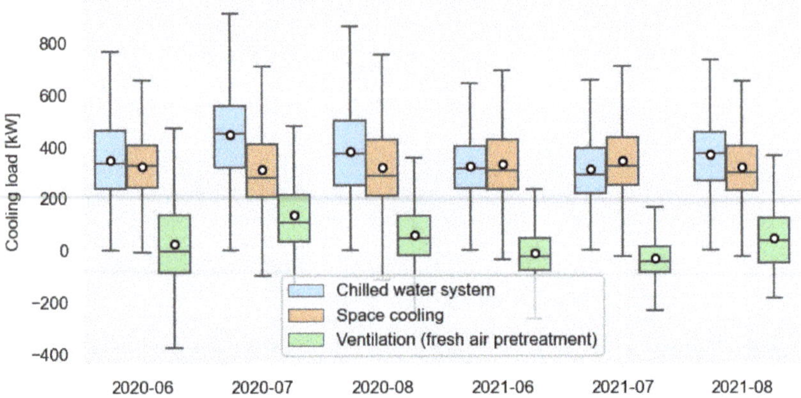

Figure 4. Cooling load overview for the case study building: the chilled water system bars show the overall cooling load incurred in the plants; the bars labelled space cooling and ventilation represent the amount of cooling load used for room conditioning and fresh air pretreatment.

3. Development of Data-Driven Measures

3.1. Chiller Sequencing

Chiller sequencing intends to take advantage of the most efficient chillers at specific cooling loads to increase the system energy efficiency. The proposed chiller sequencing is based on performance curves derived directly from the operational data. Developing performance curves based on operation data allows for capturing the performance difference between identical chillers in real-world scenarios. Manufacturer data in the chiller specifications will not be able to reflect those differences [37], whereas such differences could be further exploited to improve chilled water system performance. A similar situation was found for heat pumps [38].

3.1.1. Chilled Water System Models at Part-Load Ratio

The performance of chillers mainly depends on three factors [23]: supply water temperature at the evaporator, supply water temperature at the condenser, and the chiller part–load ratio, i.e., the ratio between the cooling load at which the chiller is operated and its nominal cooling capacity. In operation, the chiller part–load ratio appears to be the main variable that affects the system performance and has been mainly used for modelling purposes [24,25]. In fact, this simplification eases the chiller sequencing and could be justified by the slight variation in the water temperature on the demand side (evaporator) in operation, while the effect of water temperature at the condenser could be relatively well-captured by the part–load ratio itself. Indeed, if the outdoor air temperature is high, the water temperature at the condenser might be relatively high as well, which coincides with a higher cooling load and, thus, a higher chiller part–load ratio.

The chiller COP has been calculated as follows:

$$COP_{ch} = \frac{\dot{Q}_{cool}}{\sum_i \dot{W}_{ch,i}} = \frac{\dot{m}_w c_{p,w}\left(T_{ret} - T_{sup}\right)}{\sum_i \dot{W}_{ch,i}}, \tag{2}$$

where $\dot{Q}_{cool}$ is the building cooling load, which is estimated from water flowrate ($\dot{m}_w$), and supply and return water temperature (T_{sup} and T_{ret}). $c_{p,w}$ is the water specific heat at constant pressure. $\dot{W}_{ch,i}$ is the electric power of the *i*th chiller.

Figure 5 shows the chiller data-driven performance curves obtained using hourly averaged data. Note that the chiller COP is shown against the thermal load, not the part–load ratio, in order to support the development of the chiller sequencing strategy. The centrifugal chiller is referred to as CH #1, while screw chillers are referred to as CH #2 and CH #3. Since the evaporator thermal power was not measured, these curves were developed only when chillers were individually in operation (i.e., one at a time) or for a given combination of chillers (e.g., both screw chillers at the same time). To generate relatively cleaned performance curves, the following data process steps were taken:

- Low differences between water supply and return temperature (<3 °C) were excluded by statistical means;
- Low water flow rates (<10 kg/s) were excluded by statistical means;
- Low hourly electric power values (<25 kW) were excluded by statistical means; this allows for avoiding possible cycling;
- To remove transient operation, for each datum point, the corresponding combination of chillers must have been in operation for one hour and must remain in operation for one hour.

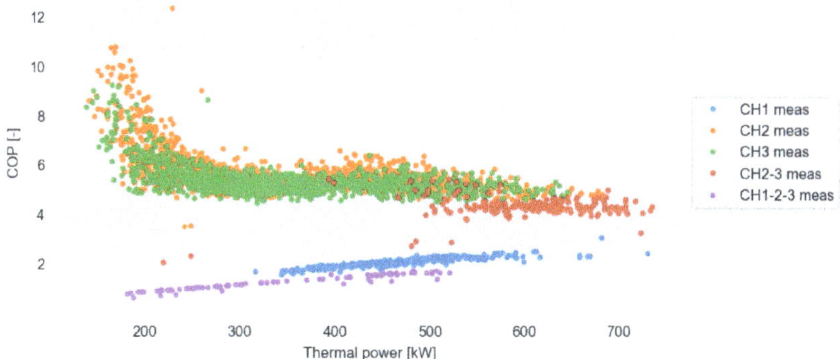

Figure 5. Chiller data-driven performance curves obtained with operational data. The electric power accounts for the chiller(s) in operation only.

The results from Figure 5 show that screw chillers (CH #2 and CH #3), although identical, have slightly different performance curves and are the most efficient units in this multi-chiller system. Both screw chillers show an unexpected COP increase at a low part–load ratio; this could be partially explained by the measurement uncertainty at a low load. When both of the screw chillers are in operation at the same time (CH #2 and #3), their combined performance is lower than when they are operated alone. This could be explained by the less-favorable operating conditions when the cooling load is high (i.e., lower inlet water temperature at the evaporator and higher inlet water temperature at the condenser). On the other hand, the centrifugal chiller (CH #1) shows an unexpected poor performance, which may indicate a technical issue. This was raised to the building technical team, which has further investigated this aspect and found that the chiller was rusted and needed in-depth cleaning. Once the chiller is cleaned, the performance curve can be easily updated to account for the new situation.

The main drawback of chiller performance curves is that they does not consider the power use of auxiliary devices, such as circulation pumps and fans from cooling towers. At a low cooling load, the chiller might be energy efficient; nonetheless, the electric power of pumps and fans might remain high. This may significantly reduce the system performance of the chilled water network. To account for the energy use from auxiliary equipment, the chilled water system COP has been calculated as follows:

$$COP_{cool} = \frac{\dot{Q}_{cool}}{\sum_i \left(\dot{W}_{ch,i} + \sum_j \dot{W}_{pp,i,j} + \sum_k \dot{W}_{fan,i,k} \right)}, \tag{3}$$

where $\dot{W}_{\mathrm{pp},i,j}$ is the electric power of the *j*th circulation pump related to the operation of the *i*th chiller; it includes pumps in both primary and secondary water loops. $\dot{W}_{\mathrm{fan},i,k}$ is the electric power of the *k*th fan related to the operation of the *i*th chiller; it includes fans in the air-cooled condensers and/or cooling towers.

Figure 6 shows the performance curves of the chilled water system. As expected, at a low cooling load, chiller electric power is relatively low and the contribution of auxiliary equipment to the total electric power becomes significant. Overall, similar trends were obtained: individual screw chillers perform the best, followed by the combination of screw chillers, the centrifugal chiller, and, last, all chillers together. Other chiller combinations were barely found in the historical data. In general, the auxiliary equipment significantly reduces the performance: the chilled water system COP (Figure 6) almost never exceeds 3 while the chiller COP (Figure 5) can easily reach 6 and beyond.

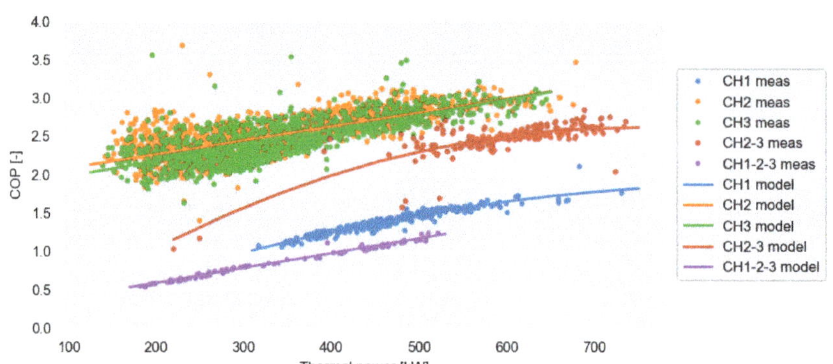

Figure 6. Chilled water system data-driven performance curves obtained with operational data. The electric power accounts for the chiller(s) in operation and the associated auxiliary equipment (i.e., circulation pumps, cooling tower fans, air-cooled condenser fans).

3.1.2. Proposed Chiller Sequencing Strategy

Chilled water system data-driven performance curves (Figure 6) were used to develop the new chiller sequencing strategy. Linear models were used to estimate the COP of the screw chillers (CH #2 and CH #3) and all of the chillers together (CH #1-3) from the thermal load; quadratic models were used for the combination of the screw chillers (CH #2-3) and the centrifugal chiller (CH #1):

$$COP_{\mathrm{CH\#1}} = -2.375e^{-6}\dot{Q}^2_{\mathrm{CH\#1}} + 4.348e^{-3}\dot{Q}_{\mathrm{CH\#1}} - 1.068, \tag{4}$$

$$COP_{\mathrm{CH\#2}} = 1.818e^{-3}\dot{Q}_{\mathrm{CH\#2}} + 1.898, \tag{5}$$

$$COP_{\mathrm{CH\#3}} = 2.004e^{-3}\dot{Q}_{\mathrm{CH\#3}} + 1.774, \tag{6}$$

$$COP_{\mathrm{CH\#2-3}} = -5.307e^{-6}\dot{Q}^2_{\mathrm{CH\#2-3}} + 7.896e^{-3}\dot{Q}_{\mathrm{CH\#2-3}} - 0.326, \tag{7}$$

$$COP_{\mathrm{CH\#1-2-3}} = 1.935e^{-3}\dot{Q}_{\mathrm{CH\#1-2-3}} + 2.009, \tag{8}$$

The model accuracy was evaluated based on the coefficient of determination (R^2), the normalized mean bias error (NMBE), and the coefficient of variation of the root mean square error (CV-RMSE) [39]. The results are given in Table 1 and show good consistency with the operational data.

Table 1. Accuracy of chilled water system COP models.

Chiller in Operation	Model	R^2	NMBE	CV-RMSE
CH #1 (centrifugal)	Quadratic	0.92	$3.3 \times 10^{-14}\%$	3.0%
CH #2 (screw)	Linear	0.53	$2.2 \times 10^{-14}\%$	7.5%
CH #3 (screw)	Linear	0.57	$-2.3 \times 10^{-14}\%$	7.5%
CH #2-3	Quadratic	0.61	$-2.1 \times 10^{-14}\%$	5.0%
CH #1-2-3	Linear	0.98	$-3.5 \times 10^{-14}\%$	2.9%

Based on the findings from Figure 6, it appears that CH #2 is the most efficient chiller, although the difference with CH #3 becomes negligible for a cooling load higher than 575 kW. At a high cooling load (above 650 kW), both screw chillers CH #2 and #3 might be required and should be used, instead of the centrifugal chiller. Switch-on and switch-off thresholds should also be considered to avoid cycling, while more advanced controls could be incorporated as well to reduce the number of switches and chiller start-ups [24]. For the present case study, the proposed chiller sequencing is described in Table 2. Switch-on and switch-off thresholds were tuned to reduce the number of chiller switches without degrading the system performance. The centrifugal chiller is not used in the proposed sequencing due to its poor performance.

Table 2. Chiller sequencing strategy: switch-on and switch-off thresholds.

Switch-On Threshold		Switch-Off Threshold	
From #2 to #3	575 kW	From #3 to #2	550 kW
From #3 to #2-3	650 kW	From #2-3 to #3	625 kW

Although more advanced controls could be used to improve the multi-chiller system by considering more complex chiller models [23], more control variables (e.g., water and air flowrates and temperatures) [23], or advanced optimization routines [40], the proposed approach requires limited operational data, only targets chiller sequencing without affecting local controls in the chilled water network (and related technical considerations), and relies on a simple part–load ratio model for the chillers, which eases the development of the sequencing strategy. Such features are intended to facilitate the implementation of the proposed strategy into actual control systems and the replicability to other multi-chiller systems.

3.2. Free Cooling

A free cooling strategy aims to cool down a building with or without the limited use of a mechanical device by introducing more outdoor air when conditions are favorable. Unlike conventional free cooling strategies which rely on natural or hybrid ventilation generally, the proposed free cooling strategy is to increase fresh air intake through the existing mechanical ventilation system in the AHUs. Increasing the ventilation flowrate might reduce the chilled water system electric power; however, it also increases the AHU fan power consumption, whose contribution can be significant to the total building electric power (see Section 2.3). The proposed free cooling strategy aims to identify outdoor conditions under which the total electric power of the chilled water system and the AHU fans is lower compared to the baseline operation, i.e., without increased fresh air flowrates.

3.2.1. Free Cooling Thermal Power

Quebec's climate shows a high potential for free cooling, which is mainly due to the large daily temperature variations, as temperatures at night can easily go below 15 °C [22]. Free cooling in these conditions especially occurs in early or late summer, or during shoulder seasons, and depends on required indoor air conditions.

To evaluate the feasibility of potential free cooling, outdoor air conditions must be lower than indoor air conditions. In this case, incorporating more fresh air cools down the building. Otherwise, more cooling from the mechanical cooling system is required to pre-condition the fresh air (see Section 2.3). This fresh-air pre-treatment can be evaluated by comparing fresh air and return (or exhaust) air conditions. In addition, the amount of potential free cooling depends on the capacity of the ventilation system to incorporate more fresh air into the system, besides occupancy needs. At night, occupancy in CI buildings is relatively low, which offers more potential for free cooling than during the daytime. Consequently, the free cooling thermal power can be estimated as follows:

$$\dot{Q}_{\text{free cooling}} = \min\left(\dot{Q}_{\text{total cooling}}, \left(\dot{m}_{\text{fre,max}} - \dot{m}_{\text{fre,occ}}\right)\left(h_{\text{OA}} - h_{\text{EA}}\right)\right), \tag{9}$$

where $\dot{Q}_{\text{free cooling}}$ is the total building cooling load (kW), $\dot{m}_{\text{fre,max}}$ and $\dot{m}_{\text{fre,occ}}$ are the maximum fresh air flowrate based on ventilation system capacity and fresh air requirements for occupancy purposes (kg/s), and h_{OA} and h_{EA} are outdoor air and exhaust air enthalpies (kJ/kg) and depend on temperature and relative humidity. From the operational data for summer 2020 and 2021, the maximum fresh air flowrate observed was 20,500 L/s; this value was used in Equation (9). In theory, Equation (9) provides the maximum free cooling thermal power by considering the maximum fresh air flowrate. However, a more sophisticated approach could be to test all possible flow rates from occupancy requirements up to the maximum; however, it would also make the approach more complex and more difficult to implement in real buildings.

Figure 7 shows the total building cooling load and the free cooling thermal power for one week in late August 2020. It is worth mentioning that this free cooling thermal power is possible due to favorable outdoor air conditions compared to indoor air conditions, and only accounts for the free cooling potential from a thermal load viewpoint. In other words, it does not necessarily mean that it is efficient to conduct free cooling from an electric power perspective (see Section 3.2.3). We can see that free cooling is possible mainly at night, although also sometimes during the day due to lower outdoor air conditions during this period of the year. Figure 7 also shows the fresh air flowrate required for occupancy needs and the flow rate available for free cooling. During the day, building occupancy is higher and reduces the potential for free cooling by decreasing the available air flowrate for free cooling (see Equation (9)), which partially explains why the potential free cooling thermal power is lower during work hours. The total fresh air flowrate is sometimes below 20,500 L/s, which occurs when the potential free cooling load exceeds the building cooling load and is adjusted to exactly match it by lowering the flowrate.

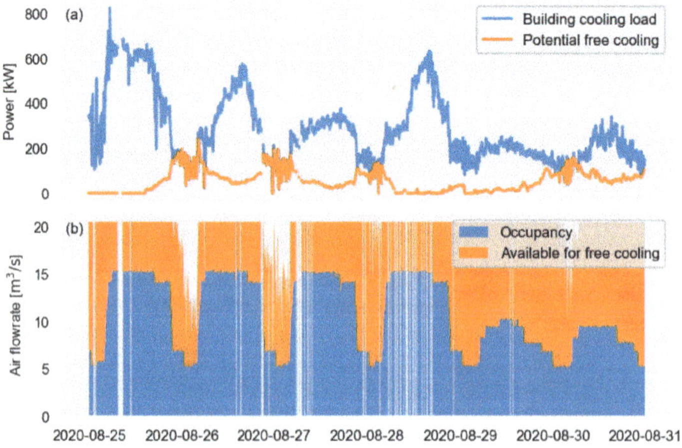

Figure 7. (**a**) Total building cooling load and free cooling thermal power, and (**b**) fresh air flowrate for occupancy needs and available for free cooling for two weeks in August 2020.

3.2.2. Chiller and AHU Fan Electric Power

To evaluate the electric power associated with the free cooling strategy, models are required to calculate the contribution of chillers and of fans located in the primary and secondary AHU systems. The chiller models developed in Section 3.1.2 were reused for this purpose.

The fan electric power was calculated as a function of flowrate by means of linear regression: the primary AHU fan power is determined from fresh air flowrate; the secondary AHU fan power is estimated from discharged air flowrate. Figure 8 shows the results for the AHU fan power.

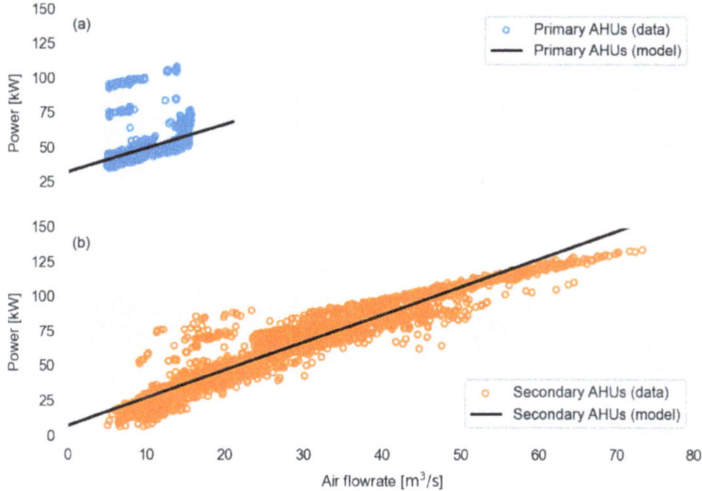

Figure 8. (**a**) Primary AHU fan power as a function of fresh air flowrate; (**b**) secondary AHU fan power as a function of discharged air flowrate.

Under free cooling, a portion of the cooling load can be satisfied by incorporating more fresh air into the ventilation system. This portion is calculated from Equation (9) and the primary AHU system operates at the maximum fresh air flowrate. The rest of the cooling load is supplied by the chillers, which operate at a lower part–load ratio. If the building cooling load is low, it might occur that the entire load can be satisfied by free cooling and the primary AHU system might need to modulate the fresh air flowrate to avoid any overcooling in the building. The total fresh air flowrate becomes the contribution of the fresh air flowrate for occupancy needs and for free cooling. In contrast, the discharged air flowrate depends on the building cooling load; Figure 9 shows this behavior under the baseline operation as well as the obtained model. Based on the model, the total discharged air flowrate under the free cooling mode was calculated as the sum of the additional fresh air flowrate and the discharged air flowrate at the cooling load covered by the chillers (i.e., total building cooling load minus free cooling thermal power). Note that the contribution of increased fan power to the building cooling load was neglected for simplicity.

Table 3 shows the accuracy of the fan power models and discharged air flowrate model based on R^2, NMBE, and CV-RMSE [39]. The results are given in Table 3 and show good consistency with the operational data.

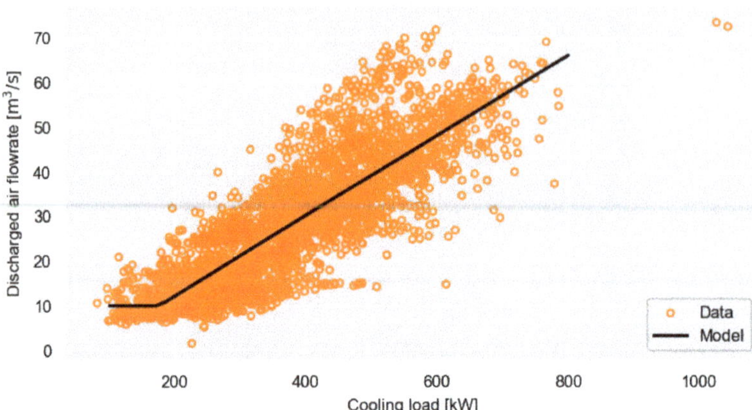

Figure 9. Discharged air flowrate as a function of the cooling load from the chilled water system under baseline operation.

Table 3. Accuracy of fan power and discharged air flowrate models.

Variable	Model	R^2	NMBE	CV-RMSE
Primary AHU fan power	Linear	0.22	$-3.6 \times 10^{-14}\%$	22.4%
Secondary AHU fan power	Linear	0.91	$3.1 \times 10^{-14}\%$	14.8%
Discharged air flowrate	Linear	0.73	0.65%	27.9%

3.2.3. Proposed Free Cooling Strategy

The free cooling thermal power given in Equation (9) mainly depends on the difference between indoor air (approximated by exhaust air) and outdoor air conditions: the larger the difference, the higher the free cooling thermal power. However, this does not guarantee that it is always energy efficient to run the building under free cooling conditions. It is therefore required to verify that the sum of the electric power of the chilled water system and that of the AHU fans is reduced compared to the baseline case. Figure 10 depicts the reduction in total electric power of the chilled water system and AHU fans as a function of the enthalpy difference between indoor and exhaust air. The negative values indicate that operating under free cooling increases the electric power. From this figure, we can clearly see a change point at 7.5 kJ/kg from which the electric power reduction becomes positive and free cooling thus becomes energy efficient. It is worth mentioning that this change point was obtained with the chiller sequencing under baseline operation; nonetheless, it does not change with the proposed chiller sequencing strategy.

The outdoor air conditions that allow for efficient free cooling were investigated and the results are shown in Figure 11. It can be clearly seen that the temperature is the main driver for efficient free cooling; however, relative humidity also plays an important role. Overall, free cooling was found to be effective when the outdoor air temperature was lower than 21 °C; however, the threshold also depends on relative humidity and a free cooling strategy based on outdoor air temperature only could have increased the electric power. Such a strategy could become quite effective during shoulder seasons when, for instance, outdoor air temperature at night can be low while cooling is still required for the building to compensate for internal gains.

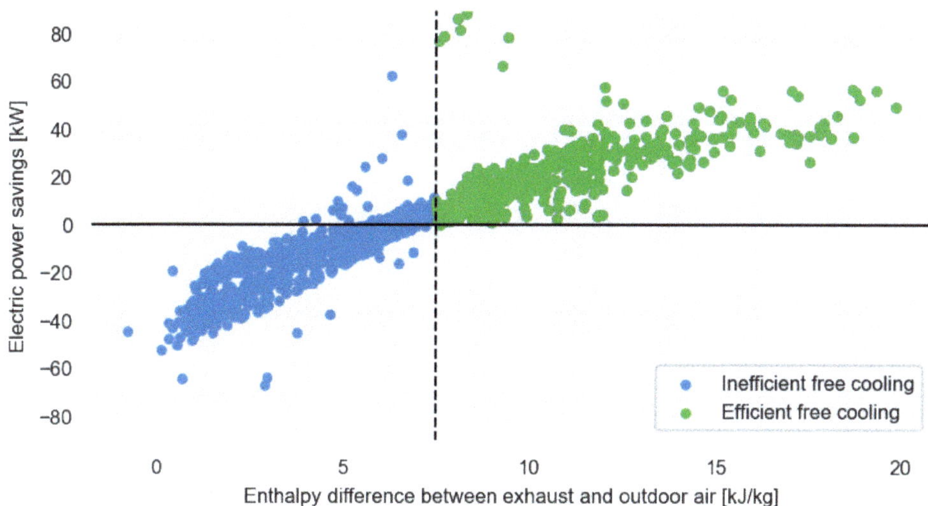

Figure 10. Reduction in total electric power of chilled water system and AHU fans by using free cooling as a function of the enthalpy difference between exhaust and outdoor air.

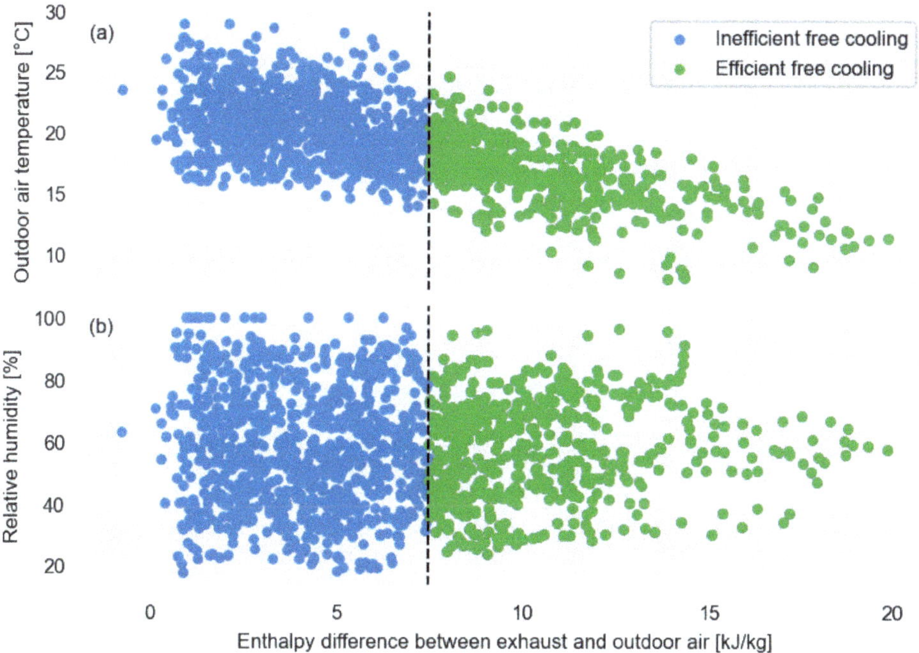

Figure 11. (**a**) Outdoor air temperature and (**b**) relative humidity for inefficient and efficient free cooling as a function of the enthalpy difference between exhaust and outdoor air.

Such a global approach not only makes the proposed free cooling strategy applicable to conventional HVAC configurations with economizers but also to more unique configurations. It requires chiller and fan electric power and fresh air flowrates, and allows for evaluation of the overall benefits of the proposed free cooling strategy. In terms of implementation, the proposed strategy is simple and could be easily implemented by using the enthalpy difference between exhaust and outdoor air, and the estimated threshold of 7.5 kJ/kg.

3.3. Supply Air Temperature Reset

3.3.1. Proposed Supply Air Temperature Reset Strategy

To improve AHU energy efficiency in a VAV system, Supply Air Temperature (SAT) setpoint is a common variable to optimize. ASHRAE G36 [31] has recommended various SAT reset strategies for AHUs. Zhang et al. [18] reported that the G36 SAT reset strategy based on outdoor dry bulb temperature saved energy in different California climates.

The existing SAT setpoints for the two AHUs in the case study are configured to be constantly equal to 13 °C. This low setpoint requests cooling load from the chillers as long as the outdoor air (close to cooling coil inlet temperature) is higher than the setpoint value due to the lack of air economizers in the AHUs (see Figure 2). This generates an unnecessary cooling load for the chillers even when the outdoor air is relatively cool and dry, especially during shoulder seasons or cold summer nights. G36 suggests resetting SAT setpoint to a higher value when the outdoor air is in favorable conditions. Specifically, it recommends adjusting the SAT setpoint proportionally from 13 °C to 18 °C when the outdoor air changes from 21 °C to 16 °C [31].

To consider the impact of humidity, this work adopts a reset strategy based on outdoor air enthalpy. The green line in Figure 12 shows the proposed SAT reset strategy based on fixed outdoor air enthalpy. The SAT setpoint range is selected to be the same as that in G36 from 13 °C to 18 °C. The low and high limits for the outdoor air enthalpy (40.5 kJ/kg and 61.6 kJ/kg, respectively) are selected by considering the same outdoor dry bulb temperature (16 and 21 °C) with a constant relative humidity at 85%. Note that this high limit of outdoor air enthalpy, 61.6 kJ/kg, is lower than the high limit, 65.1 kJ/kg (equivalent to 24 °C outdoor air at 85% relative humidity), specified in ASHRAE Standard 90.1 [34] for an air economizer based on fixed enthalpy with a fixed dry bulb temperature of outdoor air for all ASHRAE climate zones. It is, therefore, considered to be safe for the proposed SAT reset strategy not to introduce extra humidity into the building.

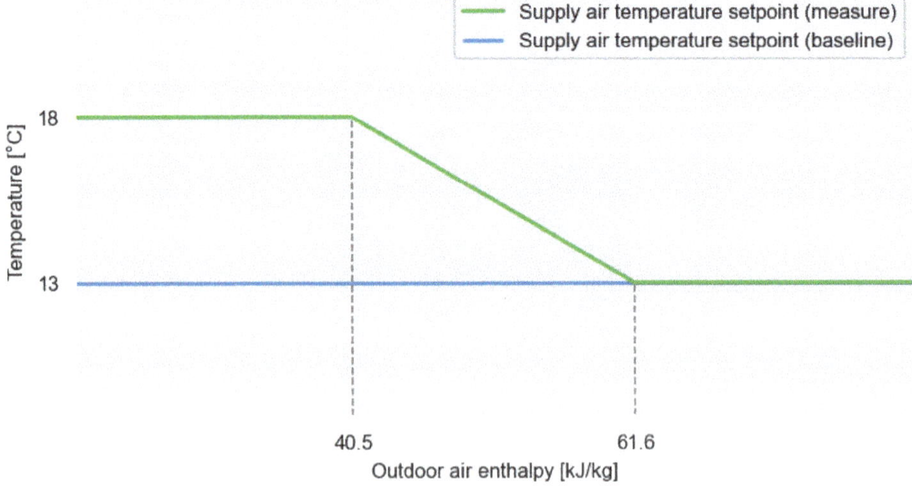

Figure 12. Proposed supply air temperature reset strategy as a function of outdoor air enthalpy while the existing supply air temperature setpoint remains constant.

3.3.2. AHU Cooling Load Calculation

To evaluate the cooling load reduction in the SAT reset strategy, it is necessary to calculate the cooling load of the coils in the AHUs. The BAS measures and stores the temperature, relative humidity, and flowrates of the fresh air, as well as the inlet and outlet temperatures of the cooling coils. Based on mass and energy balance equations [36,41], the cooling coil load of an AHU can be calculated using Equation (10):

$$q_{\text{AHU}} = \dot{m}_{a,AHU_i} \left(h_a \left(T_{\text{coil,i,in}}, \omega_{\text{coil,i,in}} \right) - h_a \left(T_{\text{coil,i,out}}, \omega_{\text{coil,i,out}} \right) \right) - \dot{m}_{w,AHU_i} h_w \left(T_{\text{coil,i,out}}, x = 0 \right), \tag{10}$$

where $\dot{m}_{a,AHU_i}$ and $\dot{m}_{w,AHU_i}$ are the mass flow rate of air and condensation water in AHU_i; indices *coil,i,in* and *coil,i,out* refer to the cooling coil inlet and outlet of AHU_i. h_a is the enthalpy of the humid air, and h_w is the water condensation from the humid air. T and ω are the temperature and humidity ratio, and x is vapor quality.

For the calculation of the baseline cooling load, the measured values were used for the outlet temperatures of the cooling coils. For the calculation of the cooling load with the reset strategy, the outlet temperatures of the cooling coils were assumed to be equal to the SAT reset setpoints. This assumption is considered to be reasonable, as the existing data of the SAT and its setpoint show negligible discrepancies. In other words, the cooling coils were always able to deliver the required cooling load within the sampling rate (5 min) of the data and there was no risk that the reset setpoint could not be reached or would be reached with significant delay.

The proposed temperature reset strategy follows a structure commonly found in BAS and generally implemented through if/then statements, which eases its implementation. Moreover, such a strategy could be replicated to other HVAC configurations with primary/secondary AHUs, as well as different AHU configurations with small modifications, as proposed by G36.

3.4. Methodology to Assess Performance

The performance of each measure was first evaluated individually, and a baseline case (reference or BAU, "Business As Usual") was defined to evaluate energy savings. It consists of: (a) the measured building cooling load (i.e., no SAT reset strategy), which is fully satisfied by the cooling system (i.e., no free cooling), whose electric power is calculated using the chiller models; (b) a baseline chiller sequencing deduced from the measured operation (i.e., which chiller in operation at a given time); and (c) measured fresh air flowrates and discharged air flowrates to calculate fan electric power (i.e., no free cooling). Once the individual assessment was performed, the impact of all measures together was evaluated. In summary, this means:

- *Chiller sequencing*: the new chiller sequencing was compared with the baseline sequencing for the same building cooling load without free cooling;
- *Free cooling strategy*: the proposed free cooling strategy was compared to the baseline case without free cooling for the same building cooling load and chiller sequencing;
- *Supply air temperature reset strategy*: the SAT reset strategy was compared with the baseline cooling load for the same chiller sequencing and without free cooling;
- *All measures together*: the air temperature reset strategy coupled with the proposed chiller sequencing and the free cooling strategy were compared with the baseline case.

To assess the energy savings, the decrease in building cooling energy ($\Delta Q_{\text{measure}}$) and the overall reduction in the cooling system electric power ($\Delta W_{\text{measure}}$) were estimated over a given period of time as shown in Equations (11) and (12). For the specific case of free cooling, the cooling system electric power reduction comes along with an increase in AHU fan power; this effect is considered in the energy savings by penalizing the cooling system electric power reduction with the fan power increase as follows:

$$\Delta Q_{\text{measure}} = \frac{\int\limits_{t} \left(\dot{Q}_{cool}^{ref} - \dot{Q}_{cool}^{meas} \right) dt}{\int\limits_{t} \dot{Q}_{cool}^{ref} dt},$$

(11)

$$\Delta W_{\text{measure}} = \frac{\int\limits_{t} \left(\left(\dot{W}_{cool,sys}^{ref} - \dot{W}_{cool,sys}^{meas} \right) + \left(\dot{W}_{fan}^{ref} - \dot{W}_{fan}^{meas} \right) \right) dt}{\int\limits_{t} \dot{W}_{cool,sys}^{ref} dt},$$

(12)

where $\dot{Q}_{cool}^{ref}$, $\dot{W}_{cool,sys}^{ref}$, and $\dot{W}_{fan}^{ref}$ are, respectively, the building cooling energy, the cooling system electric power, and the AHU fan power for the baseline case. $\dot{Q}_{cool}^{meas}$, $\dot{W}_{cool,sys}^{meas}$, and $\dot{W}_{fan}^{meas}$ are the same variables for the evaluated measure(s).

4. Performance Evaluation of Data-Driven Measures

The performances of the proposed data-driven measures are given in Table 4 with their associated thermal and electric energy savings. These results are discussed in detail in the sub-sections.

Table 4. Performance of proposed data-driven measures.

Measure	Thermal Energy (Baseline)	Thermal Energy (Measure)	Electric Energy (Baseline)	Electric Energy (Measure)
Chiller sequencing only	1429 MWh	1429 MWh (-)	676 MWh	547 MWh (−19.1%)
Free cooling only	128 MWh	128 MWh (-)	73 MWh	53 MWh (−27.2%)
Supply air temperature reset only	1264 MWh	1103 MWh (−12.7%)	614 MWh	558 MWh (−9.1%)
All measures together	1429 MWh	1261 MWh (−11.8%)	676 MWh	456 MWh (−32.5%)

4.1. Performance of Individual Measures

4.1.1. Chiller Sequencing

The performance of the proposed chiller sequencing was evaluated when one of the chiller combinations described in Section 3.1 was in operation (i.e., for which models were developed) and when data were available (i.e., no missing values). It corresponds to 89% of the whole period.

The new strategy allowed for reducing the cooling system electric power by 19% compared to the baseline case. This was mainly due to the low performance of chiller #1, which was used alone or in combination with the other chillers. From the performance curves shown in Figure 6, operating chillers #2 or #3 was also more energy efficient than the operation of both at the same time. The proposed chiller sequencing extended the operation of individual chillers, which further improved the performance. Figure 13 shows the chiller sequencing for the baseline case and the proposed strategy for two weeks in August 2020. We can see that chiller #1 is not operated anymore while chiller #2 is the most used, especially when the building cooling load is lower than 550–575 kW.

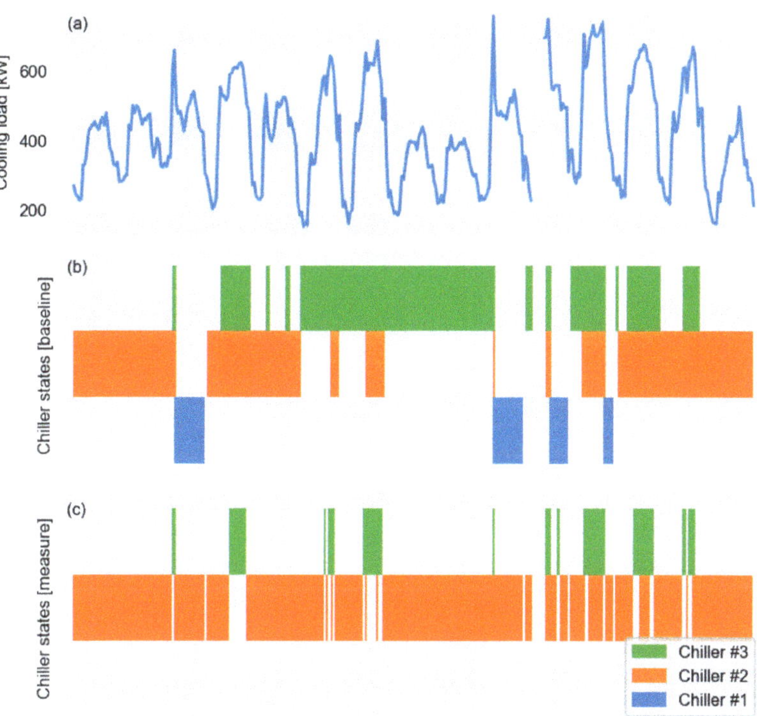

Figure 13. (**a**) Building cooling load, (**b**) baseline chiller sequencing, and (**c**) proposed chiller sequencing for a two-week period.

4.1.2. Free Cooling

To operate under free cooling mode, the building must satisfy two conditions: (a) there must be potential for free cooling (i.e., $\dot{Q}_{\text{free cooling}}$ in Equation (9) higher than zero), and (b) free cooling must be energy efficient (i.e., enthalpy difference between exhaust and outdoor air higher than 7.5 kJ/kg). During the six-month period, free cooling was possible for 33% of the time and efficient free cooling represents 37% of the period when free cooling was possible. Since outdoor air conditions must be favorable, applying the free cooling measure during shoulder seasons (e.g., May and September) would further increase its use.

Although it was not often used during the whole period (12%), Table 4 shows that efficient free cooling can provide significant savings by reducing chilled water system electric energy by 27%. Note that this number includes the increased fan power when the building operates under free cooling mode. Figure 14 shows the results for a one-week period. We can see that free cooling is generally used at night, when outdoor air temperature is cooler and occupancy fresh air requirements are lower; however, it can also be used during the day (e.g., 10 or 16 June). Figure 14b shows the electric power reduction when the building is under free cooling mode. We can clearly see that free cooling helps reduce cooling system electric power when the chilled water system is not required anymore (i.e., at night). When free cooling complements the chilled water system, interesting savings are still achieved.

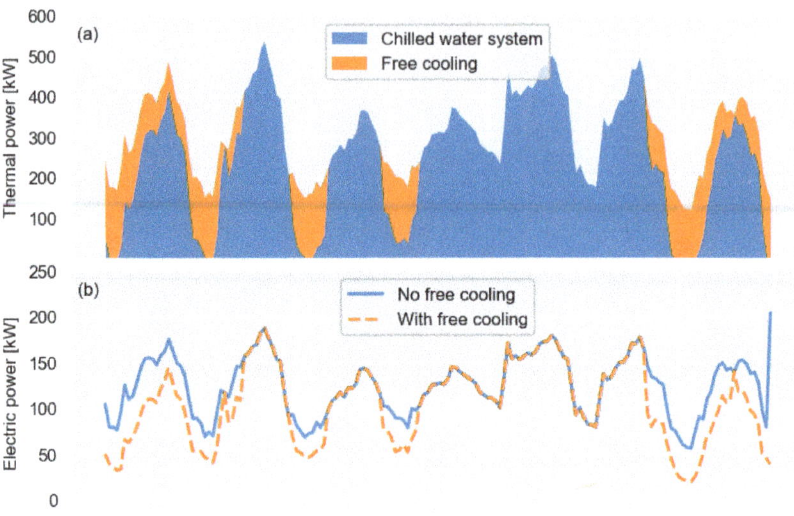

Figure 14. (**a**) Building cooling load satisfied by chilled water system or free cooling and (**b**) electric power with and without free cooling for a one-week period. The electric power is the contribution of the chilled water system power and the increased fan power, as given in Equation (12).

4.1.3. Supply Air Temperature Reset

Figure 15 shows the cooling load of the baseline and proposed reset strategy for a typical week in summer 2021. We can see that the proposed strategy requires less cooling when the outdoor air enthalpy is low, which is especially effective during 1–5 July. Low enthalpy means that the outdoor air is cool and dry during those periods, which leads to less need for the AHU system to further cool and dehumidify the fresh air entering the system. Unlike the existing control, the proposed measure resets the SAT setpoint to a higher value in those conditions and thus reduces the cooling load.

It is worth mentioning that the cooling load was estimated using air-side measurements at the AHU level and might lead to slightly different cooling load results compared to using chilled water measurements in Equation (2), which were used for the chiller sequencing measure. When combined measures are assessed, this cooling load reduction (absolute value) is used along with Equation (2). For more information about the differences and uncertainties regarding the two cooling load calculation methods, see [36,41].

Figure 16 summarizes the average cooling load of the baseline and proposed SAT measure for the two summers. We can see that the measure was able to reduce cooling load for all summer months by 7–22%, with slightly more reductions in June. It was expected that the measure could further reduce the cooling load in shoulder seasons such as in the months of May and September. The overall cooling load reduction from the SAT measure is 12.7% against the baseline for the investigated six months as shown in Table 4. The decrease in cooling energy comes along with a reduced utilization of the chilled water system and a reduction in electric power, estimated at 9.1%.

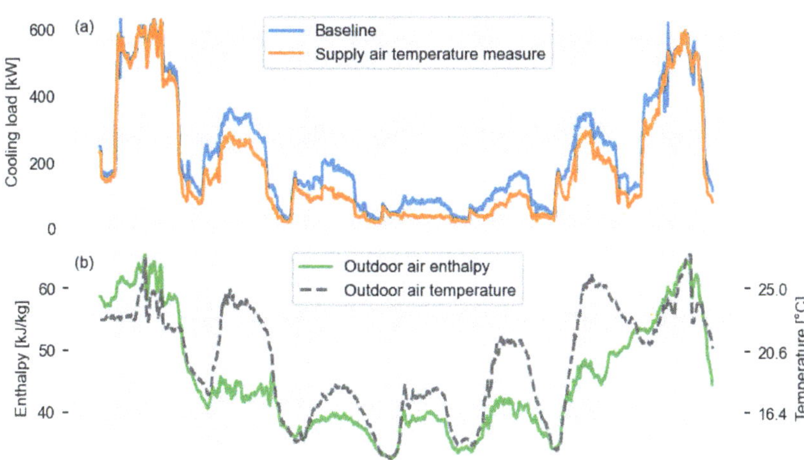

Figure 15. (**a**) Cooling load of a typical week in 2021 for the baseline and proposed supply air temperature reset strategy; (**b**) outdoor air enthalpy and dry bulb temperature for the same week.

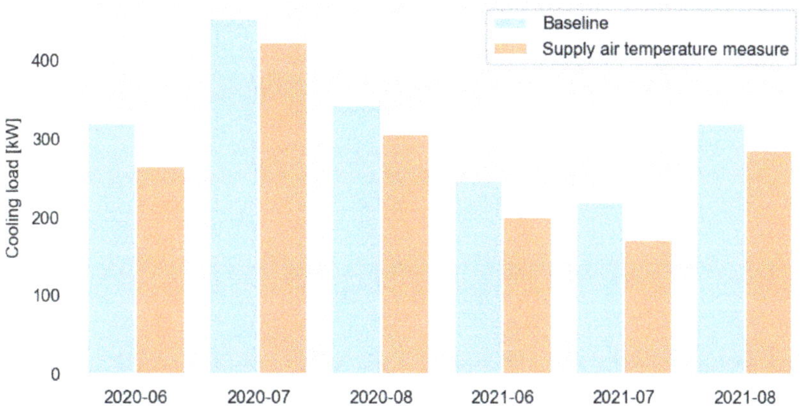

Figure 16. Average cooling load per month for the baseline and proposed supply air temperature reset strategy during 2020 and 2021 summer.

4.2. Performance of Combined Measures

The combined effect of all of the measures applied together is given in Table 4. It shows an overall reduction in thermal energy of 11.8% and in electric energy of 32.5%. The SAT reset strategy allows for reducing both the thermal and electric energy. The new chiller sequencing takes advantage of the most energy-efficient chillers and directly tackles chilled water system electric power. Finally, when effective, the free cooling strategy permits further reducing electric energy by increasing fresh air intake into the building, thus reducing the usage of the chilled water system.

Figure 17 shows the cooling load and electric power for the baseline and the combined measures for a typical week in 2021. Both thermal and electric power are consistently reduced due to the combined effects of chiller sequencing, free cooling, and SAT reset strategies.

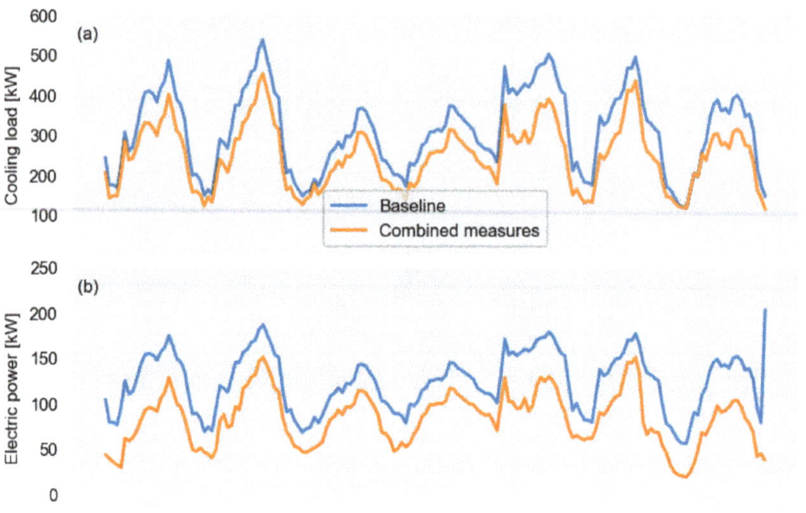

Figure 17. (**a**) Building cooling load; (**b**) electric power for the baseline and the combined measures for a typical week. The electric power is the contribution of the chilled water system power and the increased fan power, as given in Equation (12).

On a monthly basis, Figure 16 shows the cooling load reduction of the SAT reset strategy; however, it also represents the results of the combined measures, since only the SAT reset strategy affects the building cooling load. For the chilled water system electric power, Figure 18 shows the average monthly results and reductions ranging between 23% and 46%. In 2020, June provideed higher savings (34%) due to the most favorable outdoor air conditions, which allowed for operating under the effective SAT reset strategy and free cooling mode. In 2021, more significant savings were achieved, especially for the months of July and August (36–46%). These savings were mainly achieved due to the new chiller sequencing, which avoids the usage of both the chiller #1 alone and the chillers #1-3, which show poor performance, as displayed in Figure 6.

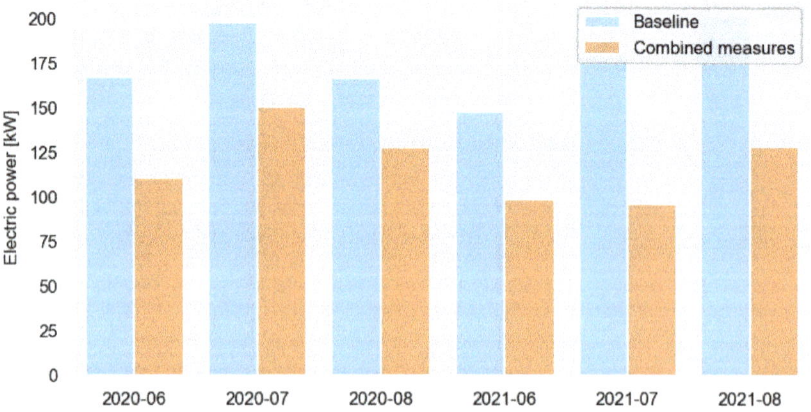

Figure 18. Average chilled water system electric power per month for the baseline and the combined measures during summer 2020 and 2021.

4.3. Generalizability of Proposed Data-Driven Measures

The proposed data-driven measures were built on models to obtain insights into the cooling system performance to further optimize building operation. To facilitate the implementation in the current control system, the proposed measures were aimed to be implemented by means of simple rule-based controls:

- *Chiller sequencing*: the sequencing can be implemented using switch-on and switch-off thresholds as given in Table 2; this sequence could act as a master controller and could be further overridden by local control rules if required (e.g., a high temperature limit can be reached, which may require a chiller switch). This strategy could be replicated to any other multi-chiller systems with the same ease of implementation.
- *Free cooling*: the free cooling mode requires the change point (7.5 kJ/kg, as shown in Figure 10) and the enthalpy calculations for both exhaust and outdoor air; the library Coolprop [42] can be used for this purpose. Additional control modifications are necessary to allow the fresh air flowrate to increase. This strategy could not only be replicated to conventional HVAC configurations with economizers, but also to more unique configurations.
- *Supply air temperature reset*: reset strategies are already commonly used in BAS for various temperatures with if/then statements. The proposed temperature reset strategy is given in Figure 12 and relies on outdoor air enthalpy estimation; the library Coolprop [42] can be used in this case as well. This strategy is inspired by ASHRAE Guideline 36 [31] and, as such, it could be replicated to many AHU systems with minor adjustments.

The proposed measures could be generalized to other buildings; however, their development depends on the available operational data while the tuning of control parameters (i.e., chiller sequencing switch-on and switch-off thresholds, free cooling change point, temperature reset strategy parameters) requires careful data analysis and model development.

5. Conclusions

This work tackled the development of three data-driven model-based control strategies to improve the cooling performance of commercial and institutional buildings:

- A new chiller sequencing based on data-driven performance curves;
- A free cooling strategy considering both chilled water system electric power and air handling unit fan power;
- A supply air temperature reset strategy based on outdoor air enthalpy.

These measures were built upon operational data and virtual energy meters and models, and were intended to be relatively simple to develop and easy to implement in actual control systems. This approach makes the proposed measures more robust and less prone to errors compared to more complex methods because the installation of additional hardware is not required, which makes the approach more easily replicable in other buildings.

These measures were developed for an existing large commercial building and were evaluated both individually and all together during summer (June–August) 2020 and 2021. The results showed that the chiller sequencing could reduce chilled water system electric energy by 19%. The free cooling strategy was efficient with a 27% reduction in electric energy; however, this mode could be activated during only a small fraction of the summer periods (12%). More energy savings would be expected by extending the analysis to shoulder seasons (e.g., May and September). Finally, the supply air temperature reset strategy helped to reduce the building cooling load by 13% and, as a side effect, reduced the electric power by 9%. All of the measures combined allowed for a reduction in building cooling load by 12% and chilled water system electric power by 33% over the six months studied.

Future work includes field implementation of the proposed measures in the case study building, replication to other existing buildings, and the development of a predictive control strategy built upon the proposed measures to further improve the building cooling performance.

Author Contributions: E.S.: Conceptualization, Methodology, Software, Validation, Formal analysis, Data Curation, Writing—original draft, Writing—review & editing, Visualization, Project administration. K.Z.: Conceptualization, Methodology, Software, Validation, Formal analysis, Data Curation, Writing—original draft, Writing—review & editing, Visualization. All authors have read and agreed to the published version of the manuscript.

Funding: The authors want to gratefully acknowledge the financial support of Natural Resources Canada through the Office of Energy Research and Development.

Data Availability Statement: Data not available due to privacy restrictions.

Acknowledgments: The authors would like to thank our external partners for sharing their experience, providing data and feedback, and internal and external reviewers for their useful comments.

Conflicts of Interest: The authors declare no conflict of interest.

References

1. Fernandez, N.; Katipamula, S.; Wang, W.; Xie, Y.; Zhao, M.; Corbin, C. *Impacts of Commercial Building Controls on Energy Savings and Peak Load Reduction*; Pacific Northwest National Laboratory, U.S. Department of Energy: Richland, WA, USA, 2017.
2. Li, X.; Wen, J. Review of Building Energy Modeling for Control and Operation. *Renew. Sustain. Energy Rev.* **2014**, *37*, 517–537. [CrossRef]
3. U.S. Green Building Council Benefits of Green Building. Available online: https://www.usgbc.org/articles/benefits-green-building (accessed on 16 September 2022).
4. Kramer, H.; Lin, G.; Curtin, C.; Crowe, E.; Granderson, J. *Proving the Business Case for Building Analytics*; Lawrence Berkeley National Laboratory: USA, 2020. [CrossRef]
5. Drgoňa, J.; Arroyo, J.; Cupeiro Figueroa, I.; Blum, D.; Arendt, K.; Kim, D.; Ollé, E.P.; Oravec, J.; Wetter, M.; Vrabie, D.L.; et al. All You Need to Know about Model Predictive Control for Buildings. *Annu. Rev. Control.* **2020**, *50*, 190–232. [CrossRef]
6. Cotrufo, N.; Saloux, E.; Hardy, J.M.; Candanedo, J.A.; Platon, R. A Practical Artificial Intelligence-Based Approach for Predictive Control in Commercial and Institutional Buildings. *Energy Build.* **2020**, *206*, 109563. [CrossRef]
7. Saloux, E.; Cotrufo, N.; Candanedo, J. A Practical Data-Driven Multi-Model Approach to Model Predictive Control: Results from Implementation in an Institutional Building. In Proceedings of the 6th International High Performance Buildings Conference, West Lafayette, IN, USA, 28 May 2021.
8. Wang, S.; Ma, Z. Supervisory and Optimal Control of Building HVAC Systems: A Review. *HVACR Res.* **2008**, *14*, 3–32. [CrossRef]
9. Afram, A.; Janabi-Sharifi, F. Theory and Applications of HVAC Control Systems–A Review of Model Predictive Control (MPC). *Build. Environ.* **2014**, *72*, 343–355. [CrossRef]
10. Taheri, S.; Hosseini, P.; Razban, A. Model Predictive Control of Heating, Ventilation, and Air Conditioning (HVAC) Systems: A State-of-the-Art Review. *J. Build. Eng.* **2022**, *60*, 105067. [CrossRef]
11. Thieblemont, H.; Haghighat, F.; Ooka, R.; Moreau, A. Predictive Control Strategies Based on Weather Forecast in Buildings with Energy Storage System: A Review of the State-of-the Art. *Energy Build.* **2017**, *153*, 485–500. [CrossRef]
12. Yu, Z.; Huang, G.; Haghighat, F.; Zhang, G. Control Strategies for Integration of Thermal Energy Storage into Buildings: State-of-the-Art Review. *Energy Build.* **2015**, *106*, 203–215. [CrossRef]
13. Péan, T.Q.; Salom, J.; Costa-Castelló, R. Review of Control Strategies for Improving the Energy Flexibility Provided by Heat Pump Systems in Buildings. *J. Process Control.* **2019**, *74*, 35–49. [CrossRef]
14. Darwazeh, D.; Duquette, J.; Gunay, B.; Wilton, I.; Shillinglaw, S. Review of Peak Load Management Strategies in Commercial Buildings. *Sustain. Cities Soc.* **2021**, *77*, 103493. [CrossRef]
15. Mirakhorli, A.; Dong, B. Occupancy Behavior Based Model Predictive Control for Building Indoor Climate—A Critical Review. *Energy Build.* **2016**, *129*, 499–513. [CrossRef]
16. Park, J.Y.; Ouf, M.M.; Gunay, B.; Peng, Y.; O'Brien, W.; Kjærgaard, M.B.; Nagy, Z. A Critical Review of Field Implementations of Occupant-Centric Building Controls. *Build. Environ.* **2019**, *165*, 106351. [CrossRef]
17. Shi, Z.; Newsham, G.; Ashouri, A.; Pardasani, A. Comparative Study of Pre-Cooling Strategies for Houses in Southern Ontario. In Proceedings of the 11th eSim Building Simulation Conference, Vancouver, BC, Canada, 14–16 June 2021.
18. Zhang, K.; Blum, D.; Cheng, H.; Paliaga, G.; Wetter, M.; Granderson, J. Estimating ASHRAE Guideline 36 Energy Savings for Multi-Zone Variable Air Volume Systems Using Spawn of EnergyPlus. *J. Build. Perform. Simul.* **2022**, *15*, 215–236. [CrossRef]
19. Raman, N.; Chen, B.; Barooah, P. A Unified MPC Formulation for Control of Commercial HVAC Systems in Multiple Climate Zones. In Proceedings of the International High Performance Buildings Conference, West Lafayette, IN, USA, 24–28 May 2021.
20. Hobson, B.W.; Gunay, H.B.; Ashouri, A.; Newsham, G.R. Occupancy-Based Predictive Control of an Outdoor Air Intake Damper: A Case Study. In Proceedings of the eSim 2020: 11th Conference of IBPSA-Canada, Vancouver, BC, Canada, 14–16 June 2020; p. 8.

21. Yuan, S.; Vallianos, C.; Athienitis, A.; Rao, J. A Study of Hybrid Ventilation in an Institutional Building for Predictive Control. *Build. Environ.* **2018**, *128*, 1–11. [CrossRef]
22. Vallianos, C.; Athienitis, A.; Rao, J. Hybrid Ventilation in an Institutional Building: Modeling and Predictive Control. *Build. Environ.* **2019**, *166*, 106405. [CrossRef]
23. Thangavelu, S.R.; Myat, A.; Khambadkone, A. Energy Optimization Methodology of Multi-Chiller Plant in Commercial Buildings. *Energy* **2017**, *123*, 64–76. [CrossRef]
24. Liao, Y.; Huang, G. A Hybrid Predictive Sequencing Control for Multi-Chiller Plant with Considerations of Indoor Environment Control, Energy Conservation and Economical Operation Cost. *Sustain. Cities Soc.* **2019**, *49*, 101616. [CrossRef]
25. Gunay, H.B.; Ashouri, A.; Shen, W. Load Forecasting and Equipment Sequencing in a Central Heating and Cooling Plant: A Case Study. *ASHRAE Trans.* **2019**, *125*, 513–523.
26. Liu, S.; Henze, G.P. Experimental Analysis of Simulated Reinforcement Learning Control for Active and Passive Building Thermal Storage Inventory: Part 1. Theoretical Foundation. *Energy Build.* **2006**, *38*, 142–147. [CrossRef]
27. Candanedo, J.A.; Dehkordi, V.R.; Stylianou, M. Model-Based Predictive Control of an Ice Storage Device in a Building Cooling System. *Appl. Energy* **2013**, *111*, 1032–1045. [CrossRef]
28. Song, X.; Zhu, T.; Liu, L.; Cao, Z. Study on Optimal Ice Storage Capacity of Ice Thermal Storage System and Its Influence Factors. *Energy Convers. Manag.* **2018**, *164*, 288–300. [CrossRef]
29. Feng, J. (Dove); Chuang, F.; Borrelli, F.; Bauman, F. Model Predictive Control of Radiant Slab Systems with Evaporative Cooling Sources. *Energy Build.* **2015**, *87*, 199–210. [CrossRef]
30. Pang, X.; Duarte, C.; Haves, P.; Chuang, F. Testing and Demonstration of Model Predictive Control Applied to a Radiant Slab Cooling System in a Building Test Facility. *Energy Build.* **2018**, *172*, 432–441. [CrossRef]
31. American Society of Heating, Refrigeration and Air Conditioning Engineers. ASHRAE Guideline 36-2021 High-Performance Sequences of Operation for HVAC Systems; ASHRAE: Atlanta, GA, USA, 2021. Available online: https://www.techstreet.com/ashrae/standards/guideline-36-2021-high-performance-sequences-of-operation-for-hvac-systems?product_id=2229690 (accessed on 25 January 2023).
32. Fan, B.; Jin, X.; Du, Z. Optimal Control Strategies for Multi-Chiller System Based on Probability Density Distribution of Cooling Load Ratio. *Energy Build.* **2011**, *43*, 2813–2821. [CrossRef]
33. Chen, Y.; Yang, C.; Pan, X.; Yan, D. Design and Operation Optimization of Multi-Chiller Plants Based on Energy Performance Simulation. *Energy Build.* **2020**, *222*, 110100. [CrossRef]
34. *90.1-2019*; American Society of Heating, Refrigeration and Air Conditioning Engineers ANSI/ASHRAE/IES Standard—Energy Standard for Buildings Except Low-Rise Residential Buildings. ASHRAE: Atlanta, GA, USA, 2019. Available online: https://www.ashrae.org/technical-resources/bookstore/standard-90-1 (accessed on 25 January 2023).
35. Stasi, R.; Ruggiero, F.; Berardi, U. The Efficiency of Hybrid Ventilation on Cooling Energy Savings in NZEBs. *J. Build. Eng.* **2022**, *53*, 104401. [CrossRef]
36. Saloux, E.; Zhang, K. Towards Integration of Virtual Meters into Building Energy Management Systems: Development and Assessment of Thermal Meters for Cooling. *J. Build. Eng.* **2022**, *65*, 105785. [CrossRef]
37. Dou, H.; Zmeureanu, R. Evidence-Based Assessment of Energy Performance of Two Large Centrifugal Chillers over Nine Cooling Seasons. *Sci. Technol. Built Environ.* **2021**, *27*, 1243–1255. [CrossRef]
38. De Coninck, R.; Helsen, L. Practical Implementation and Evaluation of Model Predictive Control for an Office Building in Brussels. *Energy Build.* **2016**, *111*, 290–298. [CrossRef]
39. Afram, A.; Janabi-Sharifi, F. Black-Box Modeling of Residential HVAC System and Comparison of Gray-Box and Black-Box Modeling Methods. *Energy Build.* **2015**, *94*, 121–149. [CrossRef]
40. Lee, W.-S.; Chen, Y.-T.; Kao, Y. Optimal Chiller Loading by Differential Evolution Algorithm for Reducing Energy Consumption. *Energy Build.* **2011**, *43*, 599–604. [CrossRef]
41. Saloux, E.; Zhang, K. Virtual Energy Metering of Whole Building Cooling Load from Both Airside and Waterside Measurements. In Proceedings of the 12th eSim Building Simulation Conference, Ottawa, ON, Canada, 21–24 June 2022.
42. Bell, I.H.; Wronski, J.; Quoilin, S.; Lemort, V. Pure and Pseudo-Pure Fluid Thermophysical Property Evaluation and the Open-Source Thermophysical Property Library CoolProp. *Ind. Eng. Chem. Res.* **2014**, *53*, 2498–2508. [CrossRef]

 *buildings*

Review

Predictive Maintenance 4.0 for Chilled Water System at Commercial Buildings: A Systematic Literature Review

Malek Almobarek [1,*], Kepa Mendibil [1] and Abdalla Alrashdan [2]

[1] Department of Design, Manufacturing, and Engineering Management, Faculty of Engineering, University of Strathclyde, Glasgow G1 1QE, UK
[2] Industrial Engineering Department, College of Engineering, Alfaisal University, Riyadh 11533, Saudi Arabia
* Correspondence: malek.almobarek@strath.ac.uk

Abstract: Predictive maintenance plays an important role in managing commercial buildings. This article provides a systematic review of the literature on predictive maintenance applications of chilled water systems that are in line with Industry 4.0/Quality 4.0. The review is based on answering two research questions about understanding the mechanism of identifying the system's faults during its operation and exploring the methods that were used to predict these faults. The research gaps are explained in this article and are related to three parts, which are faults description and handling, data collection and frequency, and the coverage of the proposed maintenance programs. This article suggests performing a mixed method study to try to fill in the aforementioned gaps.

Keywords: predictive maintenance; faults detection and diagnosis; chilled water system; commercial buildings; Industry 4.0; Quality 4.0; data-driven analysis

1. Introduction

1.1. Background

At commercial buildings (CBs)/large facilities, business work fills most people's time and occupies most employees or workforces, who spend most of their workday inside these buildings, so CBs make up a sizable portion of the built environment for the people. Common-sense drives the organizations/owners to take care of these buildings in order to avoid any negative impact on the surrounding or the internal environment of these buildings.

CBs are different from city to city and could be massive or regular ones such as universities, offices buildings, shopping malls, hotels, factories, compounds, hypermarkets, etc., and cover most of the land areas in the cities. The University of Michigan reported that CBs' floor spaces are foreseen to encompass 124.7 billion square feet by 2050, which is a 34 percent increase from 2019 [1]. Moreover, they are obviously playing a significant role in the communities, as a great CB can enhance people's more social life and can generate more jobs. However, they are approximately consuming up to 40 percent of the total global energy demand [2]. Moreover, one of the main challenges that CBs are facing is climate change. Monge-Barrio and Gutierrez indicated that climate change has a significant impact on such buildings [3]. Furthermore, climate change is predicted to have strong effects on the energy requirements of CBs, as their heating and cooling needs are highly related to temperature conditions and weather variations [4]. In addition, activities in buildings contribute to a major share of global environmental concerns [5]. These challenges motivate any facility manager or engineer to take valid actions toward building performance improvement and maintenance, as well as looking after the associated operation and maintenance (O&M) costs. This should be performed, as CBs are increasingly equipped with sophisticated engineering facilities, as well, such as Heat, Ventilation, and Air Conditioning (HVAC) equipment/machines [6]. By doing so, the facility manager will fulfil the sustainability of his/her CB [7].

Citation: Almobarek, M.; Mendibil, K.; Alrashdan, A. Predictive Maintenance 4.0 for Chilled Water System at Commercial Buildings: A Systematic Literature Review. *Buildings* **2022**, *12*, 1229. https://doi.org/10.3390/buildings12081229

Academic Editor: Etienne Saloux

Received: 30 June 2022
Accepted: 10 August 2022
Published: 13 August 2022

Publisher's Note: MDPI stays neutral with regard to jurisdictional claims in published maps and institutional affiliations.

Maintaining a particular building requires managing all the systems within it. Generally, it includes either mechanical or electrical systems. There are five major disciplines, which are (1) HVAC, (2) plumbing and fire protection, (3) electrical power and telecommunications, (4) illumination, and (5) noise and vibration control [8]. The HVAC system is a technology of internal environmental ambience that supplies thermal comfort and agreeable indoor air quality (IAQ) [9]. It is based on the contrivances and the findings from William Rankin, Nikolay Lvov, Willis Carrier, James Joule, and others [10]. It is a critical system and is playing a big role in consuming a high percentage of energy in CBs, and accordingly, there will be an assertion on the electricity bill [11]. It consumes more than thirty percent of the total energy used in CBs [12]. Cho and others argued that the energy consumption of an HVAC system for a large office building can take forty to fifty percent of the building's total energy use [13]. It is, generally, worrying the organizers at CBs about the difficulty of replacing its components, when needed, so caring and making a planned control arrangement about that will save energy, with minimal infrastructure investments [14]. However, conducting a proper and well-organized maintenance program for this system is totally required, as many researchers have found that the factor most often embroiling IAQ is maintenance related [15].

The importance of HVAC existed even before operating a particular CB, where selecting the appropriate system with its components at the beginning of its project time covers a significant part of its design. In this regard, Hassanain et al. argued that the HVAC system is one of the most convoluted systems in buildings projects [16]. The aforementioned HVAC components were listed by Sugarman, such as water chillers, cooling towers, etc. [17]. Naturally, the selection of the said system is made based on three concepts, which are the configuration of that CB, the climate conditions, and the inclination of the organization that owns them [18]. The standards that HVAC building design are held to when being created, selected, or studied come from The American Society of Heating, Refrigerating, and Air-Conditioning Engineers (ASHRAE) [19]. Furthermore, it is an important system from well-being and safety points of view, as it monitors the environment related to occupant health, such as the level of colorless odorless gas (CO_2) and humidity margins, as well as occupant thermal comfort, including ambient temperature and airflow [20]. This system, especially its ventilation and cooling part, plays a big role in reducing the infection inside CBs during the recent global pandemic (COVID-19) if a proper maintenance management exists in monitoring airflow [21]. Aebischer and others underlined that due to the impact of climate change, the need for cooling comfort inside CBs will be increased even in Europe until 2030 as the increment in temperature would be two-degree centigrade over time [22].

The sub system highlighted in this article is the chilled water system (CWS). It is considered as one of the major functions in HVAC system and it usually consumes a significant amount of the total energy amortization used in the main system [23]. ASHRAE handbook listed the components of CWS onto four as follows [24]:

1. Chillers.
2. Cooling towers.
3. Primary/secondary/condenser water pumps.
4. Terminal units: air-handling units (AHUs) and fan coil units (FCUs).

Per ASHRAE [24], the operation of CWS starts with chillers producing the chilled water required to operate the AHUs/FCUs and thereby to achieve the designed room conditions. Chillers, primary chilled water pumps, are operated and sequenced to produce chilled water at a set temperature, whereas a specified temperature of water required by the condenser component of chillers is produced by the cooling towers through the condenser water pumps. The produced chilled water is then pumped by the secondary water pumps to all the terminal units, such as AHUs and FCUs, and in case of a variable flow system, their speed is controlled to maintain a set differential pressure in the pipe network. Finally, the terminal units receive the chilled water and control their respective valve actuators to achieve the desired temperatures inside the rooms they are serving. Figure 1 shows a schematic drawing of a CWS.

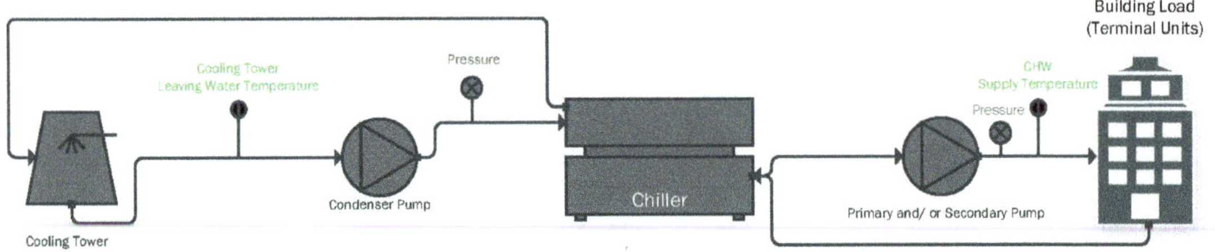

Figure 1. CWS schematic.

1.2. Predictive Maintenance Paradigm

Predictive maintenance (PdM) was first devised back in the late 1940s [25] and is basically used to assist in determining the status of an operated equipment in order to estimate the time of performing the maintenance actions [26]. According to Selcuk, it can be defined as an exercise of pre-empting failures depending on historical data in order to optimize the maintenance efforts [27]. Moreover, it is considered to be conditioned-based maintenance (CBM) to predict the likelihood of the failure time of a particular equipment and advise which maintenance task should be performed accordingly [28]. Figure 2 illustrates the position of PdM, along with other maintenance strategies. Since PdM is under the preventive maintenance (PM) category, it allows for convenient scheduling of reactive maintenance (RM) and prevents the equipment at a particular CB from any unexpected failure, where its principle is to evaluate the actual operating condition of a certain system and its components in order to optimize the O&M costs [29]. So, PdM can be considered as an enhancement of PM and RM; Figure 3 visualizes this argument. Furthermore, this research believes that PdM is significant for CB's maintenance program, as it counts on the current operational situation of an equipment and leads the concerned party to identify the expected issue immediately rather than average or expected life statistics, and also to predict when a maintenance activity will be needed. Verbert and others have assured that the routine maintenance does not usually identify the faults but can be sorted by implementing a PdM program [30].

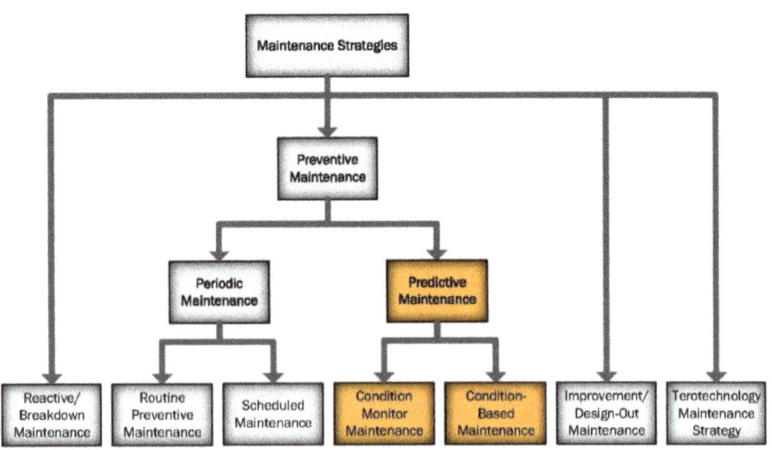

Figure 2. Structure of maintenance management strategies.

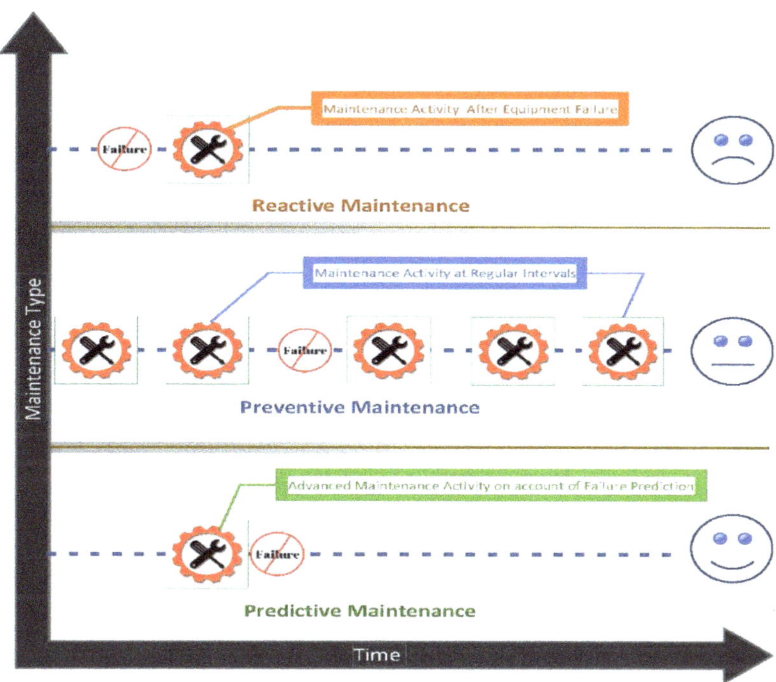

Figure 3. Effectiveness of PdM.

Having said it is a significant paradigm, well-known and key industrial manufacturers have invested in PdM to maximize machine parts and their uptime and disseminate maintenance to be more cost-effective [31]. Wang and others have argued that scheduled and unscheduled shutdowns; astronomical O&M costs; avoidable inventory; and undue maintenance activities performed on a particular equipment, machine, or system can be dwindled with PdM [32]. However, any technique has its own pros and cons; the main advantages of PdM are making the repairs based on the equipment condition, and this will sometimes lead to twenty-percent savings, as well as enriching safety aspects of the equipment and its surrounding; meanwhile, the disadvantages of PdM come from the organization's culture of hesitating to assign a sufficient budget for it [33].

PdM uses data analytics to detect equipment faults and to try to rectify operational inefficiencies with a goal of eliminating the root cause of potential system flops [34]. Amruthnath and Gupta did mention that observing equipment performance and monitoring the critical parameter of a particular system are one of the main PdM techniques [35]. Moreover, Huang and Wang considered components' monitoring of a particular system as one of PM's themes, which is the derived category of PdM [36]. Nguyen and Medjaher plus Yu and others have indicated that fault detection and diagnosis (FDD) and condition monitoring are critical components of PdM [37]. To perform automatic fault detection, PdM requires a big data collection, analytics platform, and data sufficiency [38]. The analytics platform must incorporate domain expertise, so that the algorithms have an intended application to the system under study [39]. According to Garg and Deshmukh, data sufficiency is the availability of data from enough sensors, actuators, meters, and control parameters so that a meaningful analysis can be performed accordingly [40].

Per Ran and others, maintenance in business industrial life is mainly RM and PM, with the PdM strategy being applied only for critical situations [41]. They believe that these maintenance strategies do not consider the vast amount of data that can be generated and the available approaches that align with Industry 4.0/Quality 4.0 principles, such as machine learning (ML), internet of things (IoT), Artificial Intelligence (AI), big data, advanced data analytics, data driven, cloud computing, and augmented reality.

Based on the thoughts of Chukwuekwe and others, PdM 4.0 is aligned with Industry 4.0, which is a paradigm shift in industrial processes impelled by intelligent information-processing approaches [42]. This shift in the maintenance paradigm has motivated this research's argument to believe in the PdM 4.0 paradigm, which can consider the operational status of CWS and shows the concerned manager, the maintenance engineer, or the system's user the health condition of the said system and make affirmative measures toward that, when required. Figure 4 explains the idea behind PdM 4.0.

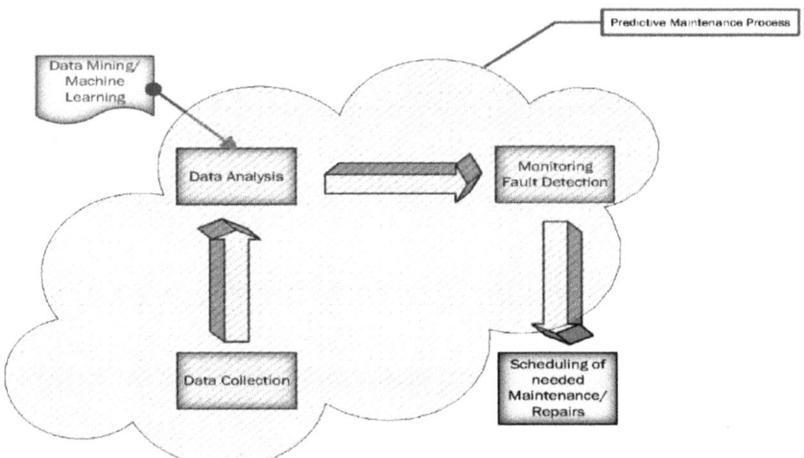

Figure 4. PdM's general plan.

2. Literature Review Methodology

2.1. Systematic Literature Review

Typically, a systematic literature review (SLR) is a kind of review that compiles varied research studies and epitomizes them in order to find the answers for a research question by using stringent methods [43]. Here, in this article, the protocol that was outlined by Kitchenham and others is followed [44]. The SLR went through four stages, as follows:

1. Determining the research questions.
2. Base of the research.
3. Criterion of the literature selection.
4. Quality assessment.

2.1.1. Stage #1

This stage is the launching of SLR. It consisted of defining the research questions (RQs) of this study. The RQs are questions that a study or research project intends to answer [45]. To make a strong RQ for all fields, especially in technology, engineering, and management, Figure 5 shows the principles that should be used [46].

As the idea of this research is to look after the studies that proposed a PdM 4.0 for CWS from an engineering management point of view, the following two RQs arose:

- RQ1: How can the faults be identified in order to predict them?
- RQ2: What are the methods that can be used to predict the faults?

2.1.2. Stage #2

This stage shows the search string and source selection. For the search string, operators called Boolean allow the researcher to use specific keywords with symbols such as "AND"

and "OR" in order to limit the relevant research papers [47]. Based on the information of previous sections, Boolean operators were exercised at the search engines as follows:

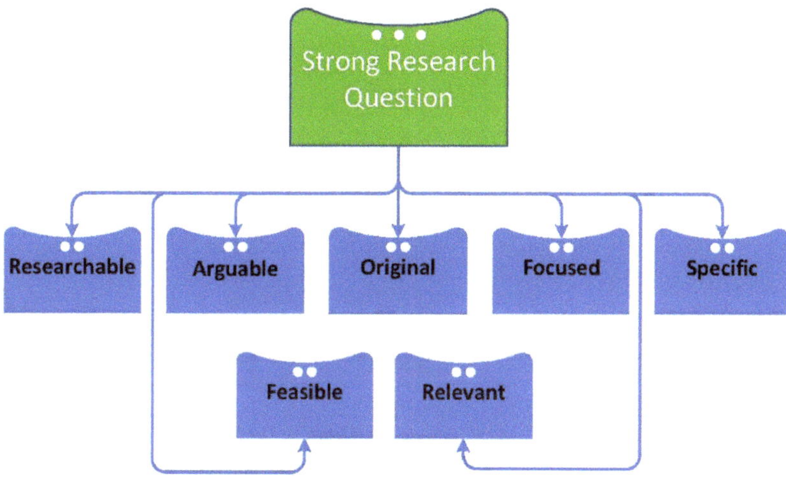

Figure 5. Principles of Strong RQ.

("Industry 4.0" OR "Quality 4.0") AND ("Machine learning" OR "Deep Learning" OR "Data Driven" OR "Artificial Intelligence") AND ("Predictive Maintenance" OR "Faults Detection" OR "Faults Diagnosis" OR "Condition Based Maintenance" OR "Condition Monitor Maintenance") AND ("Architecture" OR "Framework" OR "Management" OR "Program") AND ("Ontology" OR "Reasoning") AND ("Chilled Water System" OR "HVAC" OR "AC" OR "Chiller" OR "Cooling Tower" OR "Primary Pump" OR "Secondary Pump" OR "Condenser Pump" OR "Terminal Unit" OR "Air Handling Units" OR "Fan Coil Unit") AND ("Commercial Buildings" OR "Large Facilities").

The search engines or database used in this article, in addition to MDPI, are Google Scholar, IEEE, Springer, ACM Digital Library, Scopus, ProQuest, Web of Science, and ScienceDirect, as they are persuasive and reliable [48,49].

2.1.3. Stage #3

Following the actions that were performed within the previous two stages, all studies that were not pertinent to the aim of this article were removed. To do so, the following exclusion criteria, which are shown in Table 1, were applied.

Table 1. Exclusion criteria.

Exclusion Criteria	Reference
Papers (journals or conferences) that are not related to predictive PdM in a beeline	[48–50]
Papers that are not related to Industry 4.0 or Quality 4.0 or data-driven analysis or data mining in a beeline	[48–50]
Grey literature	[51]
Non-English publications	[51]
Pre-1999 publications	[49]
Papers that are not peer-reviewed	[52]

After that, filtering process have been implemented. Duplicate papers to be removed, thereafter, titles, and abstracts to be analyzed, and then the entire text to be analyzed [48].

2.1.4. Stage #4

Following SLR's procedure [44], the remaining articles were subjected to four questions; at least two questions out of these four questions should be fulfilled with a "yes" answer. The said four questions are as follows:

- Is the purpose of the research clearly presented?
- Does the research showed a framework/an architectural proposal or a research methodology?
- Does/do the author(s) present and discuss the results of the research?
- Does the paper used an ontology or reasoning?

2.2. Search Results

Starting from the second stage of SLR up to the fourth one, 168 studies are the total number of considered research papers in this article. Table 2 shows the papers' selection journey and how many papers are left after each stage.

Table 2. Journey of SLR.

Action	Stage Number	Number of Studies
Initial Search	2	1094
Exclusion Criteria	3	483
Duplicates Removal	3	422
Filtering Process by Title	3	328
Filtering Process by Abstract	3	244
Filtering Process by Entire Text	3	179
Quality Assessment	4	168

3. Applications

This section covers the considered studies that were mentioned in the previous section. It has four subsections—one for each of CWS components.

3.1. Chillers

PdM for chillers was presented in many ways either by a general maintenance framework or by FDD protocol in order to keep pace with the rapid industrial development. Rueda and others reported the development of FDD for liquid chillers based on AI techniques at one of the laboratory test facilities [53]. By using an artificial neural network (ANN), they predicted the temperature increment of the water-cooled condenser with almost ninety-nine per cent prediction accuracy. A similar valuable study was performed in the United Kingdom by Tassou and Grace to predict the refrigeration leak fault of a particular liquid chiller at one of the large CBs [54]. This fault was also predicted by using the Kalman Filter (KF) algorithm [55]. Han and others integrated k-nearest neighbors (KNN), support victor machine (SVM), and random forest (RF) into an ensemble diagnostic model to predict the said fault and achieved around ninety-nine per cent accuracy [56]. Liu and others stated that the leakage faults are seriously affecting the reliability of chillers, and therefore, they proposed an excellent timely and accurate method based on the adaptive moment estimation algorithm with multilayer feedforward neural networks trained with the error backpropagation neural network (Adam-BPNN) [57]. In Hong Kong and China, seven studies applied principal-component analysis (PCA) to predict several faults of sensors that are reading operational parameters, such as chilled-water flow rate, condenser water flow rate, and evaporating pressure [58–64]. Furthermore, Hu and others applied self-adaptive PCA to enhance sensors' FDD efficiency [65]. In contrast, Li and others reported that support vector data description (SVDD) is better than PCA, as PCA is not very efficient when it comes to predicting complex sensor faults, due to the weakness of Q-statistic plot, which is part of it [66]. Choi and others utilized data from one of ASHRAE projects to predict multiple sensors faults of parameters such as the evaporator water entering temperature [67]. They applied three data-driven techniques, which are multiway dynamic PCA, multiway partial least squares (PLS), and deep-learning SVM. Based on their results, they found that the first two techniques, which employed generalized likelihood ratio test, are more accurate than the neural network one (SVM). This finding emerged with another study performed by Namburu and others to predict eight different faults of chillers by using the same three techniques [68]. From another ASHRAE project, Schein and Bushby applied a hierarchical rule-based FDD to predict the scheduling fault during three different weather seasons but with no broaching to the data sample of their study [69].

Sensors faults were not usually considered in the previous studies. For example, at one of CBs in Hong Kong, performance indices (PIs) proposed to predict evaporator fouling using regression model [70]. PIs were again proposed to predict the other seven faults, such as condenser fouling using fuzzy modeling and ANN technique [71]. Both previous studies concluded that PI may not be effective in fault diagnosis. In this regard, it would be interesting if the data of one of the ASHRAE projects which were mentioned by Comstock and others were utilized for proposing a new FDD, as the sensitivity of eight common faults were already tested [72]. Han and others applied FDD for multiple simultaneous faults of two chillers using combined SVM and multi-label (MLB) techniques [73]. These combined techniques showed high accuracy detection of the chillers' performance, although the experimental data were limited. On a separate note, such techniques require sufficient training data for high-quality outputs [74,75]. Per Ma and Wang, chiller performance degradation can be detected significantly by using a hybrid quick search (HQS) method through characterizing the PIs of multiple operational parameters, such as the temperature of the condenser water supply [76].

A high chiller's load affects the performance and leads to the appearance of faults such as condenser fouling. Yu with Chan discussed that via two studies, the first one on how to improve chiller management using regression model and the other one proposed an assessment strategy of chiller's performance using clustering analysis [77,78]. Zhao and others indicated that early identification of the said fault (condenser fouling) is essential to highly maintain chiller performance and developed a virtual sensor for that fault [79]. Moreover, Magoules and others proposed a significant FDD strategy using a recursive deterministic perception neural network (RDPNN) to predict faults related to chiller's load [80]. Data from one ASHRAE project were utilized in twelve different studies to predict condenser fouling, along with other faults [81–92]. The first study applied exponentially weighted moving average (EWMA) control charts; the second one applied Bayesian belief network (BBN); the third one applied SVDD; the fourth one used SVM; the fifth one applied conditional Wasserstein generative antagonistic networks (CWGANs); the sixth one combined extended KF (EKF) and recursive one-class SVM (ROSVM); the seventh one derived a tree-structured fault dependence kernel (TFDK); the eighth one used PCA, along with SVDD; and the ninth one adopted Linear Discriminant Analysis (LDA). With regards to the tenth one, One-Dimensional Convolutional Neural Network (1D-CNN) and Gated Recurrent Unit (GRU) were applied while the eleventh one conjoined a distance rejection (DR) technique with Bayesian network (BN) via transforming the chiller FDD problem into a single-class classification problem. The last one in that group predicted seven different faults by using the large margin information fusion (LMIF) method and found that this method is more accurate than others, such as multi-class SVM (MSVM), ANN, decision tree (DT), quadratic discriminant analysis (QDA), Ada Boost (AB), and logistic regression (LR). All of these studies showed significant accuracies but did not include fault-free situation in their data training. Moreover, three more studies used the same ASHRAE project just to compare different models for the same purpose of the previous twelve studies [93–95]. The first study presented two models, one by SVM and the second by combining nonlinear least squares support vector regression (SVR) based on the differential evolution (DE) algorithm with EWMA control charts, and it was found that the second one has better prediction. The second study applied multiple linear regression (MLReg), kriging algorithm, and radial basis function (RBF) and concluded that RBF is the best. The outcome of the third study showed that ANN is more accurate than KNN and bagged tree (BT) algorithms. The impact of condenser fouling was discussed and, accordingly, a decoupling-based FDD method was proposed to predict this fault [96]. This method was applied by observing the cooling capacity and suggested to clean the condenser water tubes before data collection. Later, this method was applied again alongside another two methods for efficiency comparison purposes in detecting multiple simultaneous chiller's faults [97]. The aforementioned other two methods were MLRrg and simple linear regression (SLReg), and it was found that these two methods (MLReg and SLReg) are not very effective. Bonvini and others argued that observing the energy consumption of chiller is considerable to predict the faults that are related to the high load [98]. They introduced the FDD approach based on unscented KF (UKF), which is an advanced Bayesian nonlinear state estimation technique, to predict three of the aforementioned faults. KF can be considered as a quite proven technique and does not require long-time focused studies when applied in individual CWS devices in different CBs [99]. This pretext came from a study that used KF to detect gradual chiller degradation based on the gray-box model at the Jinmao tower of China. The said model is based on measuring and analyzing the variations of chilled water flow rate and supplied chilled water temperature through statistical process control (SPC). Moreover, Karami and Wang integrated the Gaussian mixture model regression (GMMR) technique with UKF to model a nonlinear system based on the measurement data of four operational parameters and found this to be efficient in detecting chiller degradation and reducing the number of detecting sensors, as well [100].

The chiller faults either that are related to the high load or from other issues can be linked to human interventions and, accordingly, can influence the occupant's satisfaction. Having said that, maintenance characteristics such as the skills, the knowledge, and the number of maintenance laborers were addressed by Au-Yong and others at one of the office buildings, using mixed methods [101]. Following a survey that was shared with the occupants, as well as with key responsible staff, they predicted eight of that maintenance characteristics via a regression model and found empirical evidence that such communications with the concerned parties can improve the maintenance management and lead to the occupant satisfaction. In regard to high performance levels, it has been noticed that some CBs are using building management system (BMS) software in relation to their maintenance activities. For example, Alonso and others suggested utilizing BMS, in addition to plant management software (PMS), to control CWS, and they successfully applied this idea by observing the coefficient of chillers' performance at one of the large hospitals [102]. Yan and others proposed chiller's FDD procedure to develop BMS via a hybrid model that integrated SVM with autoregressive exogenous variables (ARX) and obtained a high prediction accuracy and minimal false-alarm rate [103]. Identical results were presented by Mclintosh and Mitcell, using statistical analysis by modeling the log-mean temperature difference and condenser water temperature difference to predict six faults of chillers [104]. In addition, two studies proposed a control strategy for chiller operation uncertainty by using the Monte-Carlo simulation (MCS) [105,106]. To curb the deterioration of chillers, Beghi and others proposed a semi-data-driven approach by using PCA in differentiating anomalies from normal operation variability and a reconstruction-based contribution approach to segregate variables related to faults [107]. To minimize faults prediction errors, Kocyigit addressed eight faults and claimed to use a fuzzy interference system (FIS) and Levenberg–Marquart-type ANN (LMANN) algorithm by evaluating several operational parameters, such as condenser pressure and evaporator pressure [108]. Additionally, Gao and others presented a novel FDD strategy in combining maximal information coefficient (MIC) with a long short-term memory (LSTM) network by using a virtual sensor [109].

It has been noted from the market that multiple providers of maintenance solutions for smart CBs are proposing building information modeling (BIM) and the building automation system (BAS) in addition to BMS. Cheng and others utilized BIM with IoT sensors to predict chillers' faults through ANN and SVM [110]. However, their approach could not be applied for other CWS components due to the differences in operational parameters. From BMS data, Escobar and others used a fuzzy logic clustering (FLC) approach for smart buildings that was called the learning algorithm for multivariable data analysis (LAMDA) and succeeded in reaching a zero-error state for chillers' control [111]. Besides BMS, Srinivasan and others have used explainable AI (XAI) for chiller FDD and showed how it is significant to acquire the trust of maintenance officers [112]. Hu and others used BAS for collecting data of chiller operational parameters, such as the condenser water flow, and then used them to detect faults by using SVM [113]. Considering the fault-free situation in their model training, their approach detected only one single fault, which was compressor overcharging. The same fault was efficaciously detected by using a PCA-based EWMA and virtual refrigerant charge (VRC) algorithm [114]. From BMS, Luo and others collected the data of chilled water supply and return temperatures in every minute frequency of six days from four different weather seasons to predict six different faults, in addition to fault free condition, using k-means clustering [115]. The said faults were not fully described, and as with other studies, the frequencies of their data sampling were not justified even after the development of this approach [116]. Theiblemont and others explored state-of-the-art control strategies [117]. The first strategy, called "Model-Free Control Strategy", does not require building a model or the use of historical data. It can be performed by programming the ambient temperature based on the weather forecast of the next day. The second strategy is an intelligent one which uses AI, along with a cold thermal energy storage (CTES) system. This strategy suggests combing a fuzzy logic controller and a feed-forward controller with weather predictions. To do so, the authors listed 27 rules for that [117]. Advanced control is the third strategy, which includes two techniques: Non-Optimal Advance Predictive Control and Model Predictive Control (MPC). The Unknown-but-Bounded method is an example of the first technique, and its implementation is costly. The concept of MPC is to optimize the variables of CWS as a function of future horizon to satisfy the relevant constraints. Arteconi and others suggested applying CTES for Demand-Side Management (DSM) strategy, which can change the chiller load profile to optimize the power system from generation to delivery [118].

Some studies used the ratio between the cooling load and the energy consumed, which is called the coefficient of performance (COP), as a data sample for scheduling PdM activities. In this regard, Wu and others proposed a method to optimize the PdM scheduling for HVAC system by mixed-integer programming (MIP) [119]. The said method has two stages: the first one is the parameter generation through historical data, and the second one is the optimization by linear programming. They conducted a case study on chillers and addressed COP. The idea of the first stage is to study the operational status and then listing the related constraints while the optimizing model (second stage) has to be solved to present a high-quality PdM schedule in order to detect the chillers' degradation. The model is a bit general, and it did not consider or discuss any precise faults or issues that lead to the chiller degradation. Li and others proposed a novel FDD method using a deep belief network (DBN) [120]. Their data were collected through an IoT agent and processed through four different stages, including optimizing them by particle swarm optimization (PSO) algorithm. Moreover, they did compare DBN with deep neural network (DNN), KNN, and SVM and obtained almost same prediction accuracy. From COP data, Motomura and others developed two outstanding simulation models to evaluate multiple chillers faults [121,122]. The first model calculated the increase amount of daily peak power, while the second one tracked the decrease rate of COP. Sulaiman and others observed chillers' COP and developed an FDD approach by using deep learning (DL), multi-layer perceptron (MLP), and SVM, and they mentioned that MLP is more accurate than others [123]. From a chiller's COP data sample, Ng and others used BN to predict sensor bias of water flow temperature, but the results were not very encouraging [124]. To obtain usual promising results, Harasty and others argued that an ANN should be used in PdM management [125].

3.2. Cooling Towers

Compared to the studies on chillers, the studies on cooling towers were limited and were either part of chillers' ones or were discussed separately. Ahn and others developed a simulation model to detect three faults of cooling towers [126]. Their model was built based on the deviation of different operational parameters such as the difference between the water temperatures that are leaving the tower and the temperatures that are entering the same. The only claim against this study is the data collection, as the authors did not clarify the source of their samples that were used in the associated experiment. Zhou and others used a regression model to detect air fan degradation fault by formulating the PI of the air-flow-rate reduction [70]. The sample size of their data was small, as it was generated from only five days in the summer season, including the fault-free condition. Hu and others collected data on fan power to detect the same fault by using SVM, and their sample size was also small [113]. From a qualitative method study, Chew and Yan suggested cleaning cooling towers' fans before applying any FDD approach [127]. Khan and Zubair discussed another fault, which is fouling of fills, and predicted it very well by using a regression model [128]. Through this model, the correlation was analyzed between the PIs of different operational parameters. Per Ma and Wang, the said two faults (fouling of fills and air fan degradation) can be detected significantly by using the HQS method through characterizing the PIs of multiple operational parameters, such as the inlet water temperature [76]. Air fan faulty was again predicted by Sulaiman and others when they compared MLP, SVM, and DL methods, and they found that MLP is more accurate than others [123].

Human and organizational factors are obviously affecting PdM costs and its scheduling. In this regard, Jain and others studied the failure conditions of a particular cooling tower by introducing a process resilience analysis framework [129]. This framework utilized a BN model to integrate two factors, which are process parameter variations as a technical factor, and human and organizational factor as a social one. It illustrated the impact of the said model on PdM management from cost and safety points of view. Melani and others insisted that making a significant investment in PdM is essential to maintaining the availability of systems that are operating CBs [130]. Having said that, they developed a generalized stochastic Petri net (GSPN) model to predict multiple faults, such as those related to fans, including the operational errors caused by humans. Furthermore, Aguilar and others proposed an autonomic cycle of data analysis tasks (ACODAT) involving BMS to manage the failures of two cooling towers of opera palace in Spain [131]. They utilized three techniques, namely MLP, KNN, and gradient boosting (GB), and reached to similar prediction accuracies. To diagnose such failures, Poit and Lancon suggested CBs to use SCANSITES 3D system and surveyed several cooling towers in France and found the said system to be very useful [132].

As was performed in chillers, the FDD of sensor faults was also studied in regard to cooling towers. At the Oak Ridge National Laboratory (ORNL) in the United States of America, the air fan degradation faults of the high flux isotope reactor (HFIR) were predicted by using wireless sensors [133]. Wang and others predicted the motor degradation by using PCA [63]. Their data samples were collected through a sensor that read one of the operational parameters, which was the inlet water temperatures, and per their results, the PCA did not always record the occurrence timings of that fault, and, accordingly, they could not evaluate the PI of the aforementioned parameter. An excellent study collected data from the same parameter to predict fan degradation fault by using the KF method [99]. Another study used the KF method to observe the cooling towers' performance at one of China's CBs [134]. To reduce the false-alarm rate, the said study analyzed and measured some chosen parameters via SPC. Motomura and others developed two superb simulation models to assess multiple cooling-tower faults [121,122]. The first model checked the water flow and the outside air wet-bulb temperature, whilst the second model focused on the inlet and the outlet condenser water temperatures. Data on air wet-bulb temperature, as well as other parameters, were collected to predict a particular cooling tower's performance and to eliminate the severity of the related faults by using the BPNN method [135]. The said method resulted in the obtainment of a very good correlation coefficient between the predicted and the experimental values.

3.3. Pumps

Following the literature on cooling towers, the number of studies on pumps is almost the same. Karim and others predicted five faults of pumps—out of which two were related to the cooling system—using ANN method, and their hypothetical data showed that such a method is capable of predicting the aforementioned faults [136]. Using a clustering method, Luo and others studied the sensors bias of primary and secondary pumps, but with no full description of the faults [115]. Through the HFIR project at ORNL, Hashemian predicted three different faults by using wireless sensors [133]. These faults are excessive noise, control switch failure, and faulty starter, and all of them are related to the secondary pump. From BAS, Hu and others collected a good data sample of differential pressure to predict the degradation of secondary pump by using SVM [113]. In order to keep control on the differential pressure of primary and secondary pumps, Ma and Wang developed a simulation model that takes the water flow rates into consideration [137]. Miyata and others used MCS to detect the operational uncertainty caused by the imponderable pressure [105]. Zhou and others used a regression model to detect partial clog fault in the secondary pump by formulating the PI of the increase in the pipeline resistance [70]. On the other hand, Wang and others predicted the same fault (partial clog) by using the PCA [63]. Furthermore, Liu and others studied the pipeline resistance and then predicted the primary pump's leakage fault by using Adam-BPNN [57]. Motomura and others developed two valuable simulation models to predict the faults of primary, secondary, and condenser pumps [121,122]. From the BMS data, their first model observed the water flow in liter per minute, while the second one focused on the sensor errors, and it also studied the impact of pumps specifications, such as the caliber.

The appearance of faults obviously affects the CWS performance, whether they are caused by human interventions or by an operational issue or unreliable sensor. Au-Yong and others focused on the pumps within their mixed-method study, which was explained in the section on chillers [101]. Per the qualitative method research of Chew and Chan, the maintenance officers and the researchers are advised to check the condenser pumps for corrosion before applying any FDD approach [127]. Moreover, Yang and others proposed the use of the FDD strategy with the ML method and counted data samples via BMS that are related to pumps, but they did not specify the associated operational parameters nor the ML method [138]. Yuan and Liu used a semi-supervised learning (SSL) technique to predict severe gear damage of a particular pump and took into consideration the fault free condition while training the model [139]. Bouabdallaoui and others introduced a PdM framework by using LSTM [140]. As part of this framework, they collected data for three pumps via BAS and IoT devices, but they did not specify the associated operational parameters nor the detected faults. With regard to the state-of-the-art control strategies, Theiblemont and others suggested applying adaptive MPC to decrease the pumps running time [117].

3.4. Terminal Units

The subject component has the largest number of studies comparing it to other CWS components. Liang and Du proposed an FDD model of the HVAC system that uses mixed methods. The under-study component was an AHU of a particular CB in Hong Kong [141]. They combined the simulation-based-model method with the SVM method. Three types of faults were addressed, which are return damper jam, cooling coil blockage, and speed reducing of the supply fan, noting that false signal fault is not considered in their study. Their method was built by collecting data of multiple parameters, such as the set temperature and the indoor cooling load. The original sample size was small; because it was generated from ten operational hours but based on the qualitative output of a related research that was reviewed by Ding, they assumed that the fault would arrive within one hour [142]. So, they did depend on this assumption when finalizing their required data and obtained a bigger sample size, which was used to build the said model. Through BIM and Modelica software, Andriamamonjy and others presented a simulation model to detect damper faults of a particular AHU [143]. Their model showed the potential of BIM for a significant reduction of the manual configuration needed to disseminate such a model, which was based on calculating the normalized root mean square error (RMSE) of multiple operational parameters, such as supply air temperature under three conditions, faulty, uncertain, and fault free. In contrast to a case study performed at one of the universities, Alavi and Forcada argued that BIM cannot constitute complete information on maintenance activities when implementing decision-making frameworks [144]. The study, which discussed the impact of human interventions in the occurrence of faults and was explained in chillers and pumps sections, also included AHUs [101]. The PdM framework of Bouabdallaoui and others, which was discussed in the pumps section, was also embedded with two AHUs, but they were not defined by the predicted faults in their case study, which was performed at one of the sport facilities in France [140].

Bruton and others discussed previous procedures and proposed a good one on how to choose the appropriate ML technique based on AHUs conditions [145]. Thereafter, they developed an automated FDD for AHUs, the contents of which are data access layer to be flexible with BMS, business layer to be flexible with any combination of sensors with operational parameters, and graphical user interface to evaluate the performance of AHUs [146]. Candanedo and others used DT technique for evaluating an early stage PdM model of terminal units [147]. In a set of buildings that are between zero and thirty years old, they obtained historical data of the indoor temperatures in order to compare them with the designed ones and then to identify any abnormal behavior. They indicated that DT showed its accuracy in covering the faults possibilities. To achieve the thermal comfort inside CBs, an experiment was performed by collecting occupant skin temperatures to predict and evaluate multiple issues, such as the air velocity of AHUs, using SVM and extreme learning machine (ELM) techniques, and obtained satisfactory results from both [148]. To get high accuracy FDD model, it is advised to clean the impeller, the fan scroll, and the blower blade of AHUs before applying that model [127]. Arteconi and others suggested a state-of-the-art control strategy using DSM to reduce the required AHU's size up to 40%, which leads to energy saving [118].

The variable air volume (VAV) of AHUs was discussed in many studies. For instance, in a multi-purpose research and test facility called an environmental chamber, Cho and others conducted two studies on a number of rooms that represent CBs' standards [149,150]. In addition to the fault-free condition, the first study used ANN to predict eight faults linked to AHU parts, including VAV, while the second study applied transient pattern analysis (TPA) to isolate the said faults to reach steady-state condition. The study of Schein and Bushby, which was mentioned above in chillers section, did predict a VAV sensor fault when reading the discharge air temperature [69]. At a large academic office building in Canada, Gunay and others developed an excellent simulation model to detect five VAV sequencing logic faults in two AHUs [151]. By using ASHRAE project's data, another excellent simulation model was developed by Norford and others to detect multiple AHU's faults that are related to VAV's damper, fan, and filter coil system [152]. Moreover, Li and others proposed a simulation model to predict eleven VA faults at a particular CB in China, and they succeeded in detecting nine of the faults, including outdoor air damper stuck and multiple sensors faults [153]. Two more valuable studies precited the damper stuck fault at two different CBs: the first one applied RF, while the second developed a simulation model [154,155]. At other different CBs, thirteen interesting studies used data from one ASHRAE project to predict some faults of AHUs and FCUs, including the ones that related to VAV, and obtained an acceptable prediction accuracy for each [156–168]. The first study applied the temporal association rules mining (TARM) algorithm, while both the second and third ones applied BN. The fourth study applied ensemble rapid centroid estimation (ERCE), the fifth one applied regression tree (RT), and the sixth one applied SVM. With regard to the seventh one, the generative adversarial network (GAN) was applied, and the eighth one combined RF with SVM. The ninth one utilized simulation software called HVACSIM+, the tenth one derived LMIF, the eleventh one applied PCA, the twelfth one applied SSL, and the last one applied SVM with ARX. In contrast, Zhao and others criticized the same ASHRAE project because its data did not cover a vast range of operating conditions [169].

Combining the FDD approach with faults-isolation approach is one of the PdM ideas. A study in Canada presented this idea by applying the PCA to detect two selected faults of AHUs faults and active functional testing (AFT) to isolate the same faults [170]. Two more studies applied the PCA, but in both detecting and isolating a number of faults on AHUs [171,172]. Ranade and others developed a simulation model to predict five selected faults of FCU and VAV, including fault free condition [173]. They argued that these faults can be isolated easily by applying DT. Using data of AHU's outlet water and supply air temperatures, Shahnazari and others applied a recurrent neural network (RNN) to detect and isolate the faults of the associated sensors [174]. Moreover, Wang and Chen conducted a case study at a particular CB, which has thirty-six floors, by applying EWMA for the same purpose [175]. Wang and others applied a genetic algorithm (GA) to predict and isolate the faults of AHU's supply fan and VAV [176]. Data from BMS were utilized to predict and isolate ten selected faults of AHUs by using BN [177]. At a green CB, an excellent experiment resulted in developing four simulation models to detect and isolate four faults of AHUs (one model for each fault) [178]. Yang and others presented a pragmatic simulation model to detect only four selected faults at forty-four buildings in Canada. Their solution relied on clustering work orders datasets, which were collected from occupants' complaints, and then computing mean time between failure (MTBF) [179]. MTBF was also computed by Sanchez-Barroso and Sanz-Calcedo [180]. To detect and isolate small bias sensors faults, a novel study advised using a hybrid-model-based FDD that combines the fractal correlation dimension (FCD) algorithm with SVR [181]. Zhang and Hong explained the background of multiple AHU faults, which will help the researchers or CBs' officers to take that into consideration while making PdM programs [182].

By recalling of what is written in the chillers section about how CBs are using BMS to control CWS performance, we note that Hosamo and others stated that BMS cannot detect many faults, including those that are related to AHUs [183]. Having said that, they conducted a case study on four AHUs at a particular university which proposed a digital twin technology that utilizes BIM and IoT's sensors, noting that this technology is an ANN-based technique. On a related note, Lee and others predicted seven faults of sensors by using a general regression neural network (GRNN) model [184]. Gao and others studied the impact of the system's water temperature difference (delta-T) on AHUs' performance [185]. They developed a worthy simulation model that generates the PI of each multiple operational parameter. Choi and Yeom introduced a thermal satisfaction prediction model that combined human factors and physiological signals [186]. Their data were collected from volunteer students through a LabVIEW-based data acquisition (DAQ) system and were analyzed by multiple statistical analysis and data-mining software called WEKA. Their study showed a significant correlation between the said factors and signals. A similar study discussed IAQ and used hypothesis tests to diagnose AHU's sensors' faults [187]. Shaw and others studied the correlation between multiple operational parameters to obtain reliable FDD results [188]. In Australia, an auto FDD model, which was developed by Guo and others in one of the large CBs, merged the hidden Markov model (HMM) and SVM [189]. Their data were collected through BMS from fifteen AHU sensors, and their model was trained based on selected faults over two business months. Unfortunately, they did not specify which parameters of the said AHU were studied, nor did they consider the sensors' false signal in their model. Holub and Macek presented a simulation model within a stochastic system by addressing the set temperature of a rooftop AHU [190]. The target of their application was to detect a diagnostic fault that links to the fan. Frankly, the data used to simulate the aforementioned model were limited where they applied a hybrid system. To obtain an active simulation model, Deshmukh and others suggested (while collecting AHU's data of fault free mode) holding three operational conditions, including closing the cooling valve [191]. Ma and others introduced a PdM framework which integrated BIM, geographic information system, and reliability-centered maintenance technologies by implementing a quantitative decision-making model, along with an MCS model [192]. Their case study was performed on a virtual university campus that includes AHUs, and they found it difficult to acquire a large data sample size. Gourabpasi and Nik-Bakht indicated that the lack of knowledge in locating the sensors is causing difficulty for either data collection or sensor's FDD [193].

Terminal units' FDD can be considered as a probabilistic approach. In the USA, Dey and Dong applied BBN in a probabilistic way to predict some AHU faults at one of the universities [194]. Du and others applied the wavelet neural network (WNN) to fix the AHU's sensor bias [195]. A subtractive clustering technique and BPNN were combined to catch the missing alarm when an AHU's fault occurred [196]. For missing alarm issues, a study suggested applying LMANN to eliminate that [197]. To enhance the thermal comfort, Dudzik and others used BAS and applied ANN to develop and examine the environmental quality management system [198]. At one of Qatar's sport facilities, Elnour and others applied a neural network that clustered the RMSE of some operational parameters, and then compared that with SVR, KNN, and DT techniques [199]. They found their approach to be more efficient than the aforementioned three techniques in controlling AHU's operation. Through virtual refrigerant mass flow sensors, Kim and Braun presented an FDD approach to predict five selected faults, such as condenser fouling [200]. Lauro and others used FLC to predict the abnormal behavior of a particular building's FCU [201]. Li and Wen used wavelet transform with the PCA technique (WPCA) to predict some AHU faults [202]. Liu and others applied the Markov chain Monte Carlo (MCMC) algorithm to drive the statistical characteristics of an AHU's faults levels [203]. On a single terminal unit, Lo and others applied fuzzy GA to eliminate sensors' false signals [204]. The study of Luo and others, as was explained in the chillers and pumps sections, also included terminal units [115]. By utilizing of ASHRAE's thermal comfort database, a novel study compared the thermal sensation vote (TSV) and the predicted mean vote (PMV) by using an RF model and resulted in around sixty-five percent accuracy in TSV prediction [19]. The study of Miyata and others, as mentioned in the chillers and pumps sections, did include AHUs, but they defined the related faults [105]. From an ASHRAE project dataset, Montazeri and Kargar applied six algorithms, namely SVM, RBF, kernel PCA (KPCA), DT, DBN, and shallow neural network (SNN), to detect the sensor and actuator faults, and they stated that DT had the biggest prediction accuracy [205].

ASHRAE datasets were not the only source in developing FDD within the literature. Novikova and others utilized VAST Challenge 2016 dataset to develop a simulation model that monitored and assessed terminal units' performance at a three floors' CB [206]. The data of residential complex building were utilized by Parzinger and others for AHU's FDD, using ARX and RF techniques [207]. Both techniques showed similar and acceptable prediction accuracy. Rafati and others performed a good review of the utilization of non-intrusive load monitoring software in terminal units' FDD [208]. In cooperation with a leading building management company, Satta and others proposed a PdM approach for the cohort of seventeen appliances that are similar to terminal units and examined it at one of Italian hospitals [209]. Using the historical data of different variables, such as the indoor temperature, they used DT to detect the abnormal behavior of these appliances. They argued that the reciprocal dissimilarities between appliances' behavior can expose an upcoming fault with enough anticipation to allow for a proactive meddling and avert breakage in operation. Tehrani and others addressed one fault related to a particular terminal unit at one of the Canadian universities [210]. The said fault was filter blockage, and they used ANN to predict the behavior of the said unit. Moreover, they determined that the performance of the unit in discussion has improved by using DT instead of

ANN. Furthermore, Shakerian and others recommended applying the synthetic minority oversampling (SMO) technique to improve the prediction accuracy [211]. Sulaiman and others developed a fuzzy fault detection model for centralized CWS, using simulation [212]. They implemented the said model in the air-supply damper of an AHU, which is linked to two specific rooms. Three cases were studied in their research to simulate the said model. Two of them were related to the damper's faults, and the third one was at normal operation, without any faults. They identified these faults by checking the room-temperature variation. They mentioned that the developed model had resulted in detecting the damper faults, but with no technical details. Another FDD approach, which was presented by them and explained in the chillers and cooling towers section, also covered AHUs [123]. Thumati and others developed a generic simulation model to detect terminal unit faults and to isolate the associated residual errors [213]. Their idea could be presented perfectly by using virtual-sensor approaches such as Verbert and et al.'s one [30]. At a residential facility, Turner and others developed a simulation model for AHUs' FDD [214]. During seven days' study time, they focused on the outdoor temperature and the set indoor temperature parameters to detect selected faults, such as compressor failure. They believe that using such data-driven approaches for tracking the said parameters can help easily detect the associated faults. Van Every and others applied Gaussian regression (GR) and SVM to estimate AHU's sensor values and to detect the associate faults, respectively [215]. Velibeyoglu and others applied directed acyclic graph (DAG) to assess the detectability of AHUs' simultaneous faults and obtained promising results [216].

The usage of one or more software or systems such as DAQ, BIM, IoT sensors, BAS, and BMS is important in controlling CWS performance. Villa and others extolled the usage of such software in AHU's FDD purposes, and, accordingly, they introduced an outstanding PdM framework, using an automatic ML platform called H2O [217]. Using FLC, Wijayasekara and others assessed BMS's performance in controlling the thermal comfort inside selected rooms [218]. Alongside BMS, DT was used by Yan and others to develop a diagnostic strategy for AHUs [219]. Nine cases were addressed for their related experiment, which are eight faults, such as duct leakage, and one case for normal operation (fault free). For this experiment, they used data that were recorded from one of the ASHRAE projects. They emphasized that data-driven methods are unique to glean the useful information from large datasets and for modeling the behavior of HVAC systems. Yu and others proposed association rule mining (ARM), which is a data-mining technique, to test the correlation between all AHUs' operational parameters at one of the complex buildings that contains offices and chemical labs [220]. It seems that they faced some difficulties in regard to the data-collection part. The absence of data sources makes any ML model weak in detecting and diagnosing the faults [108].

4. Discussion

From the previous section, it is observed that chillers and terminal units were mostly researched, while there was not much research focused on cooling towers and pumps. Following SLR, the maximum number of research studies on chillers was carried out in the years 2016 and 2019, whereas on terminal units, it was performed in the year 2020. Regarding cooling towers and pumps, the year 2019 recorded the maximum number of research studies. Figure 6 highlights the research trends from the year 1999 onward.

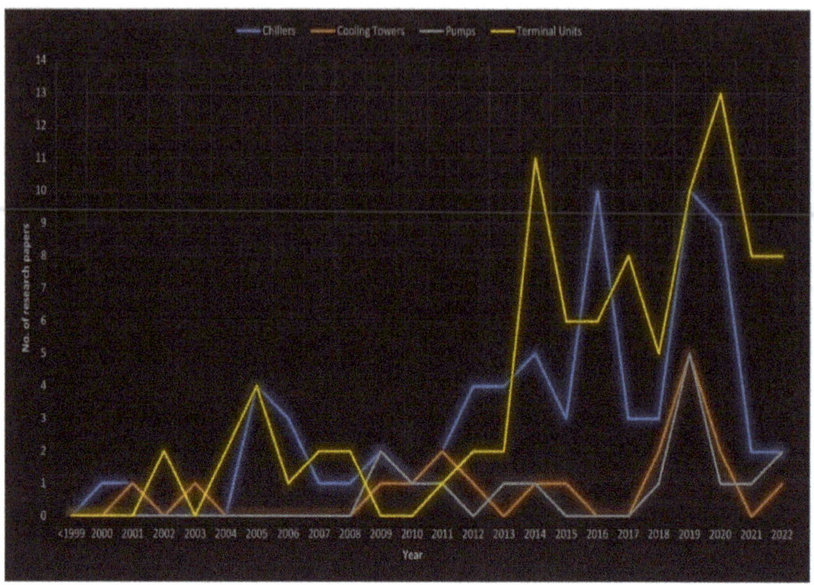

Figure 6. Research trends.

The considered studies sometimes addressed CWS components independently and other times in combinations. For combinations, some of them addressed either two components or three components in total; no research study addressed the whole system (i.e., four components) at once. Table 3 shows the number of considered studies that addressed either a single component or addressed more than one within the same research.

Table 3. Breakdown of considered studies.

Components	Number of Considered Studies
Just Chillers	57
Just Cooling Towers	6
Just Pumps	4
Just Terminal Units	82
Chiller and Cooling Towers	5
Chillers and Pumps	1
Chillers and Terminal Units	3
Cooling Towers and Pumps	1
Pumps and Terminal Units	1
Chillers, Cooling Towers, and Pumps	3
Chillers, Pumps, and Terminal Units	3
Chillers, Cooling Towers, and Terminal Units	1
Cooling Towers, Pumps, and Terminal Units	1

This research is intended to answer the aforementioned two RQs where the idea behind both of them is to explore the ways of implementing a PdM program on a CWS. The aim of RQ1 is to understand how the researchers are preparing or arranging for the PdM program. This includes checking the way of identifying and studying the system faults, which are considered as a base of implementing the PdM program. Moreover, it includes identifying the operational parameters, which allow us to observe the system and lead to the faults, as well as knowing their data sample size and their source. With regard to the aim of RQ2, it is a further action after RQ1 and intends to explore the tools, methods, frameworks, or control strategies that were applied to make the PdM program. So, the third section of this article (Application) is presented based on the aims of these RQs. Each considered study has gone through these activities unless it has missing information.

The following four subsections highlight the findings of this research. The first one addresses the faults-related information, and the second subsection focuses on the operational parameters and their data samples. Both of these subsections are related to RQ1. The third subsection is related to RQ2, while the last subsection lists and explains the research gaps.

4.1. System Faults

The following points are the major findings with regard to system faults:

- The fault is defined as any failure that may lead to a CWS breakdown over time.
- Some of the considered studies were focused on only one fault, such as condenser fouling of chillers.
- Many of the considered studies addressed refrigeration leaks, which can be considered as the most popular fault in chillers.
- Fouling of fills and air-fan degradation are the most addressed faults of cooling towers.
- The most addressed fault of pumps is the partial clogging, whereas the full clogging, which is more critical, was not addressed at all.
- For terminal units, the most addressed faults were return damper jam, cooling coil blockage, and speed reducing of the supply fan including the VAV related faults.
- There are obvious differences between the studies in the number of the addressed faults for all CWS components.
- Some of the studies addressed the same faults, while the others addressed different faults.
- Some studies did not state the addressed faults, and even if they stated the faults, they were found to be not fully described, stating "abnormal behavior" as a fault, for example.
- Some studies stated multiple faults, but they did not address or predict all of them in their case studies.
- Faults such as sensors' bias and controller false alarms were not usually considered in the aforementioned studies.
- High chiller's load is affecting the performance and leading to faults such as condenser fouling and compressor overcharging.
- Condenser fouling was found to be the fault with the most negative impact on CWS reliability, as well as on the occupants' satisfaction.
- Human factors, such as the skills of the maintenance officer who manages the system, have a significant impact on faults appearance.
- Fault-free mode has to be considered in any research in order to increase the prediction reliability.

4.2. Operational Parameters and Data Collection

Operational parameters are the measurable factors that provide numerical data of the system performance [26]. Hereunder are the major findings in this regard:

- Any operational parameter can give a glimpse of the health condition of the related CWS component.
- Water leaving temperature is the most operational parameter used for both chillers and cooling towers.
- The most operational parameter used for pumps is the differential pressure, while the space temperature is the most common one for terminal units.
- Some studies did not specify which operational parameter their studies were built on. In addition, some studies stated the operational parameters, but they did not provide detailed information about the associated data, including the sample size, source, and frequencies.
- The way of collecting the data is different for different the studies. The samples size and the frequencies were not the same for all considered studies. With regard to the data source, some studies used ASHRAE projects, part of them counted on the sensors, and others used historical records of the same buildings that were under

study. Moreover, some studies utilized IoT sensors, BMS, or BAS to obtain their data, as well as to control the system.

- In the literature considered that used sensors as a data source, it has been noticed that no study has suggested management or technical procedures in the case of unavailability of that source at a particular building.
- Using the PI of operational parameters may not be effective in fault diagnosis.
- The buildings management software, such as BMS, cannot detect all the faults. In contrast, faults predictions and system control can be improved by using them alongside ML methods.

4.3. Predictive Tools and Control

This subsection gives a swift overview on PdM tools that were applied within the considered studies and highlights the major findings of the same as per the below points:

- The methods mostly used are a simulation model, SVM, DT, and ANN. They are furnished in Table 4, along with others which are not used so commonly. The said table showcases which method is applied for which CWS component.
- There is a clear difference of views on predicting sensors' faults by using PCA. Some studies indicated that the sensors' FDD can be enhanced by using the said technique, while some other studies argued that PCA is not very efficient at predicting complex sensors' faults, due to the feebleness of one of the PCA parts, which is called Q-statistic plot; however, these studies did not present valid justifications to support this argument.
- Some studies include a comparison between several ML tools from the accuracy point of view. In most of them, DT and ANN scored the highest accuracy percentage in predicting the faults. Generally, all ML techniques showed good accuracies in predicting and diagnosing the faults.
- Combining ML techniques in PdM applications such as ARX with SVM showed positive outcomes in predicting CWS faults.
- DSM, which is one of the state-of-the-art control strategies, showed a significant impact on energy savings.
- To ensure an excellent ML model, it is advised to clean some CWS's critical parts such as the AHU's fan scroll and chiller's condenser water tubes before collecting the data.
- Having an excellent ML model would be challenging if the data sample size is inadequate.
- The literature showed that the proposed PdM programs ended with testing the ML model. Moreover, no management solutions were provided for the addressed faults.

4.4. Research Gaps

Based on the RQs, the research gaps found in this literature review from engineering management point of view can be summarized into three parts. The first part is on how the faults are identified and described, the second part is on data collection and the frequency of training the chosen model, and the third part is to what extent the proposed PdM programs are made. Hereunder is the description of the said gaps:

- The literature did not have the same faults and was concentrated only on selected faults, as some faults were either not stated/mentioned or were not fully described.
- The current literature did not specify how the data were collected or justify the period or the frequency of the collected data, and it was limited to testing the model and not controlling it.
- The suggested programs/frameworks/models contained either no or inconclusive solutions for the said faults from the management point of view, as they ended at how to detect/predict the faults. Moreover, the said programs did not study/cover the whole system comprehensively.

Table 4. Summary of CWS's PdM tools across the literature (chiller, CH; cooling tower, CT; pump, P; terminal unit, TU).

Technique	Component(s)	Technique	Component(s)	Technique	Component(s)
ANN	CH, P, and TU	DR	CH	FLC	CH and TU
KF	CH and CT	BN	CH, CT, and TU	XAI	CH
KNN	CH, CT, and TU	LMIF	CH and TU	VRC	CH
SVM	CH, CT, P, and TU	MSVM	CH	GRU	CH
RF	CH and TU	DT	CH and TU	MIP	CH
Adam-BPNN	CH and P	QDA	CH	DBN	CH and TU
PCA	CH, CT, P, and TU	AB	CH	DNN	CH
PLS	CH	LR	CH	MLP	CH and CT
Fuzzy Modeling	CH and TU	SVR	CH and TU	Simulation	CH, CT, P, and TU
MLB	CH	DE	CH	GB	CT
HQS	CH and CT	MLReg	CH	GSPN	CT
Regression	CH, CT, and P	Kriging	CH	BPNN	CT and TU
Clustering	CH and TU	RBF	CH and TU	PSO	CH
RDPNN	CH	BT	CH	DL	CH and CT
EWMA	CH and TU	SLReg	CH	SNN	TU
BBN	CH and TU	UKF	CH	SSL	P and TU
SVDD	CH	SPC	CH and CT	WPCA	TU
CWGAN	CH	GMMR	CH	MCMC	TU
EKF	CH	ARX	CH and TU	GA	TU
ROSVM	CH	MCS	CH, P, and TU	GAN	TU
TFDK	CH	Decoupling	CH	AFT	TU
LDA	CH	FIS	CH	GRNN	TU
1D-CNN	CH	LMANN	CH and TU	WEKA	TU
k-means clustering	CH	Virtual Sensor	CH and TU	HMM	TU
WNN	TU	H2O	CH and TU	RT	TU
KPCA	TU	ARM	TU	RNN	TU
SMO	TU	TARM	TU	FCD	TU
DAG	TU	ERCE	TU	DSM	CH and TU

5. Conclusions

This article aimed to answer two RQs, which are addressing the literature from an engineering management point of view. The first RQ asked about the arrangements or the preparations to identify the faults, and the second RQ is considered as a further activity of the first RQ, as it is asking about the tools that were used in line with Industry 4.0/Quality 4.0. This article implemented an SLR that includes four stages, and then it highlighted the studies performed post-1999 on PdM in CBs and explored many frameworks, programs, and methods. It also highlighted the gaps found in the literature from an engineering management point of view. Following SLR, especially the second stage, it was found that there is no research covering the entire CWS. The considered studies are covering either one, two, or three components only. Therefore, this article was made to focus on all four components. From a maintenance management point of view, it is recommended for the researchers to study the whole system or at least to give more research attention to the cooling towers and pumps.

As per the first research gap, which is mentioned in the previous section, it is concluded that CWS may have some other different faults than those that were studied in the literature, and, therefore, it is recommended that researchers perform further studies to explore more faults. With regard to the second research gap, it is recommended that the researchers verify the required sample size, as well as the frequencies of data readings/record. For data collection, it is recommended that the researchers create their own dataset from the same building in order to obtain more accurate data about the current operational situation and to avoid depending on historical record from different building or project. In addition, it is highly recommended to make a control plan after testing the ML model in order to keep continuous tracking on the building's O&M.

The suggested future course of action of this article is to make a PdM 4.0 program by using mixed methods. The first step would be to design a survey and share it with the facility management professionals at CBs located in the study area. The content of the said survey should list the faults found in the literature and ask the participants to select the ones they find in their site and add more faults, if any. This will enrich the knowledge of the research community with more faults about CWS. Moreover, it should include asking the participant about the frequency of such faults' occurrence (minimum and maximum duration). The survey should also let the participants prescribe solutions to the faults. This article believes the operational parameters that are reflecting the health condition of CWS components are leaving temperatures for chillers, as well as for cooling towers, the pressure for pumps, and space temperature for terminal units. This argument has been supported by the literature, along with practical experience. So, the survey should include a verification of this belief from the participants. Furthermore, it is recommended to conduct a pilot study to ensure that the said survey is valid.

The second step would be to make a methodological framework that contains three parts, as follows:

1. Setup Part: This part should include the following points:
   - Showing how to understand the drawings of CWS at a particular building.
   - Proposing a format to extract/identify the number of components of CWS from the said drawings (number of chillers, number of cooling towers, number of primary/secondary pumps, and number of terminal units).
   - Introducing a management procedure on checking the availability of reading tools, such as sensors, building management software, etc., and how to deal with their unavailability situation.
   - Proposing a way of formulating a team that will be responsible for data collection.
2. ML Part: This part should explain how the proposed predictive maintenance program is in line with Industry 4.0/Quality 4.0. The first task in the ML part is data collection. The readings should be from the operational parameters and should be taken based on the minimum and maximum frequencies that should come from the said survey. A check sheet is proposed to be used in recording the data at the building under study. All of these points should be formatted in a methodological manner from a management point of view. Python is suggested to be used in the next step of the ML part. For the said step, a procedure should be proposed on how to insert the data, train the ML model, and test the same model for each component of CWS.
3. Quality-Control Part: Here, a control plan for CWS through a tabulated format should be presented. The outcomes of the survey (frequencies) should again be used here as part of the said control plan. The other survey's outcome (solutions for each fault) should be summarized here, as well, in order to create a comprehensive management program for PdM 4.0 for CWS.

Author Contributions: Conceptualization, M.A. and K.M.; methodology, M.A. and A.A.; investigation, M.A.; writing—review and editing, M.A. and A.A.; supervision, K.M. All authors have read and agreed to the published version of the manuscript.

Funding: This research received no external funding.

Institutional Review Board Statement: Not applicable.

Informed Consent Statement: Not applicable.

Acknowledgments: Many thanks to the DMEM department at the University of Strathclyde, especially Anja Maier, Jorn Mehnen, Xichun Luo, Ian Whitfield, Laura Hay, and Gillian Eadie for their guidance, advice, and support. A special thanks also goes to Alfaisal University President (H.E. Mohammed Alhayaza), VP External relation and advancement (H.R.H. Princess Maha Bint Mishari Bin Abdulaziz AlSaud), VP Admin and Finance (Khaled AlKattan), and Acting Dean of College of Engineering (Muhammad Anan) for their unlimited support. In addition, a big thanks goes to O&M Manager at Alfaisal University (Eng. Kaleemoddin Ahmed) for his great assistance during this research.

Conflicts of Interest: The authors declare no conflict of interest.

References

1. Center for Sustainable Systems, University of Michigan. Commercial Buildings Factsheet. Pub. No. CSS05-05. 2020. Available online: https://css.umich.edu/publications/factsheets/built-environment/commercial-buildings-factsheet (accessed on 23 March 2022).
2. Kumar, P.; Martani, C.; Morawska, L.; Norford, L.; Choudhary, R.; Bell, M.; Leach, M. Indoor air quality and energy management through real-time sensing in commercial buildings. *Energy Build.* **2016**, *111*, 145–153. [CrossRef]
3. Monge-Barrio, A.; Gutiérrez, A.S.O. *Passive Energy Strategies for Mediterranean Residential Buildings: Facing the Challenges of Climate Change and Vulnerable Populations*; Springer: Berlin/Heidelberg, Germany, 2018.
4. Yau, Y.; Hasbi, S. A review of climate change impacts on commercial buildings and their technical services in the tropics. *Renew. Sustain. Energy Rev.* **2013**, *18*, 430–441. [CrossRef]
5. Urge-Vorsatz, D.; Petrichenko, K.; Staniec, M.; Eom, J. Energy use in buildings in a long-term perspective. *Curr. Opin. Environ. Sustain.* **2013**, *5*, 141–151. [CrossRef]
6. Lai, J.H.; Man, C.S. Developing a performance evaluation scheme for engineering facilities in commercial buildings: State-of-the-art review. *Int. J. Strateg. Prop. Manag.* **2017**, *21*, 41–57. [CrossRef]
7. Gálvez, A.; Diez-Olivan, A.; Seneviratne, D.; Galar, D. Fault detection and RUL estimation for railway HVAC systems using a hybrid model-based approach. *Sustainability* **2021**, *13*, 6828. [CrossRef]
8. Janis, R.R.; Tao, W.K. *Mechanical and Electrical Systems in Buildings*, 6th ed.; Pearson/Prentice Hall: New York, NY, USA, 2019.
9. Porges, F. *HVAC Engineer's Handbook*, 11th ed.; Routledge: London, UK, 2020.
10. Swenson, S.D. *HVAC: Heating, Ventilating, and Air Conditioning*, 3rd ed.; American Technical Publishers: Orland Park, IL, USA, 2004.
11. Aswani, A.; Master, N.; Taneja, J.; Smith, V.; Krioukov, A.; Culler, D.; Tomlin, C. Identifying models of HVAC systems using semiparametric regression. In Proceedings of the 2012 American Control Conference (ACC), Montreal, QC, Canada, 27–29 June 2012; pp. 3675–3680.
12. Li, X.; Li, Y.; Seem, J.E.; Li, P. Dynamic modeling and self-optimizing operation of chilled water systems using extremum seeking control. *Energy Build.* **2013**, *58*, 172–182. [CrossRef]
13. Cho, J.; Kim, Y.; Koo, J.; Park, W. Energy-cost analysis of HVAC system for office buildings: Development of a multiple prediction methodology for HVAC system cost estimation. *Energy Build.* **2018**, *173*, 562–576. [CrossRef]
14. Dawson-Haggerty, S.; Ortiz, J.; Jiang, X.; Hsu, J.; Shankar, S.; Culler, D. Enabling green building applications. In Proceedings of the 6th Workshop on Hot Topics in Embedded Networked Sensors, Killarney, Ireland, 28–29 June 2010; pp. 1–5.
15. Greene, R.E.; Casey, J.M.; Williams, P.L. A proactive approach for managing indoor air quality. *J. Environ. Health* **1997**, *60*, 15–22.
16. Hassanain, M.A.; Al-Hammad, A.-M.; Fatayer, F. Assessment of defects in HVAC systems caused by lack of maintenance feedback to the design team. *Arch. Sci. Rev.* **2014**, *57*, 188–195. [CrossRef]
17. Sugarman, S.C. *HVAC Fundamentals*, 3rd ed.; River Publishers: New York, NY, USA, 2016.
18. Seyam, S. Types of HVAC systems. In *HVAC System*, 1st ed.; Kandelousi, M., Ed.; IntechOpen: London, UK, 2018; pp. 49–66.
19. Luo, M.; Xie, J.; Yan, Y.; Ke, Z.; Yu, P.; Wang, Z.; Zhang, J. Comparing machine learning algorithms in predicting thermal sensation using ASHRAE Comfort Database II. *Energy Build.* **2020**, *210*, 109776. [CrossRef]
20. Schiavon, S.; Melikov, A.K.; Sekhar, C. Energy analysis of the personalized ventilation system in hot and humid climates. *Energy Build.* **2009**, *42*, 699–707. [CrossRef]
21. Ding, J.; Yu, C.W.; Cao, S.-J. HVAC systems for environmental control to minimize the COVID-19 infection. *Indoor Built Environ.* **2020**, *29*. [CrossRef]
22. Aebischer, B.; Catenazzi, G.; Jakob, M. Impact of climate change on thermal comfort, heating and cooling energy demand in Europe. In Proceedings of the ECEEE Summer Study, La Colle-sur-Loup, France, 4–9 June 2007; pp. 859–870.
23. Colmenar-Santos, A.; de Lober LN, T.; Borge-Diez, D.; Castro-Gil, M. Solutions to reduce energy consumption in the management of large buildings. *Energy Build.* **2013**, *56*, 66–77. [CrossRef]
24. ASHRAE Handbook. Available online: www.ashrae.org (accessed on 13 July 2021).
25. Prajapati, A.; Bechtel, J.; Ganesan, S. Condition based maintenance: A survey. *J. Qual. Maint. Eng.* **2012**, *18*, 384–400. [CrossRef]
26. Levitt, J. *Complete Guide to Preventive and Predictive Maintenance*; Industrial Press Inc.: New York, NY, USA, 2011.
27. Selcuk, S. Predictive maintenance, its implementation and latest trends. *Proc. Inst. Mech. Eng. Part B J. Eng. Manuf.* **2017**, *231*, 1670–1679. [CrossRef]
28. Goriveau, R.; Medjaher, K.; Zerhouni, N. *From Prognostics and Health Systems Management to Predictive Maintenance 1. Monitoring and Prognostics*; ISTE Ltd. and Wiley: London, UK, 2016.
29. Mobley, R.K. *An Introduction to Predictive Maintenance*; Elsevier: Amsterdam, The Netherlands, 2002.
30. Verbert, K.; De Schutter, B.; Babuska, R. Timely condition-based maintenance planning for multi-component systems. *Reliab. Eng. Syst. Saf.* **2017**, *159*, 310–321. [CrossRef]

31. He, Y.; Gu, C.; Chen, Z.; Han, X. Integrated predictive maintenance strategy for manufacturing systems by combining quality control and mission reliability analysis. *Int. J. Prod. Res.* **2017**, *55*, 5841–5862. [CrossRef]
32. Wang, J.; Li, C.; Han, S.; Sarkar, S.; Zhou, X. Predictive maintenance based on eventlog analysis: A case study. *IBM J. Res. Dev.* **2017**, *61*, 11:121–11:132. [CrossRef]
33. Hemmerdinger, R. *Predictive Maintenance Strategy for Building Operations: A Better Approach*; Schneider Electric: Rueil-Malmaison, France, 2014.
34. Tiddens, W.; Braaksma, J.; Tinga, T. Exploring predictive maintenance applications in industry. *J. Qual. Maint. Eng.* **2020**, *28*, 68–85. [CrossRef]
35. Amruthnath, N.; Gupta, T. A Research Study on Unsupervised Machine Learning Algorithms for early Fault Detection in Predictive Maintenance. In Proceedings of the 2018 5th International Conference on Industrial Engineering and Applications (ICIEA), Singapore, 26–28 April 2018; IEEE: Piscataway, NJ, USA, 2018; pp. 355–361.
36. Huang, C.H.; Wang, C.H. Optimization of preventive maintenance for a multi-state degraded system by monitoring component performance. *J. Intell. Manuf.* **2016**, *27*, 1151–1170. [CrossRef]
37. Nguyen, K.T.; Medjaher, K. A new dynamic predictive maintenance framework using deep learning for failure prognostics. *Reliab. Eng. Syst. Saf.* **2019**, *188*, 251–262. [CrossRef]
38. Sharma, A.; Yadava, G.S.; Deshmukh, S.G. A literature review and future perspectives on maintenance optimization. *J. Qual. Maint. Eng.* **2011**, *17*, 5–25. [CrossRef]
39. Faccio, M.; Persona, A.; Sgarbossa, F.; Zanin, G. Industrial maintenance policy development: A quantitative framework. *Int. J. Prod. Econ.* **2014**, *147*, 85–93. [CrossRef]
40. Garg, A.; Deshmukh, S. Maintenance management: Literature review and directions. *J. Qual. Maint. Eng.* **2006**, *12*, 205–238. [CrossRef]
41. Ran, Y.; Zhou, X.; Lin, P.; Wen, Y.; Deng, R. A survey of predictive maintenance: Systems, purposes and approaches. *arXiv* **2019**, arXiv:1912.07383.
42. Chukwuekwe, D.O.; Schjoelberg, P.; Roedseth, H.; Stuber, A. Reliable, robust and resilient systems: Towards development of a predictive maintenance concept within the Industry 4.0 environment. In Proceedings of the EFNMS Euro Maintenance Conference, Athens, Greece, 31 May 2016.
43. Briner, R.B.; Denyer, D. Systematic review and evidence synthesis as a practice and scholarship tool. In *Oxford Handbook of Evidence-Based Management: Companies, Classrooms and Research*; Oxford University Press: Oxford, UK, 2012.
44. Kitchenham, B. *Procedures for Performing Systematic Reviews*; Keele University: Keele, UK, 2004; Volume 33, pp. 1–26.
45. Qiu, M.; Qiu, H.; Zeng, Y. *Research and Technical Writing for Science and Engineering*, 1st ed.; CRC Press: Boca Raton, FL, USA, 2022.
46. Leong, E.C.; Heah, C.L.H.; Ong, K.K.W. *Guide to Research Projects for Engineering Students: Planning, Writing and Presenting*; CRC Press: Boca Raton, FL, USA, 2015.
47. Aliyu, M.B. Efficiency of Boolean search strings for Information retrieval. *Am. J. Eng. Res.* **2017**, *6*, 216–222.
48. Zonta, T.; da Costa, C.A.; da Rosa Righi, R.; de Lima, M.J.; da Trindade, E.S.; Li, G.P. Predictive maintenance in the Industry 4.0: A systematic literature review. *Comput. Ind. Eng.* **2020**, *150*, 106889. [CrossRef]
49. Dalzochio, J.; Kunst, R.; Pignaton, E.; Binotto, A.; Sanyal, S.; Favilla, J.; Barbosa, J. Machine learning and reasoning for predictive maintenance in Industry 4.0: Current status and challenges. *Comput. Ind.* **2020**, *123*, 103298. [CrossRef]
50. Sajid, S.; Haleem, A.; Bahl, S.; Javaid, M.; Goyal, T.; Mittal, M. Data science applications for predictive maintenance and materials science in context to Industry 4.0. *Mater. Today Proc.* **2021**, *45*, 4898–4905. [CrossRef]
51. Dolatabadi, S.H.; Budinska, I. Systematic Literature Review Predictive Maintenance Solutions for SMEs from the Last Decade. *Machines* **2021**, *9*, 191. [CrossRef]
52. Inayat, I.; Salim, S.S.; Marczak, S.; Daneva, M.; Shamshirband, S. A systematic literature review on agile requirements engineering practices and challenges. *Comput. Hum. Behav.* **2015**, *51*, 915–929. [CrossRef]
53. Rueda, E.; Tassou, S.A.; Grace, I.N. Fault detection and diagnosis in liquid chillers. *Proc. Inst. Mech. Eng. Part E J. Process Mech. Eng.* **2005**, *219*, 117–125. [CrossRef]
54. Tassou, S.A.; Grace, I.N. Fault diagnosis and refrigerant leak detection in vapour compression refrigeration systems. *Int. J. Refrig.* **2005**, *28*, 680–688. [CrossRef]
55. Navarro-Esbri, J.; Torrella, E.; Cabello, R. A vapour compression chiller fault detection technique based on adaptative algorithms. Application to on-line refrigerant leakage detection. *Int. J. Refrig.* **2006**, *29*, 716–723. [CrossRef]
56. Han, H.; Zhang, Z.; Cui, X.; Meng, Q. Ensemble learning with member optimization for fault diagnosis of a building energy system. *Energy Build.* **2020**, *226*, 110351. [CrossRef]
57. Liu, R.; Zhang, Y.; Li, Z. Leakage Diagnosis of Air Conditioning Water System Networks Based on an Improved BP Neural Network Algorithm. *Buildings* **2022**, *12*, 610. [CrossRef]
58. Hu, Y.; Chen, H.; Li, G.; Li, H.; Xu, R.; Li, J. A statistical training data cleaning strategy for the PCA-based chiller sensor fault detection, diagnosis and data reconstruction method. *Energy Build.* **2016**, *112*, 270–278. [CrossRef]
59. Xiao, F.; Wang, S.; Zhang, J. A diagnostic tool for online sensor health monitoring in air-conditioning systems. *Autom. Constr.* **2006**, *15*, 489–503. [CrossRef]
60. Mao, Q.; Fang, X.; Hu, Y.; Li, G. Chiller sensor fault detection based on empirical mode decomposition threshold denoising and principal component analysis. *Appl. Therm. Eng.* **2018**, *144*, 21–30. [CrossRef]

61. Wang, S.; Cui, J. Sensor-fault detection, diagnosis and estimation for centrifugal chiller systems using principal-component analysis method. *Appl. Energy* **2005**, *82*, 197–213. [CrossRef]
62. Xu, X.; Xiao, F.; Wang, S. Enhanced chiller sensor fault detection, diagnosis and estimation using wavelet analysis and principal component analysis methods. *Appl. Therm. Eng.* **2008**, *28*, 226–237. [CrossRef]
63. Wang, S.; Zhou, Q.; Xiao, F. A system-level fault detection and diagnosis strategy for HVAC systems involving sensor faults. *Energy Build.* **2010**, *42*, 477–490. [CrossRef]
64. Hu, Y.; Li, G.; Chen, H.; Li, H.; Liu, J. Sensitivity analysis for PCA-based chiller sensor fault detection. *Int. J. Refrig.* **2016**, *63*, 133–143. [CrossRef]
65. Hu, Y.; Chen, H.; Xie, J.; Yang, X.; Zhou, C. Chiller sensor fault detection using a self-adaptive principal component analysis method. *Energy Build.* **2012**, *54*, 252–258. [CrossRef]
66. Li, G.; Hu, Y.; Chen, H.; Li, H.; Hu, M.; Guo, Y.; Shi, S.; Hu, W. A sensor fault detection and diagnosis strategy for screw chiller system using support vector data description-based D-statistic and DV-contribution plots. *Energy Build.* **2016**, *133*, 230–245. [CrossRef]
67. Choi, K.; Namburu, S.M.; Azam, M.S.; Luo, J.; Pattipati, K.R.; Patterson-Hine, A. Fault diagnosis in HVAC chillers. *IEEE Instrum. Meas. Mag.* **2005**, *8*, 24–32. [CrossRef]
68. Namburu, S.M.; Azam, M.S.; Luo, J.; Choi, K.; Pattipati, K.R. Data-Driven Modeling, Fault Diagnosis and Optimal Sensor Selection for HVAC Chillers. *IEEE Trans. Autom. Sci. Eng.* **2007**, *4*, 469–473. [CrossRef]
69. Schein, J.; Bushby, S.T. A hierarchical rule-based fault detection and diagnostic method for HVAC systems. *HVACR Res.* **2006**, *12*, 111–125. [CrossRef]
70. Zhou, Q.; Wang, S.; Ma, Z. A model-based fault detection and diagnosis strategy for HVAC systems. *Int. J. Energy Res.* **2009**, *33*, 903–918. [CrossRef]
71. Zhou, Q.; Wang, S.; Xiao, F. A novel strategy for the fault detection and diagnosis of centrifugal chiller systems. *HVACR Res.* **2009**, *15*, 57–75. [CrossRef]
72. Comstock, M.C.; Braun, J.E.; Groll, E.A. The Sensitivity of Chiller Performance to Common Faults. *HVACR Res.* **2001**, *7*, 263–279. [CrossRef]
73. Han, H.; Gu, B.; Hong, Y.; Kang, J. Automated FDD of multiple-simultaneous faults (MSF) and the application to building chillers. *Energy Build.* **2011**, *43*, 2524–2532. [CrossRef]
74. Shi, Z.; O'Brien, W. Development and implementation of automated fault detection and diagnostics for building systems: A review. *Autom. Constr.* **2019**, *104*, 215–229. [CrossRef]
75. Mirnaghi, M.S.; Haghighat, F. Fault detection and diagnosis of large-scale HVAC systems in buildings using data-driven methods: A comprehensive review. *Energy Build.* **2020**, *229*, 110492. [CrossRef]
76. Ma, Z.; Wang, S. Online fault detection and robust control of condenser cooling water systems in building central chiller plants. *Energy Build.* **2011**, *43*, 153–165. [CrossRef]
77. Yu, F.; Chan, K. Improved energy management of chiller systems by multivariate and data envelopment analyses. *Appl. Energy* **2012**, *92*, 168–174. [CrossRef]
78. Yu, F.; Chan, K. Assessment of operating performance of chiller systems using cluster analysis. *Int. J. Therm. Sci.* **2012**, *53*, 148–155. [CrossRef]
79. Zhao, X.; Yang, M.; Li, H. A virtual condenser fouling sensor for chillers. *Energy Build.* **2012**, *52*, 68–76. [CrossRef]
80. Magoulès, F.; Zhao, H.-X.; Elizondo, D. Development of an RDP neural network for building energy consumption fault detection and diagnosis. *Energy Build.* **2013**, *62*, 133–138. [CrossRef]
81. Zhao, Y.; Wang, S.; Xiao, F. A statistical fault detection and diagnosis method for centrifugal chillers based on exponentially-weighted moving average control charts and support vector regression. *Appl. Therm. Eng.* **2013**, *51*, 560–572. [CrossRef]
82. Zhao, Y.; Xiao, F.; Wang, S. An intelligent chiller fault detection and diagnosis methodology using Bayesian belief network. *Energy Build.* **2013**, *57*, 278–288. [CrossRef]
83. Zhao, Y.; Wang, S.; Xiao, F. Pattern recognition-based chillers fault detection method using Support Vector Data Description (SVDD). *Appl. Energy* **2013**, *112*, 1041–1048. [CrossRef]
84. Yan, K.; Ma, L.; Dai, Y.; Shen, W.; Ji, Z.; Xie, D. Cost-sensitive and sequential feature selection for chiller fault detection and diagnosis. *Int. J. Refrig.* **2018**, *86*, 401–409. [CrossRef]
85. Yan, K.; Chong, A.; Mo, Y. Generative adversarial network for fault detection diagnosis of chillers. *Build. Environ.* **2020**, *172*, 106698. [CrossRef]
86. Yan, K.; Ji, Z.; Shen, W. Online fault detection methods for chillers combining extended kalman filter and recursive one-class SVM. *Neurocomputing* **2017**, *228*, 205–212. [CrossRef]
87. Li, D.; Zhou, Y.; Hu, G.; Spanos, C.J. Fault detection and diagnosis for building cooling system with a tree-structured learning method. *Energy Build.* **2016**, *127*, 540–551. [CrossRef]
88. Li, G.; Hu, Y.; Chen, H.; Shen, L.; Li, H.; Hu, M.; Liu, J.; Sun, K. An improved fault detection method for incipient centrifugal chiller faults using the PCA-R-SVDD algorithm. *Energy Build.* **2016**, *116*, 104–113. [CrossRef]
89. Li, D.; Hu, G.; Spanos, C.J. A data-driven strategy for detection and diagnosis of building chiller faults using linear discriminant analysis. *Energy Build.* **2016**, *128*, 519–529. [CrossRef]

90. Wang, Z.; Dong, Y.; Liu, W.; Ma, Z. A novel fault diagnosis approach for chillers based on 1-D convolutional neural network and gated recurrent unit. *Sensors* **2020**, *20*, 2458. [CrossRef]
91. Wang, Z.; Wang, Z.; He, S.; Gu, X.; Yan, Z.F. Fault detection and diagnosis of chillers using Bayesian network merged distance rejection and multi-source non-sensor information. *Appl. Energy* **2017**, *188*, 200–214. [CrossRef]
92. Li, D.; Zhou, Y.; Hu, G.; Spanos, C.J. Fusing system configuration information for building cooling plant Fault Detection and severity level identification. In Proceedings of the 2016 IEEE International Conference on Automation Science and Engineering (CASE), Fort Worth, TX, USA, 21–25 August 2016; IEEE: Piscataway, NJ, USA, 2016; pp. 1319–1325.
93. Tran, D.A.T.; Chen, Y.; Ao, H.L.; Cam, H.N.T. An enhanced chiller FDD strategy based on the combination of the LSSVR-DE model and EWMA control charts. *Int. J. Refrig.* **2016**, *72*, 81–96. [CrossRef]
94. Tran, D.A.T.; Chen, Y.; Jiang, C. Comparative investigations on reference models for fault detection and diagnosis in centrifugal chiller systems. *Energy Build.* **2016**, *133*, 246–256. [CrossRef]
95. Tahmasebi, M.; Eaton, K.; Nabil Nassif, P.E.; Talib, R. Integrated Machine Learning Modeling and Fault Detection Approach for Chilled Water Systems. In Proceedings of the International Conference on Artificial Intelligence (ICAI), The Steering Committee of The World Congress in Computer Science, Computer Engineering and Applied Computing (WorldComp), Madrid, Spain, 6–9 November 2019; pp. 67–71.
96. Zhao, X.; Yang, M.; Li, H. Field implementation and evaluation of a decoupling-based fault detection and diagnostic method for chillers. *Energy Build.* **2014**, *72*, 419–430. [CrossRef]
97. Zhao, X. Lab test of three fault detection and diagnostic methods' capability of diagnosing multiple simultaneous faults in chillers. *Energy Build.* **2015**, *94*, 43–51. [CrossRef]
98. Bonvini, M.; Sohn, M.D.; Granderson, J.; Wetter, M.; Piette, M.A. Robust on-line fault detection diagnosis for HVAC components based on nonlinear state estimation techniques. *Appl. Energy* **2014**, *124*, 156–166. [CrossRef]
99. Sun, B.; Luh, P.B.; Jia, Q.-S.; O'Neill, Z.; Song, F. Building Energy Doctors: An SPC and Kalman Filter-Based Method for System-Level Fault Detection in HVAC Systems. *IEEE Trans. Autom. Sci. Eng.* **2013**, *11*, 215–229. [CrossRef]
100. Karami, M.; Wang, L. Fault detection and diagnosis for nonlinear systems: A new adaptive Gaussian mixture modeling approach. *Energy Build.* **2018**, *166*, 477–488. [CrossRef]
101. Au-Yong, C.P.; Ali, A.S.; Ahmad, F. Improving occupants' satisfaction with effective maintenance management of HVAC system in office buildings. *Autom. Constr.* **2014**, *43*, 31–37. [CrossRef]
102. Alonso, S.; Morán, A.; Prada, M.Á.; Reguera, P.; Fuertes, J.J.; Domínguez, M. A data-driven approach for enhancing the efficiency in chiller plants: A hospital case study. *Energies* **2019**, *12*, 827. [CrossRef]
103. Yan, K.; Shen, W.; Mulumba, T.; Afshari, A. ARX model based fault detection and diagnosis for chillers using support vector machines. *Energy Build.* **2014**, *81*, 287–295. [CrossRef]
104. McIntosh, I.B.; Mitchell, J.W.; Beckman, W.A. Fault detection and diagnosis in chillers—Part I: Model development and application/Discussion. *ASHRAE Trans.* **2000**, *106*, 268.
105. Miyata, S.; Akashi, Y.; Lim, J.; Kuwahara, Y. Fault detection in HVAC systems using a distribution considering uncertainties. *E3S Web Conf.* **2019**, *111*, 05010. [CrossRef]
106. Liao, Y.; Sun, Y.; Huang, G. Robustness analysis of chiller sequencing control. *Energy Convers. Manag.* **2015**, *103*, 180–190. [CrossRef]
107. Beghi, A.; Brignoli, R.; Cecchinato, L.; Menegazzo, G.; Rampazzo, M.; Simmini, F. Data-driven Fault Detection and Diagnosis for HVAC water chillers. *Control Eng. Pr.* **2016**, *53*, 79–91. [CrossRef]
108. Kocyigit, N. Fault and sensor error diagnostic strategies for a vapor compression refrigeration system by using fuzzy inference systems and artificial neural network. *Int. J. Refrig.* **2015**, *50*, 69–79. [CrossRef]
109. Gao, L.; Li, D.; Li, D.; Yao, L.; Liang, L.; Gao, Y. A Novel Chiller Sensors Fault Diagnosis Method Based on Virtual Sensors. *Sensors* **2019**, *19*, 3013. [CrossRef]
110. Cheng, J.C.; Chen, W.; Chen, K.; Wang, Q. Data-driven predictive maintenance planning framework for MEP components based on BIM and IoT using machine learning algorithms. *Autom. Constr.* **2020**, *112*, 103087. [CrossRef]
111. Escobar, L.M.; Aguilar, J.; Garces-Jimenez, A.; De Mesa, J.A.G.; Gomez-Pulido, J.M. Advanced Fuzzy-Logic-Based Context-Driven Control for HVAC Management Systems in Buildings. *IEEE Access* **2020**, *8*, 16111–16126. [CrossRef]
112. Srinivasan, S.; Arjunan, P.; Jin, B.; Sangiovanni-Vincentelli, A.L.; Sultan, Z.; Poolla, K. Explainable AI for Chiller Fault-Detection Systems: Gaining Human Trust. *Computer* **2021**, *54*, 60–68. [CrossRef]
113. Hu, R.; Granderson, J.; Auslander, D.; Agogino, A. Design of machine learning models with domain experts for automated sensor selection for energy fault detection. *Appl. Energy* **2018**, *235*, 117–128. [CrossRef]
114. Liu, J.; Li, G.; Chen, H.; Wang, J.; Guo, Y.; Li, J. A robust online refrigerant charge fault diagnosis strategy for VRF systems based on virtual sensor technique and PCA-EWMA method. *Appl. Therm. Eng.* **2017**, *119*, 233–243. [CrossRef]
115. Luo, X.J.; Fong, K.F.; Sun, Y.J.; Leung, M.K.H. Development of clustering-based sensor fault detection and diagnosis strategy for chilled water system. *Energy Build.* **2019**, *186*, 17–36. [CrossRef]
116. Luo, X.; Fong, K. Novel pattern recognition-enhanced sensor fault detection and diagnosis for chiller plant. *Energy Build.* **2020**, *228*, 110443. [CrossRef]
117. Thieblemont, H.; Haghighat, F.; Ooka, R.; Moreau, A. Predictive control strategies based on weather forecast in buildings with energy storage system: A review of the state-of-the art. *Energy Build.* **2017**, *153*, 485–500. [CrossRef]

118. Arteconi, A.; Hewitt, N.; Polonara, F. State of the art of thermal storage for demand-side management. *Appl. Energy* **2012**, *93*, 371–389. [CrossRef]

119. Wu, Y.; Maravelias, C.T.; Wenzel, M.J.; ElBsat, M.N.; Turney, R.T. Predictive maintenance scheduling optimization of building heating, ventilation, and air conditioning systems. *Energy Build.* **2021**, *231*, 110487. [CrossRef]

120. Li, P.; Anduv, B.; Zhu, X.; Jin, X.; Du, Z. Across working conditions fault diagnosis for chillers based on IoT intelligent agent with deep learning model. *Energy Build.* **2022**, *268*, 112188. [CrossRef]

121. Motomura, A.; Miyata, S.; Adachi, S.; Akashi, Y.; Lim, J.; Tanaka, K.; Kuwahara, Y. Fault evaluation process in HVAC system for decision making of how to respond to system faults. *IOP Conf. Ser. Earth Environ. Sci.* **2019**, *294*, 012054. [CrossRef]

122. Motomura, A.; Miyata, S.; Akashi, Y.; Lim, J.; Tanaka, K.; Tanaka, S.; Kuwahara, Y. Model-based analysis and evaluation of sensor faults in heat source system. *IOP Conf. Ser. Earth Environ. Sci.* **2019**, *238*, 012036. [CrossRef]

123. Sulaiman, N.A.; Abdullah, M.P.; Abdullah, H.; Zainudin, M.N.S.; Yusop, A.M. Fault detection for air conditioning system using machine learning. *IAES Int. J. Artif. Intell.* **2020**, *9*, 109. [CrossRef]

124. Ng, K.H.; Yik, F.W.H.; Lee, P.; Lee, K.K.Y.; Chan, D.C.H. Bayesian method for HVAC plant sensor fault detection and diagnosis. *Energy Build.* **2020**, *228*, 110476. [CrossRef]

125. Harasty, S.; Böttcher, A.D.; Lambeck, S. Using Artificial Neural Networks for Indoor Climate Control in the Field of Preventive Conservation. *E3S Web Conf.* **2019**, *111*, 04054. [CrossRef]

126. Ahn, B.C.; Mitchell, J.W.; McIntosh, I.B. Model-based fault detection and diagnosis for cooling towers/Discussion. *ASHRAE Trans.* **2001**, *107*, 839.

127. Chew, M.Y.L.; Yan, K. Enhancing Interpretability of Data-Driven Fault Detection and Diagnosis Methodology with Maintainability Rules in Smart Building Management. *J. Sens.* **2022**, *2022*, 1–48. [CrossRef]

128. Khan, J.-U.; Zubair, S.M. A study of fouling and its effects on the performance of counter flow wet cooling towers. *Proc. Inst. Mech. Eng. Part E J. Process. Mech. Eng.* **2004**, *218*, 43–51. [CrossRef]

129. Jain, P.; Pistikopoulos, E.N.; Mannan, M.S. Process resilience analysis based data-driven maintenance optimization: Application to cooling tower operations. *Comput. Chem. Eng.* **2019**, *121*, 27–45. [CrossRef]

130. Melani, A.H.; Murad, C.A.; Netto, A.C.; Souza, G.F.; Nabeta, S.I. Maintenance Strategy Optimization of a Coal-Fired Power Plant Cooling Tower through Generalized Stochastic Petri Nets. *Energies* **2019**, *12*, 1951. [CrossRef]

131. Aguilar, J.; Garces, A.; Avendaño, A.; Macias, F.; White, C.; Gomez-Pulido, J.; De Mesa, J.G.; Garces-Jimenez, A. An Autonomic Cycle of Data Analysis Tasks for the Supervision of HVAC Systems of Smart Building. *Energies* **2020**, *13*, 3103. [CrossRef]

132. Piot, S.; Lancon, H. New Tools for the Monitoring of Cooling Towers. In Proceedings of the 6th European Workshop on Structural Health Monitoring-Tu, Dresden, Germany, 3–6 July 2012; Volume 4.

133. Hashemian, H. Wireless sensors for predictive maintenance of rotating equipment in research reactors. *Ann. Nucl. Energy* **2011**, *38*, 665–680. [CrossRef]

134. Sun, L.; Jia, H.; Jin, H.; Li, Y.; Hu, J.; Li, C. Partial Fault Detection of Cooling Tower in Building HVAC System. In Proceedings of the International Conference on Smart City and Intelligent Building, Hefei, China, 15–16 September 2018; Springer: Singapore, 2018; pp. 231–240.

135. Xu, Z.T.; Mao, J.C.; Pan, Y.Q.; Huang, Z.Z. Performance Prediction of Mechanical Draft Wet Cooling Tower Using Artificial Neural Network. *Adv. Mater. Res.* **2014**, *1070–1072*, 1994–1997. [CrossRef]

136. Karim, J.A.; Lanyau, T.; Maskin, M.; Anuar, M.A.S.; Soh, A.C.; Rahman, R.Z.A. Preliminary study on fault detection using artificial neural network for water-cooled reactors. *J. Mech. Eng. Sci.* **2020**, *14*, 7469–7480. [CrossRef]

137. Ma, Z.; Wang, S. An optimal control strategy for complex building central chilled water systems for practical and real-time applications. *Build. Environ.* **2009**, *44*, 1188–1198. [CrossRef]

138. Yang, C.; Shen, W.; Chen, Q.; Gunay, B. Toward Failure mode and effect analysis for heating, ventilation and air-conditioning. In Proceedings of the 20th IEEE International Conference on Computer Supported Cooperative Work in Design (CSCWD 2017), Wellington, New Zealand, 26–28 April 2017; pp. 409–413.

139. Yuan, J.; Liu, X. Semi-supervised learning and condition fusion for fault diagnosis. *Mech. Syst. Signal Process.* **2013**, *38*, 615–627. [CrossRef]

140. Bouabdallaoui, Y.; Lafhaj, Z.; Yim, P.; Ducoulombier, L.; Bennadji, B. Predictive Maintenance in Building Facilities: A Machine Learning-Based Approach. *Sensors* **2021**, *21*, 1044. [CrossRef] [PubMed]

141. Liang, J.; Du, R. Model-based fault detection and diagnosis of HVAC systems using support vector machine method. *Int. J. Refrig.* **2007**, *30*, 1104–1114. [CrossRef]

142. Ding, S. *(2005) Model-Based Fault Diagnosis in Dynamic Systems Using Identification Techniques*; Simani, S., Fantuzzi, C., Patton, R.J., Eds.; Springer: London, UK, 2003; 282p, ISBN 1-85233-685-4.

143. Andriamamonjy, A.; Saelens, D.; Klein, R. An auto-deployed model-based fault detection and diagnosis approach for Air Handling Units using BIM and Modelica. *Autom. Constr.* **2018**, *96*, 508–526. [CrossRef]

144. Alavi, H.; Forcada, N. User-Centric BIM-Based Framework for HVAC Root-Cause Detection. *Energies* **2022**, *15*, 3674. [CrossRef]

145. Bruton, K.; Raftery, P.; Kennedy, B.; Keane, M.; Sullivan, D.O. Review of automated fault detection and diagnostic tools in air handling units. *Energy Effic.* **2013**, *7*, 335–351. [CrossRef]

146. Bruton, K.; Raftery, P.; O'Donovan, P.; Aughney, N.; Keane, M.; Sullivan, D.O. Development and alpha testing of a cloud based automated fault detection and diagnosis tool for Air Handling Units. *Autom. Constr.* **2014**, *39*, 70–83. [CrossRef]

147. Sittón-Candanedo, I.; Hernández-Nieves, E.; Rodríguez-González, S.; de la Prieta-Pintado, F. Fault predictive model for HVAC Systems in the context of Industry 4.0. *Inf. Netw.* **2018**, *9*, 10.
148. Chaudhuri, T.; Zhai, D.; Soh, Y.C.; Li, H.; Xie, L. Thermal comfort prediction using normalized skin temperature in a uniform built environment. *Energy Build.* **2017**, *159*, 426–440. [CrossRef]
149. Cho, S.-H.; Hong, Y.-J.; Kim, W.-T.; Zaheer-Uddin, M. Multi-fault detection and diagnosis of HVAC systems: An experimental study. *Int. J. Energy Res.* **2005**, *29*, 471–483. [CrossRef]
150. Cho, S.-H.; Yang, H.-C.; Zaheer-Uddin, M.; Ahn, B.-C. Transient pattern analysis for fault detection and diagnosis of HVAC systems. *Energy Convers. Manag.* **2005**, *46*, 3103–3116. [CrossRef]
151. Gunay, B.; Torabi, N.; Schillinglaw, S. Classification of sequencing logic faults in multiple zone air handling units: A review and case study. *Sci. Technol. Built Environ.* **2022**, *28*, 577–594. [CrossRef]
152. Norford, L.K.; Wright, J.A.; Buswell, R.A.; Luo, D.; Klaassen, C.J.; Suby, A. Demonstration of Fault Detection and Diagnosis Methods for Air-Handling Units. *HVACR Res.* **2002**, *8*, 41–71. [CrossRef]
153. Li, T.; Deng, M.; Zhao, Y.; Zhang, X.; Zhang, C. An air handling unit fault isolation method by producing additional diagnostic information proactively. *Sustain. Energy Technol. Assess.* **2021**, *43*, 100953. [CrossRef]
154. Gao, T.; Boguslawski, B.; Marié, S.; Béguery, P.; Thebault, S.; Lecoeuche, S. Data mining and data-driven modelling for Air Handling Unit fault detection. *E3S Web Conf.* **2019**, *111*, 05009. [CrossRef]
155. Deshmukh, S.; Samouhos, S.; Glicksman, L.; Norford, L. Fault detection in commercial building VAV AHU: A case study of an academic building. *Energy Build.* **2019**, *201*, 163–173. [CrossRef]
156. Piscitelli, M.S.; Mazzarelli, D.M.; Capozzoli, A. Enhancing operational performance of AHUs through an advanced fault detection and diagnosis process based on temporal association and decision rules. *Energy Build.* **2020**, *226*, 110369. [CrossRef]
157. Zhao, Y.; Wen, J.; Wang, S. Diagnostic Bayesian networks for diagnosing air handling units faults—Part II: Faults in coils and sensors. *Appl. Therm. Eng.* **2015**, *90*, 145–157. [CrossRef]
158. Zhao, Y.; Wen, J.; Xiao, F.; Yang, X.; Wang, S. Diagnostic Bayesian networks for diagnosing air handling units faults–part I: Faults in dampers, fans, filters and sensors. *Appl. Therm. Eng.* **2017**, *111*, 1272–1286. [CrossRef]
159. Yuwono, M.; Guo, Y.; Wall, J.; Li, J.; West, S.; Platt, G.; Su, S.W. Unsupervised feature selection using swarm intelligence and consensus clustering for automatic fault detection and diagnosis in heating ventilation and air conditioning systems. *Appl. Soft Comput.* **2015**, *34*, 402–425. [CrossRef]
160. Yan, R.; Ma, Z.; Zhao, Y.; Kokogiannakis, G. A decision tree based data-driven diagnostic strategy for air handling units. *Energy Build.* **2016**, *133*, 37–45. [CrossRef]
161. Yan, K.; Zhong, C.; Ji, Z.; Huang, J. Semi-supervised learning for early detection and diagnosis of various air handling unit faults. *Energy Build.* **2018**, *181*, 75–83. [CrossRef]
162. Yan, K.; Huang, J.; Shen, W.; Ji, Z. Unsupervised learning for fault detection and diagnosis of air handling units. *Energy Build.* **2019**, *210*, 109689. [CrossRef]
163. Tun, W.; Wong, J.K.-W.; Ling, S.-H. Hybrid Random Forest and Support Vector Machine Modeling for HVAC Fault Detection and Diagnosis. *Sensors* **2021**, *21*, 8163. [CrossRef]
164. Pourarian, S.; Wen, J.; Veronica, D.; Pertzborn, A.; Zhou, X.; Liu, R. A tool for evaluating fault detection and di-agnostic methods for fan coil units. *Energy Build.* **2017**, *136*, 151–160. [CrossRef]
165. Li, S.; Wen, J.; Zhou, X.; Klaassen, C.J. Development and validation of a dynamic air handling unit model, Part 2. *ASHRAE Transact.* **2010**, *116*, 57–73.
166. Li, S.; Wen, J. Application of pattern matching method for detecting faults in air handling unit system. *Autom. Constr.* **2014**, *43*, 49–58. [CrossRef]
167. Fan, C.; Liu, X.; Xue, P.; Wang, J. Statistical characterization of semi-supervised neural networks for fault detection and diagnosis of air handling units. *Energy Build.* **2021**, *234*, 110733. [CrossRef]
168. Mulumba, T.; Afshari, A.; Yan, K.; Shen, W.; Norford, L.K. Robust model-based fault diagnosis for air handling units. *Energy Build.* **2015**, *86*, 698–707. [CrossRef]
169. Zhao, Y.; Li, T.; Zhang, X.; Zhang, C. Artificial intelligence-based fault detection and diagnosis methods for building energy systems: Advantages, challenges and the future. *Renew. Sustain. Energy Rev.* **2019**, *109*, 85–101. [CrossRef]
170. Padilla, M.; Choinière, D. A combined passive-active sensor fault detection and isolation approach for air handling units. *Energy Build.* **2015**, *99*, 214–219. [CrossRef]
171. Qin, J.; Wang, S. A fault detection and diagnosis strategy of VAV air-conditioning systems for improved energy and control performances. *Energy Build.* **2005**, *37*, 1035–1048. [CrossRef]
172. Wang, S.; Xiao, F. AHU sensor fault diagnosis using principal component analysis method. *Energy Build.* **2004**, *36*, 147–160. [CrossRef]
173. Ranade, A.; Provan, G.; Mady, A.E.-D.; O'Sullivan, D. A computationally efficient method for fault diagnosis of fan-coil unit terminals in building Heating Ventilation and Air Conditioning systems. *J. Build. Eng.* **2019**, *27*, 100955. [CrossRef]
174. Shahnazari, H.; Mhaskar, P.; House, J.M.; Salsbury, T.I. Modeling and fault diagnosis design for HVAC systems using recurrent neural networks. *Comput. Chem. Eng.* **2019**, *126*, 189–203. [CrossRef]
175. Wang, H.; Chen, Y. A robust fault detection and diagnosis strategy for multiple faults of VAV air handling units. *Energy Build.* **2016**, *127*, 442–451. [CrossRef]

176. Wang, H.; Chen, Y.; Chan, C.W.; Qin, J.; Wang, J. Online model-based fault detection and diagnosis strategy for VAV air handling units. *Energy Build.* **2012**, *55*, 252–263. [CrossRef]

177. Xiao, F.; Zhao, Y.; Wen, J.; Wang, S. Bayesian network based FDD strategy for variable air volume terminals. *Autom. Constr.* **2014**, *41*, 106–118. [CrossRef]

178. Yang, H.; Cho, S.; Tae, C.S.; Zaheeruddin, M. Sequential rule based algorithms for temperature sensor fault detection in air handling units. *Energy Convers. Manag.* **2008**, *49*, 2291–2306. [CrossRef]

179. Yang, C.; Shen, W.; Chen, Q.; Gunay, B. A practical solution for HVAC prognostics: Failure mode and effects analysis in building maintenance. *J. Build. Eng.* **2018**, *15*, 26–32. [CrossRef]

180. Sánchez-Barroso, G.; Sanz-Calcedo, J.G. Application of Predictive Maintenance in Hospital Heating, Ventilation and Air Conditioning Facilities. *Emerg. Sci. J.* **2019**, *3*, 337–343. [CrossRef]

181. Yang, X.B.; Jin, X.Q.; Du, Z.M.; Zhu, Y.H.; Guo, Y.B. A hybrid model-based fault detection strategy for air handling unit sensors. *Energy Build.* **2013**, *57*, 132–143. [CrossRef]

182. Zhang, R.; Hong, T. Modeling of HVAC operational faults in building performance simulation. *Appl. Energy* **2017**, *202*, 178–188. [CrossRef]

183. Hosamo, H.H.; Svennevig, P.R.; Svidt, K.; Han, D.; Nielsen, H.K. A Digital Twin predictive maintenance framework of air handling units based on automatic fault detection and diagnostics. *Energy Build.* **2022**, *261*, 111988. [CrossRef]

184. Lee, W.Y.; House, J.M.; Kyong, N.H. Subsystem level fault diagnosis of a building's air-handling unit using general regression neural networks. *Appl. Energy* **2004**, *77*, 153–170. [CrossRef]

185. Gao, D.C.; Wang, S.; Shan, K.; Yan, C. A system-level fault detection and diagnosis method for low delta-T syndrome in the complex HVAC systems. *Appl. Energy* **2016**, *164*, 1028–1038. [CrossRef]

186. Choi, J.H.; Yeom, D. Development of the data-driven thermal satisfaction prediction model as a function of human physiological responses in a built environment. *Build. Environ.* **2019**, *150*, 206–218. [CrossRef]

187. Najeh, H.; Sing, M.; Ploix, S. Faults and Failures in Smart Buildings: A New Tool for Diagnosis. In *Towards Energy Smart Homes*; Ploix, S., Amayri, M., Bouguila, N., Eds.; Springer International Publishing: Midtown Manhattan, NY, USA, 2021; pp. 433–462.

188. Shaw, S.R.; Norford, L.K.; Luo, D.; Leeb, S.B. Detection and diagnosis of HVAC faults via electrical load monitoring. *HVACR Res.* **2002**, *8*, 13–40. [CrossRef]

189. Guo, Y.; Wall, J.; Li, J.; West, S. Real-Time HVAC Sensor Monitoring and Automatic Fault Detection System. In *Sensors for Everyday Life*; Mukhopadhyay, S., Ed.; Springer: Cham, Switzerland, 2017; Volume 23, pp. 39–54.

190. Holub, O.; Macek, K. HVAC simulation model for advanced diagnostics. In Proceedings of the 2013 IEEE 8th International Symposium on Intelligent Signal Processing, Funchal, Portugal, 16–18 September 2013; IEEE: Piscataway, NJ, USA, 2013; pp. 93–96.

191. Deshmukh, S.; Glicksman, L.; Norford, L. Case study results: Fault detection in air-handling units in buildings. *Adv. Build. Energy Res.* **2020**, *14*, 305–321. [CrossRef]

192. Ma, Z.; Ren, Y.; Xiang, X.; Turk, Z. Data-driven decision-making for equipment maintenance. *Autom. Constr.* **2020**, *112*, 103103. [CrossRef]

193. Gourabpasi, A.H.; Nik-Bakht, M. Knowledge Discovery by Analyzing the State of the Art of Data-Driven Fault Detection and Diagnostics of Building HVAC. *CivilEng* **2021**, *2*, 53. [CrossRef]

194. Dey, D.; Dong, B. A probabilistic approach to diagnose faults of air handling units in buildings. *Energy Build.* **2016**, *130*, 177–187. [CrossRef]

195. Du, Z.; Jin, X.; Yang, Y. Wavelet Neural Network-Based Fault Diagnosis in Air-Handling Units. *HVACR Res.* **2008**, *14*, 959–973. [CrossRef]

196. Du, Z.; Fan, B.; Jin, X.; Chi, J. Fault detection and diagnosis for buildings and HVAC systems using combined neural networks and subtractive clustering analysis. *Build. Environ.* **2014**, *73*, 1–11. [CrossRef]

197. Du, Z.; Fan, B.; Chi, J.; Jin, X. Sensor fault detection and its efficiency analysis in air handling unit using the combined neural networks. *Energy Build.* **2014**, *72*, 157–166. [CrossRef]

198. Dudzik, M.; Romanska-Zapala, A.; Bomberg, M. A neural network for monitoring and characterization of buildings with Environmental Quality Management, Part 1: Verification under steady state conditions. *Energies* **2020**, *13*, 3469. [CrossRef]

199. Elnour, M.; Himeur, Y.; Fadli, F.; Mohammedsherif, H.; Meskin, N.; Ahmad, A.M.; Hodorog, A. Neural net-work-based model predictive control system for optimizing building automation and management systems of sports facilities. *Appl. Energy* **2022**, *318*, 119153. [CrossRef]

200. Kim, W.; Braun, J.E. Development, implementation, and evaluation of a fault detection and diagnostics system based on integrated virtual sensors and fault impact models. *Energy Build.* **2020**, *228*, 110368. [CrossRef]

201. Lauro, F.; Moretti, F.; Capozzoli, A.; Khan, I.; Pizzuti, S.; Macas, M.; Panzieri, S. Building fan coil electric consumption analysis with fuzzy approaches for fault detection and diagnosis. *Energy Procedia* **2014**, *62*, 411–420. [CrossRef]

202. Li, S.; Wen, J. A model-based fault detection and diagnostic methodology based on PCA method and wavelet transform. *Energy Build.* **2014**, *68*, 63–71. [CrossRef]

203. Liu, Z.; Huang, Z.; Wang, J.; Yue, C.; Yoon, S. A novel fault diagnosis and self-calibration method for air-handling units using Bayesian Inference and virtual sensing. *Energy Build.* **2021**, *250*, 111293. [CrossRef]

204. Lo, C.H.; Chan, P.T.; Wong, Y.K.; Rad, A.B.; Cheung, K.L. Fuzzy-genetic algorithm for automatic fault detection in HVAC systems. *Appl. Soft Comput.* **2007**, *7*, 554–560. [CrossRef]
205. Montazeri, A.; Kargar, S.M. Fault detection and diagnosis in air handling using data-driven methods. *J. Build. Eng.* **2020**, *31*, 101388. [CrossRef]
206. Novikova, E.; Bestuzhev, M.; Shorov, A. The visualization-driven approach to the analysis of the HVAC data. In Proceedings of the International Symposium on Intelligent and Distributed Computing, Saint-Petersburg, Russia, 7–10 October 2019; Springer: Cham, Swizterland, 2019; pp. 547–552.
207. Parzinger, M.; Hanfstaengl, L.; Sigg, F.; Spindler, U.; Wellisch, U.; Wirnsberger, M. Residual Analysis of Predictive Modelling Data for Automated Fault Detection in Building's Heating, Ventilation and Air Conditioning Systems. *Sustainability* **2020**, *12*, 6758. [CrossRef]
208. Rafati, A.; Shaker, H.R.; Ghahghahzadeh, S. Fault Detection and Efficiency Assessment for HVAC Systems Using Non-Intrusive Load Monitoring: A Review. *Energies* **2022**, *15*, 341. [CrossRef]
209. Satta, R.; Cavallari, S.; Pomponi, E.; Grasselli, D.; Picheo, D.; Annis, C. A dissimilarity-based approach to predictive maintenance with application to HVAC systems. *arXiv* **2017**, arXiv:1701.03633.
210. Tehrani, M.M.; Beauregard, Y.; Rioux, M.; Kenne, J.P.; Ouellet, R. A Predictive Preference Model for Maintenance of a Heating Ventilating and Air Conditioning System. *IFAC-Pap.* **2015**, *48*, 130–135. [CrossRef]
211. Shakerian, S.; Jebelli, H.; Sitzabee, W.E. Improving the Prediction Accuracy of Data-Driven Fault Diagnosis for HVAC Systems by Applying the Synthetic Minority Oversampling Technique. In *Computing in Civil Engineering*; ASCE Library: Reston, VA, USA, 2021; pp. 90–97.
212. Sulaiman, N.A.; Othman, M.F.; Abdullah, H. Fuzzy Logic Control and Fault Detection in Centralized Chilled Water System. In Proceedings of the 2015 IEEE Symposium Series on Computational Intelligence, Cape Town, South Africa, 7–10 December 2015; pp. 8–13.
213. Thumati, B.T.; Feinstein, M.A.; Fonda, J.W.; Turnbull, A.; Weaver, F.J.; Calkins, M.E.; Jagannathan, S. An online model-based fault diagnosis scheme for HVAC systems. In Proceedings of the 2011 IEEE International Conference on Control Applications (CCA), Denver, CO, USA, 28–30 September 2011; IEEE: Piscataway, NJ, USA, 2011; pp. 70–75.
214. Turner, W.J.N.; Staino, A.; Basu, B. Residential HVAC fault detection using a system identification approach. *Energy Build.* **2017**, *151*, 1–17. [CrossRef]
215. Van Every, P.M.; Rodriguez, M.; Jones, C.B.; Mammoli, A.A.; Martínez-Ramón, M. Advanced detection of HVAC faults using unsupervised SVM novelty detection and Gaussian process models. *Energy Build.* **2017**, *149*, 216–224. [CrossRef]
216. Velibeyoglu, I.; Noh, H.Y.; Pozzi, M. A graphical approach to assess the detectability of multiple simultaneous faults in air handling units. *Energy Build.* **2018**, *184*, 275–288. [CrossRef]
217. Villa, V.; Bruno, G.; Aliev, K.; Piantanida, P.; Corneli, A.; Antonelli, D. Machine Learning Framework for the Sustainable Maintenance of Building Facilities. *Sustainability* **2022**, *14*, 681. [CrossRef]
218. Wijayasekara, D.; Linda, O.; Manic, M.; Rieger, C. Mining Building Energy Management System Data Using Fuzzy Anomaly Detection and Linguistic Descriptions. *IEEE Trans. Ind. Inform.* **2014**, *10*, 1829–1840. [CrossRef]
219. Yan, R.; Ma, Z.; Kokogiannakis, G.; Zhao, Y. A sensor fault detection strategy for air handling units using cluster analysis. *Autom. Constr.* **2016**, *70*, 77–88. [CrossRef]
220. Yu, Z.J.; Haghighat, F.; Fung, B.C.; Zhou, L. A novel methodology for knowledge discovery through mining associations between building operational data. *Energy Build.* **2012**, *47*, 430–440. [CrossRef]

Article

Fault Types and Frequencies in Predictive Maintenance 4.0 for Chilled Water System at Commercial Buildings: An Industry Survey

Malek Almobarek [1,*], Kepa Mendibil [1], Abdalla Alrashdan [2] and Sobhi Mejjaouli [2]

[1] Department of Design, Manufacturing, and Engineering Management, Faculty of Engineering, University of Strathclyde, Glasgow G1 1XQ, UK
[2] Industrial Engineering Department, College of Engineering, Alfaisal University, Riyadh 50927, Saudi Arabia
* Correspondence: malek.almobarek@strath.ac.uk

Abstract: Predictive Maintenance 4.0 (PdM 4.0) showed a highly positive impact on chilled water system (CWS) maintenance. This research followed the recommendations of a systematic literature review (SLR), which was performed on PdM 4.0 applications for CWS at commercial buildings. Per the SLR, and to start making an excellent PdM 4.0 program, the faults and their frequencies must be identified. Therefore, this research constructed an industry survey, which went through a pilot study, and then shared it with 761 maintenance officers in different commercial buildings. The first goal of this survey is to verify the faults reported by SLR, explore more faults, and suggest a managerial solution for each fault. The second goal is to determine the minimum and maximum frequencies of faults occurrence, while the third goal is to verify selected operational parameters, in which their data can be used in smart buildings applications. A total of 304 responses are considered in this study, which identified additional faults and provided faults solutions for all CWS components. Based on the survey outcomes, justifiable frequencies are proposed, which can be used in creating the dataset of any machine learning model, and then to control the CWS performance.

Keywords: predictive maintenance; chilled water system; commercial buildings; industry 4.0; quality 4.0; survey; faults; frequencies

Citation: Almobarek, M.; Mendibil, K.; Alrashdan, A.; Mejjaouli, S. Fault Types and Frequencies in Predictive Maintenance 4.0 for Chilled Water System at Commercial Buildings: An Industry Survey. *Buildings* **2022**, *12*, 1995. https://doi.org/10.3390/buildings12111995

Academic Editor: Etienne Saloux

Received: 18 October 2022
Accepted: 15 November 2022
Published: 16 November 2022

Publisher's Note: MDPI stays neutral with regard to jurisdictional claims in published maps and institutional affiliations.

1. Introduction

The commercial building (CB) is defined as a large facility that contains sophisticated systems such as a chilled water system (CWS) and power plants [1]. Globally, the number of CBs have rapidly increased and require significant attention from a maintenance management point of view [2]. Wireman defined maintenance as managing any assets that are owned by the organization [3]. Duffuaa and others pointed out that maintenance can be looked such as a system with input and an associated output [4]. The input part contains workforces, management, tools, equipment, and machines, while the output part comprises the equipment or machines that are working perfectly, fulfilling the reliability concepts, and well-configured to reach the scheduled operational time. According to the European Standard, maintenance connects all managerial actions that are required during the life cycle of a particular CB's equipment [5,6].

Previously, the attention provided to the maintenance was not recognized as it was considered as a "Cinderella Function" due to some historical reasons and it can be surmounted by new information technologies [7]. Up to around 1940, maintenance was considered as an inescapable cost and once the failure of a particular equipment happened, the maintenance technician should be servicing the same equipment based on a call request [8]. In 1968, it was predestined that better maintenance practices in the United Kingdom (UK) could have economized approximately GBP 300 million per year of lost production; because of the unavailability of a particular equipment [9]. In 1972, the significance of building maintenance was first recognized by the responsible authorities in the UK [10]. Maintenance became one of the important managerial departments or functions that should be included in the company's organizational hierarchy [11]. Maintenance, in the 21st century is a huge business where operating and maintaining CBs is significantly taking more time than designing and constructing the same building during its project time, also the life cycle cost of operating and maintaining the same building is about sixty percent to eighty-five percent of the total cost, whereas its design and its construction are about five percent to ten percent [12]. Moreover, alongside energy costs, maintenance costs can be the biggest portion of the operational budget [13]. Many researchers have argued that implementing good and effective maintenance management increases the equipment performance and this is definitely maximizing the revenues, minimizing the operation and maintenance (O and M) costs, and then growing the organisations' profits [14–17]. In this regard, Cholasuke and others explained how to maximise organisations' profit through implementing maintenance management [18]. He listed some factors such as trying to minimise the accidents or failures. Dhillon presented an approach containing steps and important principles for maintenance management in a cost-effective manner such as measurement comes before control [19].

Maintenance can be actioned in many ways, depending on the operational status and the strategy of the organisation. However, Seeley categorized maintenance types for buildings [20]. He mentioned that it can be considered as a planned activity which can be organized by scheduling the building operation and tracking its performance, so it can be considered as a scheduled activity as well. In contrast, it can be also considered as unplanned activity [20,21]. In addition, it can be performed as a preventive task by controlling the building operation to reduce the probability of destruction, to avoid the failure of a mechanical or an electrical system (s), or to maintain an item performance from any unexpected breakdown [20,22]. Furthermore, this task can be considered as a planned, scheduled, or predictive activity [20,23]. In case of failure, it can be considered as a corrective workorder to return a system to its standard operation. For immediate action, maintenance can be known as an emergency task such as big water leakage or power outage. Furthermore, Kanisuru classified maintenance into four major types [24]. The first one is reactive (RM), corrective, or breakdown maintenance. The second one is preventive maintenance (PM), which he categorised it into two major types, predictive maintenance (PdM), which is considered in this article, and periodic one. The third major type of maintenance is the improvement or design maintenance while the fourth one is technology maintenance.

Applying quality engineering and quality control concepts in maintenance management and processes is one of the keys of making it a successful program. Having said that, Marquez and others presented a modelled framework containing eight phases, which are linked to four blocks (effectiveness, efficiency, assessment, and improvement) [25]. The philosophy behind this framework was to make a strategic plan for the organisations to improve the outcome of maintenance program. Applying the maintenance control function is another valuable technique for improving the maintenance management [26]. This function contains four phases, which are planning, organising, implementing, and controlling. Specifically, this research believes in the paradigm of the last phase (controlling), which is measuring the performance of the maintained equipment, taking predictive and corrective actions, and reviewing the associated policies and procedures. Quality control is one of the main domains of evaluating the risks of maintenance activities [27]. Chanter and Swallow argued that constructing industrial quality control procedures in maintenance projects prevents the organisation from any managerial failure with any contractual party [28]. One of the concepts of maintenance management is the process improvement, which is part of industrial quality control and was performed in many applications using Six-Sigma techniques [29–32]. With a specific regard to this research, the quality engineering approach is a data driven in PdM filed. It is one of the important features of the fourth industrial revolution, which is called Industry 4.0 [33]. Many research studies have proposed a data driven PdM framework for multiple purposes as part of quality control [34–38]. Zonnenshan and Kenett indicated that quality engineering is a crucial dimension of the processes in maintenance, and it should be a data driven, and accordingly they proposed a framework called Quality 4.0 [39]. According to Zonta and others, PdM is a rife area in the locomotion across Quality 4.0 and its related data are the clef to produce acquaintance that can partake to predict decisions [40].

This research follows a systematic literature review (SLR) study, which was looked into in the studies that proposed PdM frameworks or fault detection and diagnosis (FDD) protocols, which are in line with Industry 4.0/ Quality 4.0 [1]. The aforementioned SLR study explained and focused on CWS, which has four components as shown in Figure 1.

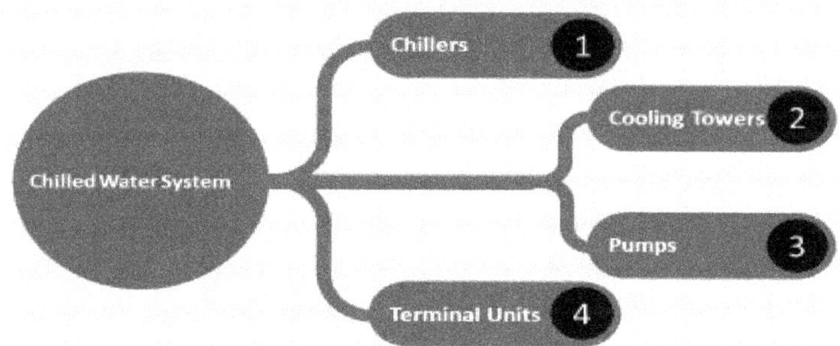

Figure 1. CWS components.

The said SLR has two research questions (RQ) from engineering management point of view as follows:

- RQ1: How can the faults be identified in order to predict them?
- RQ2: What are the methods that can be used to predict the faults?

Based on these RQs, the SLR considered 168 studies that applied different machine learning (ML) algorithms to predict CWS faults [1]. In addition, two more review studies have been considered in this regard [41,42]. The fault here is defined as any operational issue that may negatively affect CWS over time [1]. Having said ML, the said studies were considered different operational parameters to collect their readings. These readings must cover two modes, which are fault existing and fault free [1,43]. The SLR defined these readings as fault frequencies, which are used to build the prediction model [1]. So, this research focuses on these two elements (faults and their frequencies), as the research gaps resulted from the said SLR were related to both. Table 1 shows the gaps that are intended to be filled by this research.

Table 1. SLR's research gaps.

Research Gap	Related Element
The literature did not have the same faults and was concentrated only on selected faults, as some faults were either not stated/mentioned or were not fully described.	Fault
The current literature did not specify how the data were collected or justify the period or the frequency of the collected data, and it was limited to testing the model and not controlling it.	Frequency
The suggested programs/frameworks/models contained either no or inconclusive solutions for the said faults from the management point of view, as they ended with how to detect/predict the faults. Moreover, the said programs did not study/cover the whole system comprehensively.	Fault

To fill these three gaps, SLR recommended an industry survey to execute the following points:

- To validate the faults that are identified by the literature. Furthermore, from the first research gap the SLR argued that CWS may have some other different faults, so the survey can explore more ones;
- To identify the minimum and maximum faults' occurrence timings. These timings help to determine the frequencies, which can be used in creating the dataset of any ML model; for more prediction accuracy;
- To understand how to solve/ fix the faults, which are either identified by the literature or by the survey.

2. Methodology

As mentioned in the previous section, this article is intended to address the aforementioned gaps, which are related to CWS's faults and their frequencies by proposing a survey. A survey is defined as a research method utilised to collect data from a predetermined group of participants to obtain information and thoughts on diverse topics of interest [44]. Collecting data that are related to any study is important in terms of harmonization with the research goals [45]. In this research, the data should allow to investigate the operational circumstances at the buildings that are managed by the respondents. As recommended by the aforementioned SLR, the survey should be looked onto the CBs in the area/ city of the building that is going to be studied [1], and for this article Riyadh city of the Kingdom of Saudi Arabia was chosen.

2.1. Survey Construction

This research followed four guidelines while constructing the survey [46,47]. They are furnished in Figure 2.

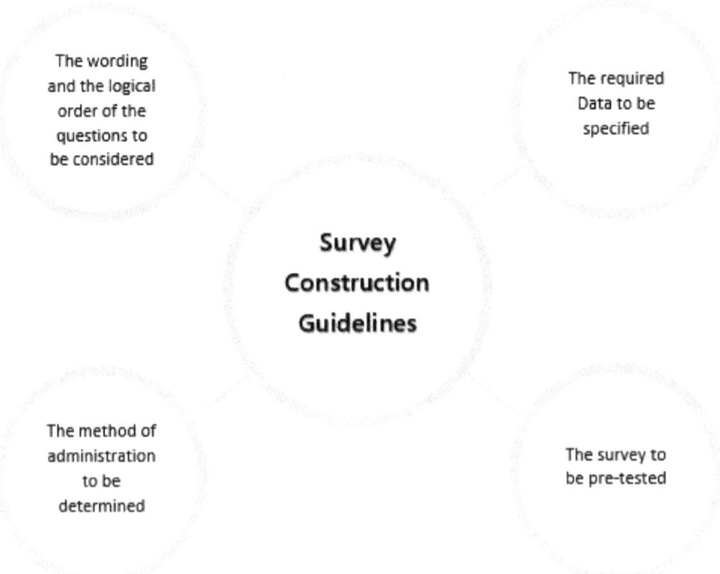

Figure 2. Survey construction guidelines.

The proposed survey contains four parts, which acquire quantitative and qualitative data. The said survey began with asking the participants about the availability of CWS in their facilities. The answer required for this enquiry is either 'yes' or 'no', and the participants were asked to complete the survey if the answer was 'yes'. The second part of the survey is related to the faults and asked the participants about the observation of listed faults that were collected from the literature for each CWS component and to provide additional faults that occurred in their buildings. Moreover, the participants were asked to suggest solution for each listed/ provided fault.

The third part of the survey is related to the faults frequencies by asking the participants which the most fault occurred so often and which most fault occurred rarely, and then to specify the frequency occurrence time for both. The further action of this part is to pick the frequencies' minimum value of the faults that occur often, and to pick the frequencies' maximum value of the faults that occur rarely. The idea here is to utilise these two values in creating the dataset that is going to be used in building any ML model. The minimum frequency is proposed to be used as a time interval in collecting the data for each component, for example, the readings should be taken every forty-five minutes. In contrast, the idea of using the maximum one is to consider it as a study period, for example, the readings should be taken over three months. Asking the participants to state those two faults is to make obtaining the frequencies information easy and to avoid any misunderstanding. As advised by the said SLR, the source of the associated data are the readings of any chosen operational parameter of each component as these operational parameters can reflect the health conditions of CWS components [1]. The fourth part of the said survey is seeking the opinion of the participants about the chosen operational parameter of each CWS component for validation purposes. The parameters are the water leaving temperature for chillers and cooling towers, pressure for pumps, and space temperature for terminal units. Up to this point, the first two guidelines, which are related to the specification of the data type and the consideration of the questions' wording and their logical order, were fulfilled. Regarding the third guideline, which is related to the method of administration, the self-completed questionnaire is selected as an instrument to collect the data. The last guideline, which is related to pre-testing the survey, was performed through a pilot study, which is explained in the next subsection.

2.2. Pilot Study

In order to adhere to the fourth guideline, the draft questionnaire was sent to ten experts from academia and industry for their review and advice. In addition to the questions, the draft included an explanation of the research goal and expectation. This is to ensure that the survey with its questions is fulfilling the needs from validity and reliability points of view and to rise the response rate [48,49]. The experts from academia were from different departments/ fields, which are industrial engineering, mechanical engineering, electrical engineering, economics, and operation management. On the other hand, a manufacturer of each CWS component as well as an O and M contractor were the experts from the industry.

The experts were provided a month to reply with their feedback. Table 2 shows the main comments from both side, academia and industry.

Table 2. Pilot study's outcomes.

Academia	Industry
The anonymity of the participants and their organisations must be protected by adding a statement in this regard.	The minimum age of CBs that are managed by the respondents should be three years.
Add a note into the solution part to make the answer short and manageable.	The CBs should have valid commercial registration with the concerned authority.
Clarify the questions that are related to the frequency part with an example.	
The form of writing the solutions for the new faults should be made united for every participant.	A separate part for each component should be fabricated.
	Avoid using abbreviations.

After addressing the experts' comments, the survey was finalised and inserted into a web-based platform, which is explained in the next sub-section. Figure 3 shows an excerpt from the survey that is related to the cooling tower part. The same is repeated with other CWS components.

Part #1:

- Do you have chilled water system at your facility?

- ○ Yes
- ○ No

If yes, then please complete the following parts:

Part #2:

- Please choose from the below furnished faults for each component of the chilled water system that you encounter at your facility and provide the details. Kindly add those that you encounter but are not listed here as new. Also, beside each fault, kindly propose a managerial solution to fix that fault in a timely manner.

Cooling Towers

Fault	Do you encounter this at your facility?		Solution
Air Fan Degradation	○ Yes	○ No	
Fills Fouling	○ Yes	○ No	
Sensor Bias	○ Yes	○ No	
New Faults and their solutions: (Please write this form: (Fault: Solution), (Fault: Solution)			

Part #3:

- What is the most fault occurring so often during your operational time? Please state the frequency of its occurrence and mention the time unit (for example, it is possibly occurred for every 45 minutes).

Fault: ....; Frequency: ....

- What is the most fault occurring rarely during your operational time? Please state the frequency of its occurrence and mention the time unit (for example, it is possibly occurred within 6 weeks).

Fault: ....; Frequency: ....

Part #4:

- Please tick appropriately if the furnished parameter best predicts the cooling tower's health condition and if not, please recommend an alternative.

Water Leaving Temperature (°C)	Is it the best to predict the health condition?		Alternative
	○ Yes	○ No	

Figure 3. Excerpt of cooling tower's survey.

2.3. Targeted Participants and Survey Distribution

Following the outcome of the pilot study, the concerned authority in Riyadh city was contacted, and accordingly the professionals' contacts of seven hundred and sixty-one (761) CBs were received, which contained e-mails and phone numbers. The professionals are facility managers, O and M managers, and support services managers. For survey distribution, the associated technique of the survey administration method, which is part of the survey construction guidelines, is the web-based questionnaire through the Survey-Monkey platform. The aforementioned platform generated e-mails to the participants, and following Andrews' and others' recommendations, an informative title of the e-mail was used as well as using careful language in the body of said e-mail in order to encourage the participants to open the e-mail, to click on the survey link, and to downplay the onus of the participants [50]. In this regard, the agreement with the participants was deal with their answers anonymously by writing a statement in the body of the generated e-mails. The duration of the survey was three months, and auto reminders were sent every seven working days. In addition to the auto reminders, a follow-up mechanism was used via phone calls for a more response rate [51].

3. Survey Results and Discussion

As the outcomes of the survey contained quantitative and qualitative data, an analysis activity was taken and then the said outcomes were summarised. The time spent on this task was three weeks.

3.1. Response Rate

From the mentioned 761 CBs that were contacted, three hundred and thirty-six (336) responses were received within the provided time, out of which three hundred and four (304) responses have CWS at their facilities, which are considered in this study. Figure 4 illustrates the response rate of the survey.

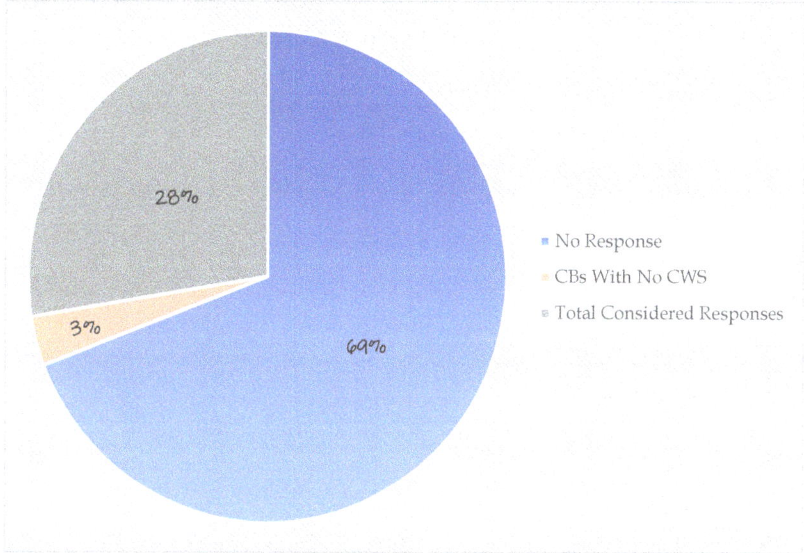

Figure 4. Survey responses.

The auto reminders and the phone calls' follow-up paid-off as the responses increased during the second and the third months, respectively. Figure 5 explains the number of responses for each month of the provided duration.

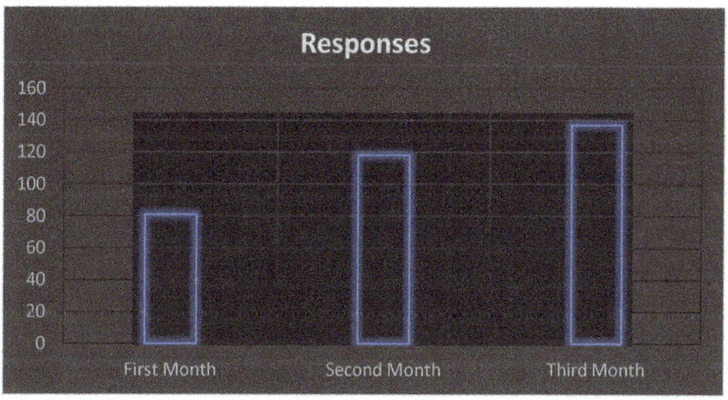

Figure 5. Monthly number of responses.

3.2. CWS Faults

To summarise the CWS faults that were addressed in the SLR, a fish-bone diagram was prepared to highlight the same as shown in Figure 6. In contrast, the argument of the aforementioned SLR was proven here as the survey came up with more faults for each CWS component. The outcome provided the research community with seventeen (17) faults for chillers, thirteen (13) faults for cooling towers, ten (10) faults for pumps, and twenty (20) faults for terminal units. In this regard, Figure 7 illustrates the difference increment between the literature and the survey outcomes.

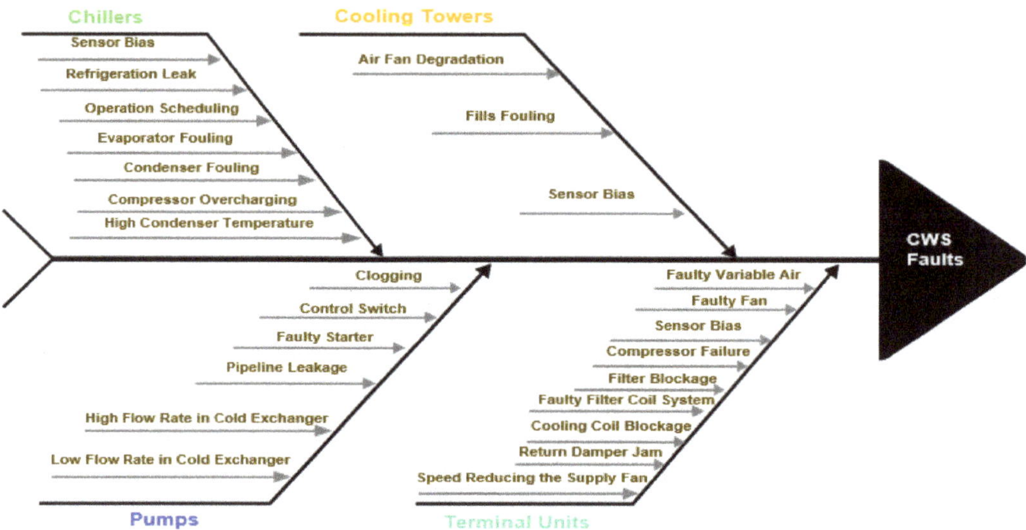

Figure 6. Faults presented in the literature.

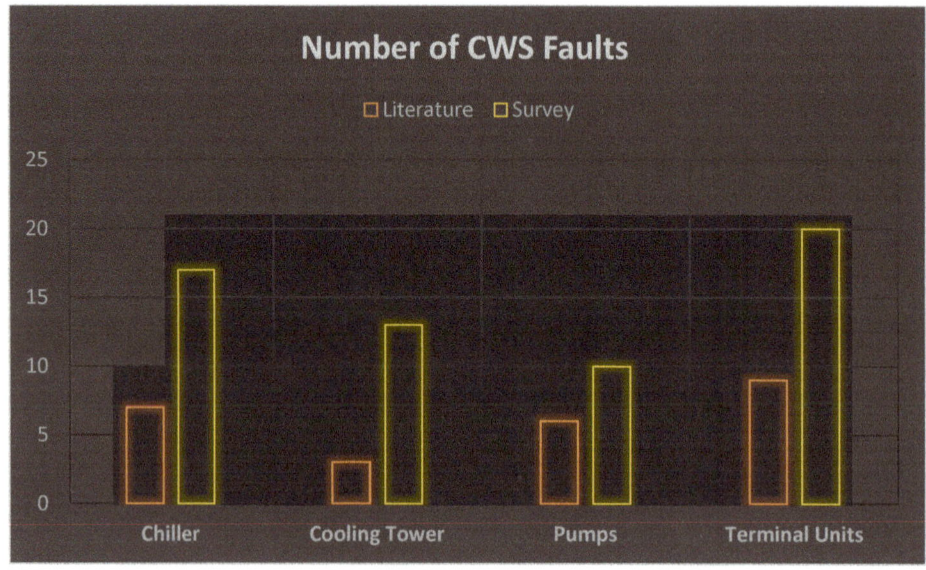

Figure 7. Difference in faults number.

The appearance of faults is different between the CBs that are managed by the participants, who repeated some faults for each component and stated some other faults a few times or only one time. Table 3 shows the most repeated fault for each CWS component in the CBs that are managed by the participants.

Table 3. Most repeated faults.

Component	Fault	Repeating Percentage
Chillers	Refrigeration Leak	100%
Cooling Towers	Malfunctioning Blowdown System	89%
Pumps	Noisy Non-Return Valve	91%
Terminal Units	Low Static Pressure	84%

From the above table, a refrigeration leak is the most common fault for chillers, and this is aligned with what was mentioned in the said SLR. For the cooling towers, the literature mostly addressed fills fouling and air fan degradation faults, while the survey showed that the malfunctioning blowdown system fault is mostly repeated and therefore, more research attention should be provided to this fault. The same is seen with pumps and terminal units where the literature mostly addressed the pumps' clogging fault and the terminal units' return damper jam fault, while the survey showed different faults that are repeated by the majority of the participants and therefore, they should be provided more focus in the upcoming research.

As stated previously, the faults that were addressed in the literature were listed in the survey for each component and the participants were asked if they observe them, and then were asked to state other faults that are not listed. Furthermore, the survey provided managerial solutions for both the faults that were studied in the literature and the new faults that are resulted from the survey, whereas the previous studies of the said literature ended their proposed PdM or FDD programs by tracing the faults with no solutions to fix the studied faults. To summarise these outcomes, a table for each component describes the faults and its source, which is either from the literature or from the survey, as well as lists a solution/action for each fault. Table 4 is for chillers, and Tables 5–7 are for cooling towers, pumps, and terminal units, respectively.

Table 4. Chillers faults and solutions.

Fault	Identified By	Solution/Action
Refrigeration Leak	Literature	All the components of refrigeration system including tube, joints, and valves should be checked, tested, and rectified as appropriate.
Evaporating Fouling	Literature	The associated parameters should be checked and the tubes descaled.
Compressor Overcharging	Literature	The factory sheet should be checked and then the charge reduced accordingly.
Operation Scheduling	Literature	The control switch should be reset.
Condenser Fouling	Literature	The associated parameters should be checked and the tubes descaled to fix the evaporating fouling.
High Condenser Temperature	Literature	The return water temperature should be checked, and then the tubes descaled.
Sensor Bias	Literature	The controller, the artificial agent, or the sensor should be checked, verified, and replaced if needed.
Low Discharge Superheat	Survey	The liquid refrigerant flow in the compressor should be checked and adjusted.
Low Evaporator Refrigerant Temperature	Survey	The expansion valve and the filter should be checked and cleaned.
Low Oil Pressure	Survey	The oil filters should be cleaned, and the oil pump with its quality should be checked and rectified.

Table 4. *Cont.*

Fault	Identified By	Solution/Action
Low Condenser Flow	Survey	The pressure of the condenser pump in operation should be checked and rectified.
Low Chilled Water Flow	Survey	The pressure of the secondary pump in operation should be checked and rectified.
Low Cooler Delta-T	Survey	The lowering efficiency of the primary pump in operation should be checked and the pressure reset.
High Cooler Delta-T	Survey	The accuracy of the water flow's control should be checked and rectified and amended.
High Compressor Lift	Survey	The water flow should be checked.
High Motor Temperature	Survey	The compressor parameters should be checked.
High Motor Ampere	Survey	The linked mechanical system and the motor winding should be checked and rectified.
High Condenser Approach	Survey	The connected tunnel of the cooling tower in operation should be checked and serviced.
High Evaporator Approach	Survey	The assigned water temperature set-point should be checked and reset if needed.
High Condenser Pressure	Survey	The strainer should be checked and cleaned.
Relief Valve Discharge	Survey	The pressure sensors should be checked and fixed.
Vibration	Survey	The supply water temperature should be checked, and the mountings should be reassembled.
Imbalanced Line Current	Survey	The loose connection at the terminals should be rectified.
Manual Guide Vane Target	Survey	The override sittings should be checked and reset.

Table 5. Cooling towers faults and solutions.

Fault	Identified By	Solution/Action
Air Fan Degradation	Literature	The fan should be checked physically, and then repaired by grinding, or replaced if needed.
Fills Fouling	Literature	The fills should be cleaned or replaced if needed.
Sensor Bias	Literature	The controller, the artificial agent, or the sensor should be checked, verified, and replaced if needed.
Unusual Sound	Survey	The bearings of the motor in operation should be checked.
Malfunctioning Blowdown System	Survey	The solenoid valves should be checked.
High Water Total Dissolved Solid	Survey	The chemical treatments should be checked.
Fills Clogging	Survey	The chemical treatments should be checked, and then the required chemicals refilled if needed.
Low Circulating Water Flow Rate	Survey	The filters should be checked and cleaned or replaced if needed.
Vibration	Survey	The motor in operation and its blade alignment should be checked.
Over Current	Survey	The phase voltage and other electrical connection should be checked.
Rise in Circulating Water Temperature	Survey	The filters should be checked and cleaned, or replaced if needed.
Damaged Fan	Survey	The associated motor should be replaced.
Faulty Water Level Valve	Survey	The valve should be replaced.
Faulty Isolation Valve	Survey	The valve should be replaced.
Motor Overheating	Survey	The voltage should be checked and adjusted.
Low Water Basin Level	Survey	The water makeup system should be checked and the water level should be increased.

Table 6. Pumps faults and solutions.

Fault	Identified By	Solution/Action
Clogging	Literature	The strainer should be checked and cleaned in case of partial clogging, and deep cleaned with chemicals and high-pressured water in case of full clogging.
Faulty Control Switch	Literature	The switch is should be troubleshooted or replaced if needed.
Faulty Starter	Literature	The electrical connection should be checked and rectified.
Pipeline Leakage	Literature	The pipe joint and its fittings should be checked and then welded or replaced if needed.
High Flow Rate in Cold Exchange	Literature	The right pump speed should be checked and adjusted.
Low Flow Rate in Cold Exchange	Literature	The right pump speed should be checked and adjusted.
Abnormal or Excessive Noise	Survey	The associated bearings and shaft should be checked and fixed.
Motor Vibration	Survey	The bearings and the foundation support should be checked and rectified.
Motor Heats-up	Survey	The bearings and the associated fan should be checked and rectified, ground, or replaced if needed.
Leakage from Pump Set	Survey	The associated gland and joints should be checked and reassembled.
Leakage from Valves	Survey	The associated joints should be checked, reassembled, or the valve to be replaced if needed.
Pumps Run but Provides No Water	Survey	The valves should be checked and made free of air.
Pumps Run at Reduced Capacity	Survey	The Stainer should be checked and cleaned.
Noisy Non-Return Valve	Survey	The valve set-up should be checked and replaced if needed.
Improper Pump Water Alignment	Survey	Realignment.
Sensor Bias	Survey	The controller, the artificial agent, or the sensor should be checked, verified, and replaced if needed.

Table 7. Terminal units faults and solutions.

Fault	Identified By	Solution/Action
Faulty Variable Air Volume	Literature	The damper connection and controller should be checked and rectified.
Faulty Fan	Literature	The fan should be checked, rectified, and replaced if needed.
Compressor Failure	Literature	The voltage and related control accessories should be checked before replacing formalities.
Filter Blockage	Literature	The filter should be cleaned or replaced if needed.
Faulty Filter Coil System	Literature	The dirt and debris should be cleared.
Cooling Coil Blockage	Literature	The fresh air damper position should be checked, and the air speed should be reduced if needed.
Return Damper Jam	Literature	The damper should be serviced and replaced if needed.
Speed Reducing the Supply Fan	Literature	The blower tips should be checked and cleaned.
Sensor Bias	Literature	The controller, the artificial agent, or the sensor should be checked, verified, and replaced if needed.
Dirty Air Flow	Survey	The bag filer section should be checked and cleaned.
Faulty Supply air Damper	Survey	The damper should be replaced.
Loose Belts	Survey	The associated pulleys, mountings, and V. belts quality should be checked and rectified.
Air Trapped in Cooling Coil	Survey	The coil should be checked and cleaned.
Faulty Control Valve	Survey	The associated voltage should be checked, and the valve should be replaced if needed.
Broken Belts	Survey	The associated pulley should be checked and rectified, and then the belt should be replaced.
Noisy Motor	Survey	The blower bearings should be checked and fixed.
Faulty Bearing	Survey	The bearing should be replaced.
Motor Overload	Survey	The power voltage and electrical accessories should be checked and rectified.
Noisy Contactors	Survey	The terminal in operation should be cleaned.
Vibration	Survey	The associated blowers should be aligned.
Motor Overheating	Survey	The voltage and rated amperes should be checked and adjusted.
Damaged Insulation on Pipe	Survey	The insulation material should be replaced and then sealed properly.
Variable Frequency Drive Soft Starter	Survey	The associated parameters should be reset.
Low Static Pressure	Survey	The air flow rate should be checked and decreased.
Damaged Insulation on Duct	Survey	The duct should be vacuumed, and the defective insulation should be replaced.
Faulty Fresh Air Damper	Survey	The air speed should be reduced, or the damper should be replaced if needed.
Faulty Exhaust Air Damper	Survey	The connected duct should be vacuumed, or the damper should be replaced if needed.
Faulty Cooling Valve Actuator	Survey	The voltage should be checked and adjusted.
Faulty Damper Actuator	Survey	The voltage should be checked and adjusted, and the air flow rate should be minimised.

Part of the research gaps of the said SLR was related to the coverage of the previous studies where the literature did not focus on the entire CWS [1]. They were either considered one, two, or three components only [1]. In return, the said SLR argued to study the whole system; in order to obtain a comprehensive PdM program. Having said that, the information in Tables 4–7 confirmed this argument as many faults in a particular component are due to the health condition of another component.

3.3. CWS Faults' Frequencies

This subsection is related to the third part of the survey, which its concept was explained in the methodology section. The participants' responses were counted by tabulating the frequencies' values of the faults that are occurring so often (x's) as well as the frequencies' values of the faults that are occurred rarely (y's). Then, the smallest value of the x's and the biggest value of the y's are identified. In addition, these two values were scored by the majority of the participants. Table 8 summarises these outcomes where each CWS component has two values (x and y) and shows the highest percentage of how many times each value is repeated by the participants.

Table 8. Frequencies outcome.

Component	X (Minutes)	Repeating Percentage	Y (Weeks)	Repeating Percentage
Chillers	30	75%	12	56%
Cooling Towers	30	68%	16	39%
Pumps	60	49%	24	34%
Terminal Units	45	40%	8	39%

From the above table, the proposed readings' time interval of the chosen operational parameters for the chiller in operation is every 30 min over a study period of 12 weeks. The same procedure is for other CWS components taking into consideration that these frequencies are convenient and applicable for the chosen city (Riyadh). These frequencies provide a justifiable thought on how to build the ML model, to predict the faults, and to control the entire CWS. Having said the operational parameters, the survey asked the participants in its fourth part about their opinion on specific operational parameters for each CWS component. This is to ensure that these operational parameters are valid and are the best to provide the health condition of the CWS components, and accordingly to find and predict the faults. Table 9 shows that the majority of the participants said the chosen operational parameters are the best to predict the health condition of CWS components, and accordingly verifies the belief of the said SLR.

Table 9. Operational parameters.

Component	Operational Parameters	Participants, Who Are with the Choice
Chillers	Water Leaving Temperature (°C)	98%
Cooling Towers	Water Leaving Temperature (°C)	96%
Pumps	Pressure (Bar)	100%
Terminal Units	Space Temperature (°C)	90%

4. Conclusions

This article is a contribution of the published SLR research on the applications of PdM 4.0 for CWS at CBs. It is intended to fill the research gaps that are listed by the said SLR, which are related to the CWS faults and their frequencies. To address these research gaps, a survey was prepared as concluded by the said SLR, and it went through constructive guidelines and a pilot study. The survey was sent to 761 CBs in Riyadh city, which have valid registration certificates with the concerned authority. As recommended by the pilot study, the CBs that were contacted have a minimum age of three years. Within three months' time, 336 responses were received, out of which 304 responses are considered in this research as they have CWS in their facilities.

Besides the outcomes of the said SLR, the survey enriched the research community with additional faults that are affecting CWS performance. It provided 17 faults for chillers, 13 faults for cooling towers, 10 faults for pumps, and 20 faults for terminal units, while the said literature only addressed 7 faults for chillers, 3 faults for cooling towers, 6 faults for pumps, and 9 faults for terminal units. Moreover, the survey provided a manageable solution or action for each fault, which either addressed in the literature or resulted from the survey, whereas the previous studies, which were mentioned in the said SLR, ended with only detecting the faults and they did not provide solutions to fix the same. Furthermore, this research showed the importance of covering the entire CWS in future studies as there are faults appearing in a particular CWS component due to another component's health condition.

The main part of any PdM 4.0 program is applying ML algorithms to predict the system faults, and these algorithms require a dataset to build the predictive model. Having said that, the said literature has not justified the frequencies of fault appearance, which were used to create their datasets, noting that the source of these frequencies was the readings of chosen operational parameters. The survey resulted in justifiable frequencies for each component, which can be used in creating the dataset of any ML model and to control the CWS after training and testing the said model. The chosen operational parameters here are the water leaving temperature for chillers and cooling towers, the pressure for pumps, and the space temperature for the terminal units, which are the best to predict the faults as verified by the survey. To create a dataset for each CWS component, the readings of water leaving temperature of a particular chiller are proposed to be taken every 30 min for a time period of 12 weeks. For a particular cooling tower, the readings of water leaving temperature are proposed to be taken every 30 min for a time period of 24 weeks. For a particular pump, the readings of pressure are proposed to be taken every 60 min over a time period of 24 weeks. For a particular terminal unit, the readings of the space temperature should be taken every 45 min for a time period of 8 weeks. The readings must be taken while operating the system and should be collected by professional technicians or users. These data are recommended to be recorded by one of the quality engineering tools such as check sheet. The check sheet should include cells for writing down the readings as well as cells for the inspection results, which should be either '1' in case of fault, or '0' in case of no fault. So, the dataset contains two columns, one is for the readings of the particular component and the other one is for the related inspection results.

The scientific contributions of this research are summarised as follows:

- The survey outcomes verified that there are more faults, which are not addressed or identified by the literature;
- Based on the survey results, faults frequencies are proposed to provide the researchers a justifiable thought on how to build the dataset for any ML model;
- A solution or action is provided for each fault to provide a comprehensive management view as the literature ended the proposed PdM programs by tracing the faults.

Though this research and the said SLR made a rigorous investigation on PdM 4.0 application for CWS, researchers are advised to perform further studies to explore more insights on these applications. Below are the suggested future research agendas:

- While making a PdM 4.0 program, the proposed frequencies can be considered to create the dataset for ML algorithms;
- This research confirmed that the appearance of the faults at a particular CWS component is due to the health condition of another component. Therefore, future research should consider covering the whole system for a more effective PdM program;
- The selected building is assumed to have sensors, for example, to collect readings of the operational parameters. In case of no reading's tools, a procedure on how to have them available should be discussed and proposed.

Author Contributions: Conceptualization, M.A. and K.M.; methodology, M.A.; investigation and analysis, M.A., A.A. and S.M.; writing—review and editing, M.A., A.A. and S.M.; supervision, K.M. All authors have read and agreed to the published version of the manuscript.

Funding: This research received no external funding.

Acknowledgments: Many thanks to the DMEM department at the University of Strathclyde, especially Anja Maier, Jorn Mehnen, Xichun Luo, Ian Whitfield, Laura Hay, and Gillian Eadie for their guidance, advice, and support. A special thanks also goes to Alfaisal University President (Mohammed Alhayaza), VP External relation and advancement (Maha Bint Mishari Bin Abdulaziz AlSaud), VP Admin and Finance (Khaled AlKattan), and Acting Dean of College of Engineering (Muhammad Anan) for their unlimited support. In addition, a big thanks goes to O and M Manager at Alfaisal University (Eng. Kaleemoddin Ahmed) as well as to Eng. Mohamed Yunus for their great assistance during this research.

Conflicts of Interest: The authors declare no conflict of interest.

References

1. Almobarek, M.; Mendibil, K.; Alrashdan, A. Predictive Maintenance 4.0 for Chilled Water System at Commercial Buildings: A Systematic Literature Review. *Buildings* **2022**, *12*, 1229. [CrossRef]
2. Hauashdh, A.; Jailani, J.; Rahman, I.A. Strategic approaches towards achieving sustainable and effective building maintenance practices in maintenance-managed buildings: A combination of expert interviews and a literature review. *J. Build. Eng.* **2022**, *45*, 103490. [CrossRef]
3. Wireman, T. *Developing Performance Indicators for Managing Maintenance*, 2nd ed.; Industrial Press Inc.: New York, NY, USA, 2005.
4. Duffuaa, S.O.; Raouf, A. *Planning and Control of Maintenance Systems*, 2nd ed.; Springer International Publishing: Zürich, Switzerland, 2015.
5. Marquez, A.C.; Gupta, J.N. Contemporary maintenance management: Process, framework and supporting pillars. *Omega* **2006**, *34*, 313–326. [CrossRef]
6. Márquez, A.C. *The Maintenance Management Framework: Models and Methods for Complex Systems Maintenance*, 1st ed.; Springer Science & Business Media: Berlin, Germany, 2007.
7. Sherwin, D. A review of overall models for maintenance management. *J. Qual. Maint. Eng.* **2000**, *6*, 138–164. [CrossRef]
8. Murthy, D.N.P.; Atrens, A.; Eccleston, J.A. Strategic maintenance management. *J. Qual. Maint. Eng.* **2002**, *8*, 287–305. [CrossRef]
9. Kelly., A. *Strategic Maintenance Planning*, 1st ed.; Elsevier Ltd.: London, UK, 2006.
10. Allen, D. What is Building Maintenance? *Facilities* **1993**, *11*, 7–12. [CrossRef]
11. Pintelon, L.M.; Gelders, L.F. Maintenance management decision making. *Eur. J. Oper. Res.* **1992**, *58*, 301–317. [CrossRef]
12. Lewis, A.; Riley, D.; Elmualim, A. Defining high performance buildings for operations and maintenance. *Int. J. Facil. Manag.* **2010**, *1*, 1–16.
13. Dekker, R. Applications of maintenance optimization models: A review and analysis. *Reliab. Eng. Syst. Saf.* **1996**, *51*, 229–240. [CrossRef]
14. Willmott, H. Business process re-engineering and human resource management. *Pers. Rev.* **1994**, *23*, 34–46. [CrossRef]
15. Alsyouf, I. The role of maintenance in improving companies' productivity and profitability. *Int. J. Prod. Econ.* **2007**, *105*, 70–78. [CrossRef]
16. Naidu, S.; Amalesh, J.; Rao, P.V.; Sawhney, R. An Empirical Model for Maintenance Strategy Selection based on Organizational Profit. In *Proceedings of the Industrial Engineering Research Conference, Norcross, GA, USA, 26–27 December 2009*; Institute of Industrial and Systems Engineers (IISE): Peachtree Corners, GA, USA, 2009; pp. 1765–1770.
17. Xia, T.; Sun, B.; Chen, Z.; Pan, E.; Wang, H.; Xi, L. Opportunistic maintenance policy integrating leasing profit and capacity balancing for serial-parallel leased systems. *Reliab. Eng. Syst. Saf.* **2021**, *205*, 107233. [CrossRef]
18. Cholasuke, C.; Bhardwa, R.; Antony, J. The status of maintenance management in UK manufacturing organisations: Results from a pilot survey. *J. Qual. Maint. Eng.* **2004**, *10*, 5–15. [CrossRef]
19. Dhillon, B.S. *Engineering Maintenance: A Modern Approach*, 1st ed.; CRC Press: Boca Raton, FL, USA, 2002.
20. Seeley, I.H. *Building Maintenance*, 2nd ed.; Macmillan International Higher Education: New York, NY, USA, 1987.
21. Tambe, P.P.; Mohite, S.; Kulkarni, M.S. Optimisation of opportunistic maintenance of a multi-component system considering the effect of failures on quality and production schedule: A case study. *Int. J. Adv. Manuf. Technol.* **2013**, *69*, 1743–1756. [CrossRef]
22. Gouiaa-Mtibaa, A.; Dellagi, S.; Achour, Z.; Erray, W. Integrated Maintenance-Quality policy with rework process under improved imperfect preventive maintenance. *Reliab. Eng. Syst. Saf.* **2018**, *173*, 1–11. [CrossRef]
23. Curcurù, G.; Galante, G.; Lombardo, A. A predictive maintenance policy with imperfect monitoring. *Reliab. Eng. Syst. Saf.* **2010**, *95*, 989–997. [CrossRef]
24. Kanisuru, A.M. Sustainable Maintenance Practices and Skills for Competitive Production System, Chapter 3. In *Skills Development for Sustainable Manufacturing*; IntechOpen: Rijeka, Croatia, 2017.

25. Márquez, A.C.; de León, P.M.; Fernández, J.G.; Márquez, C.P.; Campos, M.L. The maintenance management framework: A practical view to maintenance management. *J. Qual. Maint. Eng.* **2009**, *15*, 167–178. [CrossRef]
26. Duffuaa, S.O.; Haroun, A.E. Maintenance control. In *Handbook of Maintenance Management and Engineering*; Springer: London, UK, 2009; pp. 93–113.
27. Kosztyan, Z.T.; Pribojszki-Németh, A.; Szalkai, I. Hybrid multimode resource-constrained maintenance project scheduling problem. *Oper. Res. Perspect.* **2019**, *6*, 100129. [CrossRef]
28. Chanter, B.; Swallow, P. *Building Maintenance Management*, 2nd ed.; Blackwell Publishing, John Wiley & Sons.: Hoboken, NJ, USA, 2008.
29. Wang, X.; Wang, Y.; Xu, D. Lean six sigma implementations in equipment maintenance process. In Proceedings of the International Conference on Quality, Reliability, Risk, Maintenance, and Safety Engineering (ICQRMS), Chengdu, China, 15–18 June 2012; IEEE: Piscataway, NJ, USA, 2012; pp. 1391–1395.
30. Zasadzien, M. Application of the Six Sigma method for improving maintenance processes-case study. In Proceedings of the 6th International Conference on Operations Research and Enterprise Systems, Porto, Portugal, 23–25 February 2017; SCITEPRESS: Setubal, Portugal, 2017; 2, pp. 314–320.
31. Schafer, F.; Schwulera, E.; Otten, H.; Franke, J. From Descriptive to Predictive Six Sigma: Machine Learning for Predictive Maintenance. In Proceedings of the 2nd International Conference on Artificial Intelligence for Industries (AI4I), Laguna Hills, CA, USA, 25–27 September 2019; IEEE: Piscataway, NJ, USA, 2019; pp. 35–38.
32. Vaidya, S.; Bhosle, S.; Ambad, P. DMAIC Approach to Improve Carbon Weighing Compliance of Banburry Machine. In *Computing in Engineering and Technology*; Springer: Singapore, 2020; pp. 803–816.
33. Kenett, R.S.; Redman, T.C. *The Real Work of Data Science: Turning Data into Information, Better Decisions, and Stronger Organizations*, 1st ed.; John Wiley & Sons.: Hoboken, NJ, USA, 2019.
34. Rodseth, H.; Schjolberg, P. Data-driven predictive maintenance for green manufacturing. In Proceedings of the 6th international workshop of advanced manufacturing and automation. Advances in Economics, Business and Management Research, University of Manchester, Manchester, UK, 10–11 November 2016; Atlantis Press: Amsterdam, The Netherlands, 2016; pp. 36–41.
35. Gerum PC, L.; Altay, A.; Baykal-Gursoy, M. Data-driven predictive maintenance scheduling policies for railways. *Transp. Res. Part C: Emerg. Technol.* **2019**, *107*, 137–154. [CrossRef]
36. Antomarioni, S.; Bevilacqua, M.; Potena, D.; Diamantini, C. Defining a data-driven maintenance policy: An application to an oil refinery plant. *Int. J. Qual. Reliab. Manag.* **2019**, *36*, 77–97. [CrossRef]
37. Cheng, J.C.; Chen, W.; Chen, K.; Wang, Q. Data-driven predictive maintenance planning framework for MEP components based on BIM and IoT using machine learning algorithms. *Autom. Constr.* **2020**, *112*, 103087. [CrossRef]
38. Pisacane, O.; Potena, D.; Antomarioni, S.; Bevilacqua, M.; Emanuele Ciarapica, F.; Diamantini, C. Data-driven predictive maintenance policy based on multi-objective optimization approaches for the component repairing problem. *Eng. Optim.* **2021**, *53*, 1752–1771. [CrossRef]
39. Zonnenshain, A.; Kenett, R.S. Quality 4.0—The challenging future of quality engineering. *Qual. Eng.* **2020**, *32*, 614–626. [CrossRef]
40. Zonta, T.; da Costa, C.A.; da Rosa Righi, R.; de Lima, M.J.; da Trindade, E.S.; Li, G.P. Predictive maintenance in the Industry 4.0: A systematic literature review. *Comput. Ind. Eng.* **2020**, *150*, 106889. [CrossRef]
41. Mirnaghi, M.S.; Haghighat, F. Fault detection and diagnosis of large-scale HVAC systems in buildings using data-driven methods: A comprehensive review. *Energy Build.* **2020**, *229*, 110492. [CrossRef]
42. Kim, W.; Katipamula, S. A review of fault detection and diagnostics methods for building systems. *Sci. Technol. Built Environ.* **2017**, *24*, 3–21. [CrossRef]
43. Zhang, W.; Yang, D.; Wang, H. Data-driven methods for predictive maintenance of industrial equipment: A survey. *IEEE Syst. J.* **2019**, *13*, 2213–2227. [CrossRef]
44. Schouten, B.; Peytchev, A.; Wagner, J. *Adaptive Survey Design*, 1st ed.; Chapman and Hall/CRC: New York, NY, USA, 2017.
45. Ramakrishnan, T.; Jones, M.C.; Sidorova, A. Factors influencing business intelligence (BI) data collection strategies: An empirical investigation. *Decis. Support Syst.* **2012**, *52*, 486–496. [CrossRef]
46. Anseel, F.; Lievens, F.; Schollaert, E.; Choragwicka, B. Response rates in organizational science, 1995–2008: A meta-analytic review and guidelines for survey researchers. *J. Bus. Psychol.* **2010**, *25*, 335–349. [CrossRef]
47. Ghauri, P.; Gronhaug, K. *Research Methods in Business Studies: A Practical Guide*, 3rd ed.; Pearson Education Limited: London, UK, 2005.
48. Easterby-Smith, M.; Thorpe, R.; Jackson, P.; Jaspersen, L. *Management and Business Research*, 6th ed.; Sage Publications Ltd.: London, UK, 2018.
49. Saunders, M.N.K.; Lewis, P.; Thornhill, A. *Research Methods for Business Students*, 8th ed.; Pearson: London, UK, 2019.
50. Andrews, D.; Nonnecke, B.; Preece, J. Electronic survey methodology: A case study in reaching hard-to-involve Internet users. *Int. J. Hum.-Comput. Interact.* **2003**, *16*, 185–210. [CrossRef]
51. Hudson, D.; Seah, L.H.; Hite, D.; Haab, T. Telephone presurveys, self-selection, and non response bias to mail and internet surveys in economic research. *Appl. Econ. Lett.* **2004**, *11*, 237–240. [CrossRef]

Article

Predictive Maintenance 4.0 for Chilled Water System at Commercial Buildings: A Methodological Framework

Malek Almobarek [1,*], Kepa Mendibil [1] and Abdalla Alrashdan [2]

[1] Department of Design, Manufacturing and Engineering Management, Faculty of Engineering, University of Strathclyde, Glasgow G1 1XQ, UK
[2] Industrial Engineering Department, College of Engineering, Alfaisal University, Riyadh 50927, Saudi Arabia
* Correspondence: malek.almobarek@strath.ac.uk

Abstract: Predictive maintenance is considered as one of the most important strategies for managing the utility systems of commercial buildings. This research focused on chilled water system (CWS) components and proposed a methodological framework to build a comprehensive predictive maintenance program in line with Industry 4.0/Quality 4.0 (PdM 4.0). This research followed a systematic literature review (SLR) study that addressed two research questions about the mechanism for handling CWS faults, as well as fault prediction methods. This research rectified the associated research gaps found in the SLR study, which were related to three points; namely fault handling, fault frequencies, and fault solutions. A framework was built based on the outcome of an industry survey study and contained three parts: setup, machine learning, and quality control. The first part explained the three arrangements required for preparing the framework. The second part proposed a decision tree (DT) model to predict CWS faults and listed the steps for building and training the model. In this part, two DT algorithms were proposed, C4.5 and CART. The last part, quality control, suggested managerial steps for controlling the maintenance program. The framework was implemented in a university, with encouraging outcomes, as the prediction accuracy of the presented prediction model was more than 98% for each CWS component. The DT model improved the fault prediction by more than 20% in all CWS components when compared to the existing control system at the university.

Keywords: predictive maintenance; Industry 4.0; Quality 4.0; decision tree algorithm; chilled water system; HVAC; commercial buildings; industrial engineering; engineering management

Citation: Almobarek, M.; Mendibil, K.; Alrashdan, A. Predictive Maintenance 4.0 for Chilled Water System at Commercial Buildings: A Methodological Framework. *Buildings* **2023**, *13*, 497. https://doi.org/10.3390/buildings13020497

Academic Editor: Etienne Saloux

Received: 12 January 2023
Revised: 7 February 2023
Accepted: 10 February 2023
Published: 12 February 2023

1. Introduction

1.1. Overview

In the past decades and especially today, the downtowns of large cities have been mainly made up of commercial buildings, and the owners or the caretakers of these buildings make efforts to develop strategies and plans for their upkeep and to control their equipment. One of the said strategies is predictive maintenance (PdM), which is defined as a strategical monitoring approach that optimizes the usability of a particular equipment/system [1]. On a related note, PdM 4.0, which is in line with Industry 4.0/Quality 4.0, can determine the best time to detect equipment/system faults using machine learning (ML) models or artificial intelligence (AI) [2]. Bousdekis and others have outlined the benefits of developing the said strategies, especially PdM 4.0, and indicated that they have shown a positive impact for improving many aspects related to the organizations, such as maintenance and operation costs, replacement costs, repair downtime and verifications, machine failures, spare part stock, part service life, production, operator safety, and overall profit [3]. Using the outputs of a novel AI approach [4], PdM 4.0 can be considered a control task that maintains buildings efficiently. Moreover, PdM 4.0 ensures the sustainability of the buildings, as it allows the human and the machine to be harmonized [5,6]. In contrast, Achouch and others discussed the challenges of PdM 4.0 regarding four aspects, which are financial and organizational limitations, the limitations of data sources, activity limitations for repairing machines, and the limitations in the deployment of industrial PdM models [7].

This article focuses on one of the most important building utility systems, which is the chilled water system (CWS). It is part of the heat, ventilation, and air conditioning (HVAC) system and contains four main components, which are chillers, cooling towers, pumps, and terminal units, which are operated in an interactive way [8]. It plays a significant role in controlling the ambient temperature, which should meet the satisfaction of the buildings' occupants [9]. Furthermore, efficiently maintaining a CWS will prevent premature replacement of its components and save energy [8]. Therefore, the goal of this research was to present a comprehensive PdM 4.0 program for CWS via a methodological framework. There are a number of studies that have presented PdM 4.0 programs for CWS, focusing on either on one, two, or three components of the system to predict its faults. These faults were predicted through ML techniques such as the decision tree (DT) algorithm [10–12], artificial neural network algorithm [13–15], and support vector machine algorithm [16–18]. Moreover, a systematic literature review (SLR) addressed PdM 4.0 applications in 168 studies on CWS [19]. This research followed this SLR study, which was underpinned by two research questions, responded to three research gaps, and proposed a route for PdM 4.0. Table 1 shows the research questions and the research gaps, while Figure 1 visualizes the proposed PdM 4.0 route.

Table 1. Research Questions and Gaps.

Research Question	Research Gap
(1) How can faults be identified, in order to predict them?	(1) The literature did not consider the same faults and only concentrated on selected faults, as some faults were either not stated/mentioned or were not fully described.
	(2) The current literature does not specify how data were collected or justify the period or the frequency of the collected data, as well as being limited to testing the model and not controlling it.
(2) What are the methods that can be used to predict the faults?	(3) The suggested programs/frameworks/models did not contain, or contained inconclusive, solutions for the mentioned faults from a management point of view, as they ended at how to detect/predict the faults. Moreover, these programs did not comprehensively study/cover the whole system.

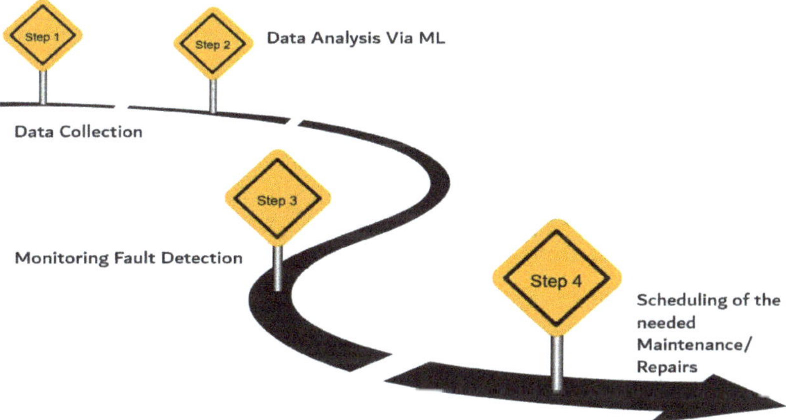

Figure 1. PdM 4.0 Route.

In addition, this research utilized the outcomes of an industry survey (IS) study [20] in building the framework. The IS study provided two tools that were used in this research, which were the fault frequencies and fault solutions. The utilization of these tools are explained in the next sections. Following Jebreen [21], and as the IS study collected quantitative and qualitative data from a number of professional participants [20], this research employed an inductive approach that proposed a methodological framework for a PdM 4.0 program for CWS in commercial buildings. From a philosophical point of view, this research follows the pragmatic paradigm, which was introduced by William James in 1898 [22]. This is defined as a philosophical tradition that considers ideologies as instruments for prediction as well as for problem solving [23]. According to Sakib and Wuest, the aforementioned research paradigm is ideal for PdM research [24]. Therefore, this research is an extension of the previously published SLR and IS studies [19,20]. It rectifies the research gaps that were identified by the SLR study and applies the recommendations of the IS study.

From an ML point of view, this research utilized the DT algorithm within the proposed framework; as the SLR study indicated that DT typically shows a high accuracy for predicting faults that affect the condition of a CWS over time [19]. The next subsection gives an overview of the mentioned algorithm.

1.2. Decision Tree Algorithm

DT is a common ML algorithm that is mainly used for classification, prediction, and regression applications. It has many benefits, and Sharma and Kumar argued that it can be used to predict continuous and discrete values [25]. They also indicated that it can capture nonlinear relationships, as well as being easier to use than other ML algorithms for understanding, interpretation, and visualization [25].

The DT has a tree-like structure, with a root node and intermediate nodes that split into branches. The last intermediate node is split into leaves and is terminated with an end node. Each node represents a classification or prediction feature. A branch or a leaf represents the possible value of the feature. The path from the root node to the end node is labeled using the predicted outcome or target classification, which is assigned using an existing training dataset. Using supervised training algorithms, the features are split recursively from top down according to certain criteria. Figure 2 depicts the general structure of the tree [26].

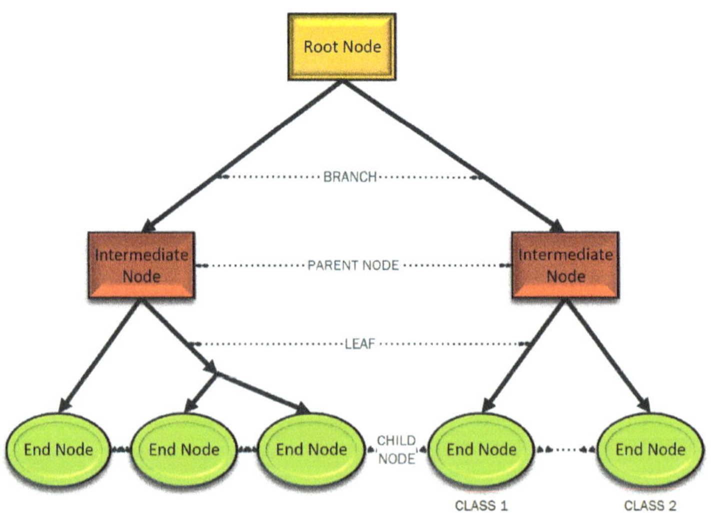

Figure 2. General Structure of the DT.

2. Methodological Framework

Constructing management frameworks for projects, continuous activities, or any other core program of building facility management gives structure to the program and allows corrective measures that can achieve related goals [27]. The framework in this research was built from an engineering management point of view, where each part of the framework contains multiple managerial steps. Table 2 describes the parts of the proposed framework, as well as the objective of each part.

Table 2. Framework Structure.

Part	Objective
Setup	• To understand the CWS at the building under study, in order to identify the numbers of each component, as well as their location at the site; • To ensure that the data reading tools are in the right locations; • To prepare the data collection plan, which includes data collection tools, determining the schedule of data collection, and formation of the team who will collect the data.
Machine Learning	• To formulate the algorithm, train the prediction model, and test it.
Quality Control	• To make a control plan for the maintenance program and evaluate the prediction model.

The above parts should be followed in the same logical order as shown and detailed in the next subsections.

2.1. Setup Part

In order to prepare the framework, three stages are suggested, to be gone through in the same order as they are listed in the following subsubsections.

2.1.1. CWS Drawing

As recommended by SLR [19], the first step in preparing the framework is to understand the as-built drawing of CWS at the building under study, in order to determine the number of CWS components installed there and to determine their locations around the site; and then to study the whole system accordingly. Such drawings show the actual building layout and are normally handed over to the facility management after completion of the building construction [28]. Following the standard in [29], this research made a simplified schematic CWS drawing; in order to easily identify the numbers of each component, as shown in Figure 3.

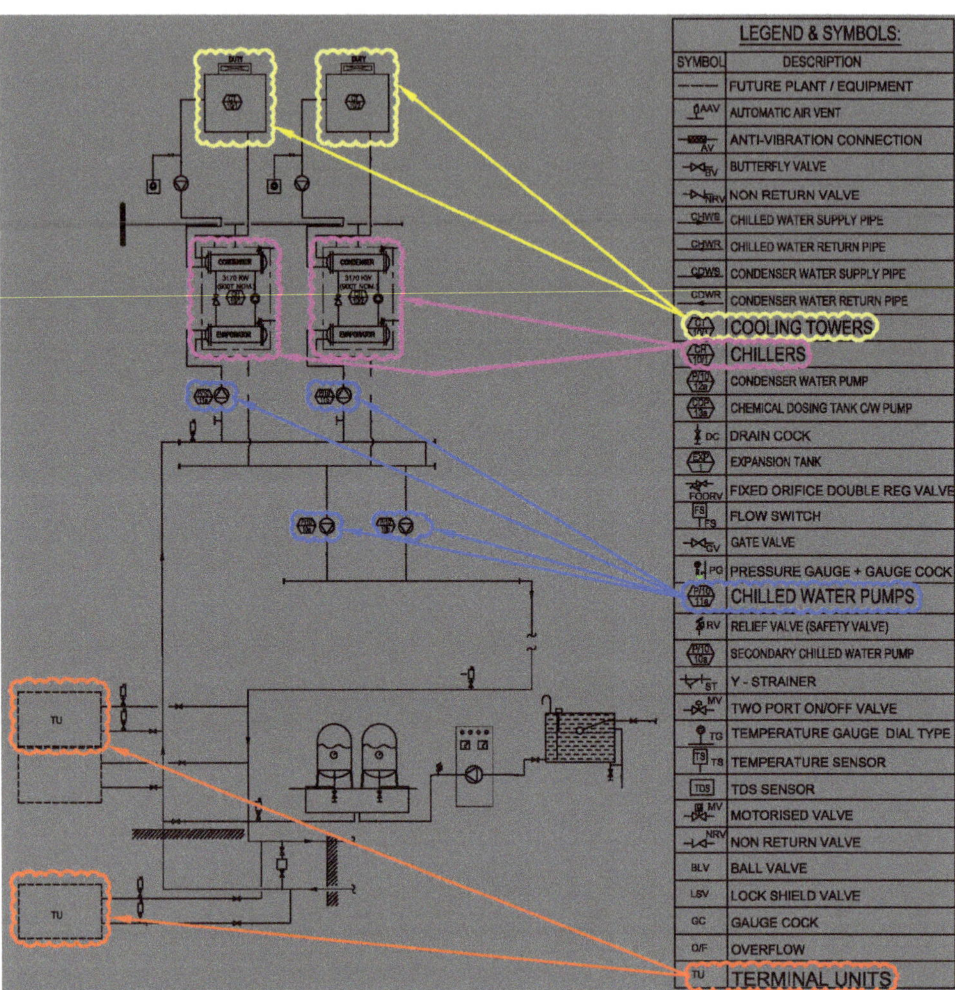

Figure 3. CWS As-Built Drawing.

2.1.2. Reading Tools for Operational Parameters

Following the SLR and IS studies [19,20], one of the fundamentals of PdM 4.0 is the datasets that contain the readings of the CWS operational parameters. Here, operational parameters are defined as quantifiable factors that give numerical data about the performance of the CWS [19]. In this research, the operational parameters chosen were the temperature of water leaving the chillers and cooling towers, pressure for pumps, and the space temperature for terminal units; as they are the best for showing the health condition of these components [20].

In order to collect the readings of these parameters, the associated tools were assumed to be available at the building under study. The measurement tools can be meters, gauges, sensors, thermostats, or any other agent, such as the building management system (BMS). In case of the unavailability of these reading tools, the authors in [30,31] outlined procedures on how to install such tools. Following the standard operating procedure in [29], Table 3 shows the best location for installing the reading tool for each CWS component in the building under study.

Table 3. Best Location for Reading Tools.

CWS Component	Location
Chiller	Chilled water supply header
Cooling Tower	Straight pipeline entering the condenser
Pump	Discharge pipeline
Terminal Unit	1.5 m above the floor level in a space or in the return air duct

Once the reading tools are installed, they have to be connected to the computer unit (CU) that will be used in the PdM program. Kayastha and others outlined a procedure for how to connect such tools to computers [32]. This course of action was utilized and is explained in the third subsection of the methodological framework section (Quality Control).

2.1.3. Data Collection

After determining the numbers of each component and finalizing the reading tools, the last stage of the setup is data collection. The IS study proposed time frequencies to collect data in a building [20]. Following these proposals, the readings of water temperature leaving a particular chiller should be taken every thirty minutes over a study period of twelve weeks. The same should be applied for cooling towers, but over a study period of sixteen weeks. With regard to pumps, the readings of pressure should be taken every hour over a study period of twenty-four weeks. For terminal units, the readings of space temperature should be taken every forty-five minutes over a study period of eight weeks. The SLR and IS studies suggested utilizing a check sheet to collect the data for each component, which should contain the readings as well as the inspection results. The inspection results will be either '1' in case of fault or '0' in case of no fault. As recommended by the IS study, the check sheet must be filled out by experienced technicians or users [20]. Each check sheet should be recorded by two team members, one for the morning and part of the afternoon shift, and one for the evening and the second part of the afternoon shift. Appendix A shows a proposed check sheet for terminal units, and the same was applied for other CWS components, taking into consideration the differences in the time intervals and the unit of the operational parameters between the components. After collecting the data, a file for each particular component should be created in Excel, and then the information from the related check sheet should be logged. Thus, each file should contain two columns, one for the readings and the another for the inspection results [20], and then it should be saved in the CU in csv format. Therefore, these files present the required datasets, and at to this point, the setup is completed, and accordingly the ML part can be started, as explained in the next subsection.

2.2. Machine Learning Part

In this part, two DT algorithms are recommended for use, which are the C4.5, a successor of the iterative dichotomiser 3 (ID3), and the classification and regression tree (CART) algorithm, as they are efficient for splitting the trees [33]. The basic principle of the splitting mechanism is to select a root node from the 'N' features and subsequently decide which attribute should be used next as the intermediate node. Different statistical criteria should be used to make these decisions, such as the Gain Ratio and the Gini Index. According to Grąbczewski [34], the Gain Ratio criterion is mainly used in the C4.5 algorithm, while the Gini Index is used in the CART algorithm. The Gain Ratio is calculated as in Equation (1):

$$Gain\ Ratio_{(A)} = \frac{Information\ Gain}{SplitInfo} = \frac{Entropy(parent) - \sum_{j=1}^{k} Entropy(j, child)}{\sum_{j=1}^{k} \frac{D_j}{D} log_2 \frac{D_j}{D}} \tag{1}$$

In information theory, entropy measures the uncertainty in data. The entropy(parent) measures the amount of randomness (impurity) in the parent node before it splits. D is the number of instances in the parent node and Dj is the number of instances in the child j, and k is the number of discrete values of an attribute A, which is tested at the parent node for splitting. The entropy at each child node is found using Equation (2):

$$Entropy = -\sum_{i=1}^{n} p_i \, log_2 \, p_i \tag{2}$$

where p_i is the probability of selecting an instance in class i, and n is the number of classes. The attribute that is selected for splitting the parent node is the one with the highest Gain Ratio. Similarly, the Gini Index for the CART algorithm can be found by Equation (3):

$$Gini \; Index_{(A)} = \sum_{j=1}^{k} \frac{D_j}{D} Gin(j, child) \tag{3}$$

Similarly to the Entropy, the Gini Index measures the impurity at the parent node. The Gini of a child node is found using Equation (4):

$$Gini = 1 - \sum_{i=1}^{n} p_i^2 \tag{4}$$

The attribute that is selected for splitting at the parent node is the one with the smallest Gini Index. In this research, this attribute is the operational parameter of each CWS component. Many programming languages/software can read collected data and train the prediction model, such as Python [35]. The software should be installed in the CU and the required codes should be written in a way that allows reading the files (datasets) for each CWS component, which were mentioned in the data collection stage of the setup part, and then to train and test the model. The next section of this article (Implementation and Discussion) gives a case study on how a prediction model is trained and tested.

2.3. Quality Control Part

This is the last part of the proposed framework, and its goal is to ensure the prediction model is working correctly, as well as to rectify the faults immediately. To do this, this research suggests making a control plan, which should contain monitoring and response actions [36]. Table 4 clarifies the descriptions of these two actions, as well as who is responsible for executing each action.

Table 4. Control Plan.

Quality Control Action	Description	Responsible
Monitoring	The prediction model should be connected to the reading tools, which were connected to the CU during the setup part. This is to ensure that the CU shows a continuous reading for each CWS component.	Information Technology (IT) Department or Programming Supplier
Response	When the prediction model shows a fault, which is a "1" as a result of a particular reading, the related component should be inspected and then to be rectified as per the solutions tabulated in the IS article [20].	Facility Department Officer/technician

After that, the response actions should be documented as follows:

- Listing the lessons learned from the proposed PdM program, such as focusing on the faults that occurred, and then brainstorming permanent solutions to avoid the reoccurrence of such faults;

- Tracking the spare part stock;
- Ensuring that the CU is working efficiently;
- Training more technicians to be familiar with the prediction model;
- Making regular reports about the performance of the proposed PdM program for future improvements.

3. Implementation and Results

This section presents a case study on the proposed framework. The case study was performed at a university in Riyadh city, Kingdom of Saudi Arabia. Implementation of the framework was carried out as per the three parts proposed in the previous section (Methodological Framework).

3.1. Implementaion of Setup Part
3.1.1. CWS Drawing

The main goal of the proposed framework is to make a PdM 4.0 program that considers the whole CWS (i.e., all CWS components). Therefore, to start implementing the framework, the CWS as-built drawing was collected and then, following Figure 3, the numbers of each CWS component were determined, as shown in Table 5, as well as their locations around the site.

Table 5. Number of CWS Components.

CWS Component	Quantity
Chiller	5
Cooling Tower	7
Pump	19
Terminal Unit	72

3.1.2. Reading Tools

At this stage, the standard shown in Table 3 was followed, and it was ensured that the reading tools for the operational parameters of each CWS component were in the best location. Figures 4–7 show the reading tool location for each CWS component. As stated in the previous section, these tools read the temperature for water leaving each chiller and cooling tower, the pressure for pumps, and the space temperature for terminal units.

Figure 4. Chiller Reading Tool.

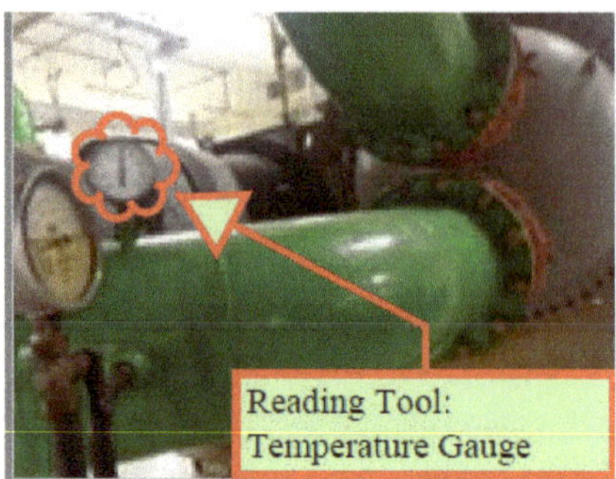

Figure 5. Cooling Tower Reading Tool.

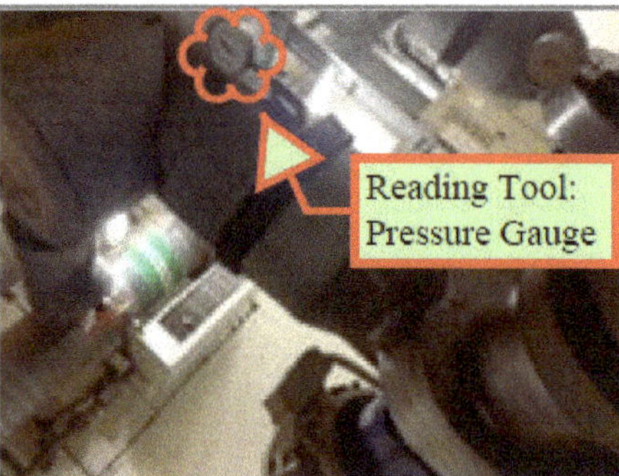

Figure 6. Pump Reading Tool.

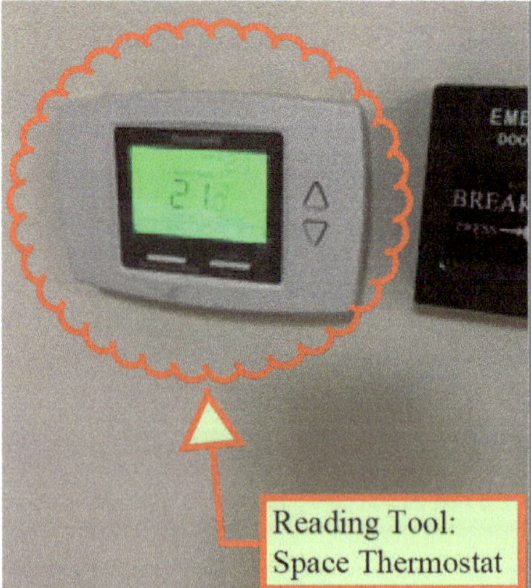

Figure 7. Terminal Unit Reading Tool.

Through the IT department of the university, these reading tools were connected via sensors to a CU, to be ready for the quality control part.

3.1.3. Data Collection

The most important stage in setting up the PdM 4.0 program was the datasets required to build the prediction model. As the previous two stages had been finalized, data collection was started as per the recommendations of the IS study [20]. This research used the recommended minimum frequencies as time intervals when collecting the data, and the recommended maximum frequencies were used as study periods for each CWS component.

Twelve qualified technicians from the university were assigned for the subject matter. Two operational units for each CWS component were selected as subjects. The readings for the water temperatures of each chiller and cooling tower were collected using check sheets. The same was performed for the pressures for each pump and the space temperatures for each terminal unit. In addition, the inspection result, which was either a fault "1" or fault free "0", was included for each check sheet. Appendix B illustrates a fully filled one day check sheet for a particular pump, and Table 6 shows the data collection plan.

Table 6. Data Collection Plan.

CWS Component	Time Interval for Reading and Inspection (Minutes)	Study Time (Weeks)	Study Period
Chiller	30	12	From 29 May 2022 to 20 August 2022
Cooling Tower	30	16	From 29 May 2022 to 17 September 2022
Pump	60	24	From 29 May 2022 to 12 November 2022
Terminal Unit	45	8	From 29 May 2022 to 23 July 2022

After that, an Excel file was created for each component, and the information in all related check sheets was transferred to the associated Excel file. Following the procedure proposed in the methodological framework section, each Excel file represented a dataset that contained two cells, one for the readings and another for the inspection results, as shown in Appendix C for a one of the cooling towers. After that, each file was named and saved in csv format. For example, for a particular pump, the file was named and saved as "pu.csv"; so it could be read when training the prediction model, as shown in the next section.

3.2. Implementation of the Machine Learning Part

A DT was built for each CWS component selected. As stated in the methodological framework section, the faults of each component were predicted using the related attributes. Table 7 shows the attribute and the training data size for each unit of the selected CWS components that were in operation.

Table 7. Main Inputs of The Prediction Model.

CWS Component	Attribute	Data Size
Chiller	Water Leaving Temperature (°C)	2688
Cooling Tower	Water Leaving Temperature (°C)	3584
Pump	Pressure (Bar)	2688
Terminal Unit	Space Temperature (°C)	1288

The C4.5 and CART algorithms were used to train the tree. Different training parameters were used to optimize the tree accuracy. The parameters included the training to testing ratio and the level of pruning. The model was executed in Python, with the script shown in Figure 8 being for a particular pump. The same was done with other CWS components, taking into consideration the changes in file reading/loading.

```python
# Load libraries
!pip install pydotplus
import csv
import numpy as np
import pandas as pd
from sklearn.tree import DecisionTreeClassifier # Import Decision Tree Classifier
from sklearn.model_selection import train_test_split # Import train_test_split function
from sklearn import metrics #Import scikit-learn metrics module for accuracy calculation
# load training dataset
load_file= pd.read_csv("pu.csv")
#split dataset in features and target variable
feature_cols = ['Pressure']
X = load_file.iloc[:, :-1].values  # Features
y = load_file.iloc[:, -1].values  # Target variable
# Split dataset into training set and test set
X_train, X_test, y_train, y_test = train_test_split(X, y, test_size=0.3, random_state=1) # 70% training and 30% test
# Create Decision Tree classifer object
clf = DecisionTreeClassifier(criterion='gini',max_depth = 1)

# Train Decision Tree Classifer
clf = clf.fit(X_train,y_train)

#Predict the response for test dataset
y_pred = clf.predict(X_test)
# Model Accuracy, how often is the classifier correct?
print("Accuracy:")
#print("Predicted values:")
out1 =metrics.accuracy_score(y_test, y_pred)
print(out1)
#write output to file CSV
f=open('results.csv', 'w')
writer=csv.writer(f, delimiter=',')

for i in range(0, len(y_pred)):
    float_list = [X_test[i],y_pred[i]]
    #print(len(y_pred),i,float_list)
    writer.writerow(float_list)

f.close()
import matplotlib.pyplot as plt
from sklearn.tree import plot_tree
plt.figure(figsize=(10,8), dpi=70)
plot_tree(clf, feature_names=feature_cols, filled=True);
```

Figure 8. DT Python Code.

The initial run of the training stage was performed without pruning, which led to prediction overfitting, as can be seen for the chiller tree in Figure 9. The same was done for the other CWS components.

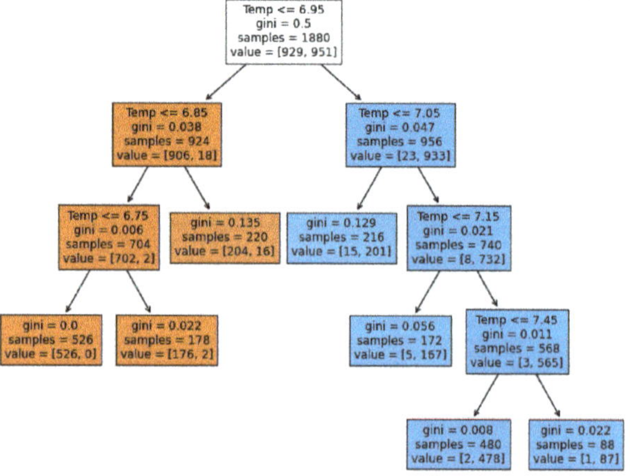

Figure 9. Chiller DT without Pruning.

Examining the different pruning methods, the optimally trained trees for each CWS component were found, as shown in Figures 10–13.

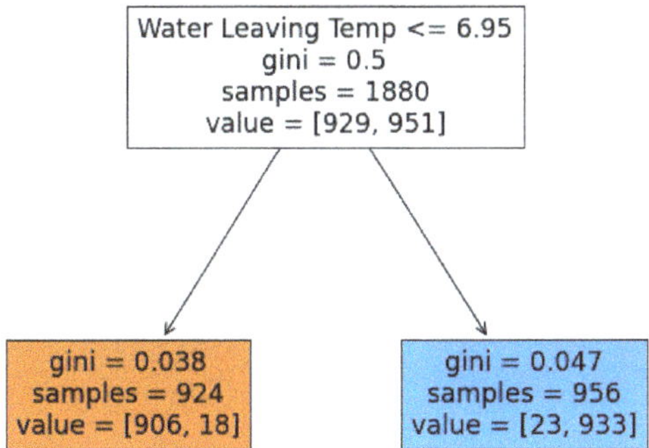

Figure 10. Chiller DT.

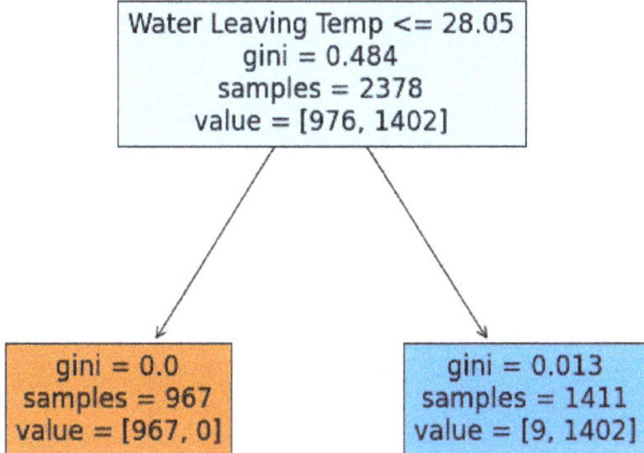

Figure 11. Cooling Tower DT.

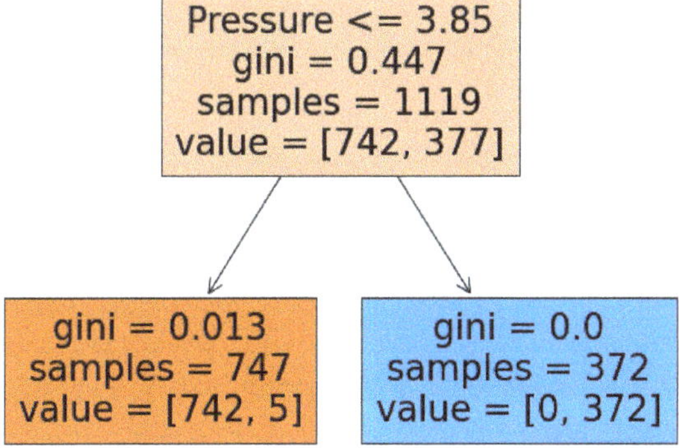

Figure 12. Pump DT.

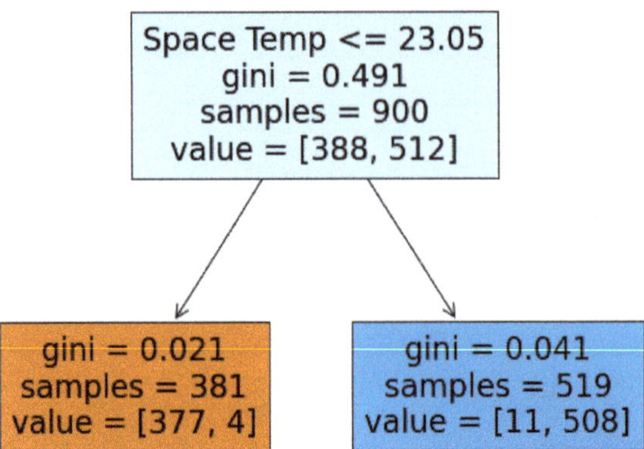

Figure 13. Terminal Unit DT.

Changing the training to testing ratio and the training algorithms had a very small impact on the prediction accuracy. A 70 to 30 percent training to testing ratio was adopted using the CART training algorithm. The prediction accuracies of each component at the optimal DT setting are presented in Table 8.

Table 8. CWS Component Prediction Accuracies.

CWS Component	Prediction Accuracy (%)
Chiller	98.50
Cooling Tower	99.60
Pump	99.80
Terminal Unit	99.20

3.3. Implementation of the Quality Control Part

After successfully building the prediction model, the control plan mentioned in Table 4 was actioned. In the monitoring of the control plan, the prediction model was connected to the CU, in order to begin the second stage of the plan (Response). After that, the readings of all CWS components at the university, which are shown in Table 5, were observed daily, as per the minimum frequencies mentioned in Table 6, for a month time, excluding weekends. For example, for a particular terminal unit, the reading of space temperature was observed every 45 min. During this period, the DT model predicted 16 different faults in the chillers, 11 different faults in the cooling towers, 12 different faults in the pumps, and 19 different faults in the terminal units, noting that the occurrence of some faults was repeated. Table 9 lists how many faults in total were predicted by the DT model within the said period, as well as the fault that occurred most for each CWS component. All fault signals from the CU, which were displayed as "1", led to real faults around the site. Thereafter, all the faults within this period were solved immediately after their appearance, by following the solutions mentioned in [20].

Table 9. Number of CWS Faults.

CWS Component	Number of Faults	Most Occurred Fault
Chiller	101	Refrigeration Leak
Cooling Tower	113	Malfunctioning Blowdown System
Pump	79	Noisy Non-Return Valve
Terminal Unit	138	Low Static Pressure

Furthermore, the facility department at the university was advised to keep observing the readings as was convenient and to inspect the site in case of a fault "1". In addition, they were advised to document the response action of the control plan, as per the steps proposed in the previous section (Methodological Framework). On a related note, the existing monitoring system implemented by the department was a BMS, and they were asked to give a report for the same period that was used for observing the whole CWS via the DT model. The report contained the total number of faults that were predicted by the BMS for each CWS component. Figure 14 shows a comparison between the DT model and the BMS in predicting the faults within the same period.

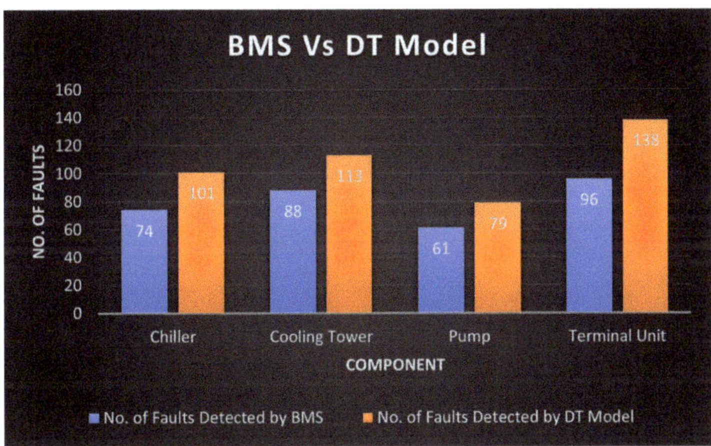

Figure 14. Comparison of Prediction Performance.

4. Discussion

The case study applied a methodological framework proposing three parts to build an efficient PdM 4.0 program. During the setup part, the first stage gave an overview of the building CWS, by determining the number of units of each component, as well as identifying their locations on site. This action was easily carried out using the schematic shown in Figure 3. The second stage was to give a clear picture about the best location for the reading tools for each CWS component. The third stage was to produce a clear map on how to collect the data, as the main objective of the setup part was to create a dataset for each CWS component. These datasets were essential to allow beginning the second part of the proposed framework. To recall what was mentioned in the previous sections, each row of a dataset contained a reading of the operational parameter in one column and its associated inspection result in another column.

In the second part, the datasets were used in building the ML model. The results were encouraging, as the DT model showed a very high prediction accuracy for each CWS component, as shown in Table 8. This confirmed that the fault frequencies proposed in [20], which were used while collecting the data of this research, are valid for tracking faults. In the third part of the proposed framework, the aforementioned control plan in Table 4 facilitated the execution of the prediction model. The empirical period of this part provided the following findings:

- The C4.5 and CART algorithms had a similar prediction accuracy for each CWS component.
- The DT model had a better performance than the BMS in predicting the faults for all CWS components, as shown in Figure 14. This fulfilled the requirements of the facility department, who manage the CWS at the university.
- The most frequent fault in chillers was refrigeration leaks. This was also confirmed by the SLR study [19], as well as the IS study [20], which reported this fault as the most common in chillers;

- A malfunctioning blowdown system was the most common fault in the cooling towers. This finding matches what was found in the IS study [20]. The IS study stated that the majority of the survey's participants suffered from this fault;
- With regard to the pumps, a noisy non-return valve occurred most often. This also matches the information provided by the IS study [20], where the majority of the survey's participants faced this fault continuously;
- Low static pressure in the terminal units occurred more than twice a day. The IS study [20] had already confirmed that the most of the survey's participants were finding this fault on a regular basis while operating the associated terminal unit;
- The solutions provided in the IS study [20] gave practical actions to rectifying the predicted faults. In this regard, one of the research gaps listed by the SLR study was that the previous 168 studies considered did not cover the whole CWS (i.e., all for components) and ended their PdM programs once detecting the faults [19]. However, the SLR study recommended having control measures, including fault solution, after completing the prediction model, which will allow a comprehensive PdM program, such as the proposed framework.

As stated in the first section of this article (Introduction), this research is a response to the mentioned SLR study [19] as well as the IS study [20]. Considering the gaps in Table 1, this study covered the first gap and prepared tools to rectify the second and the third gaps [20]. The first tool in the IS study was the frequencies, which were used in collecting the data and in controlling the whole CWS. The second tool was the solutions to faults, which were used in the quality control part. Therefore, this research has contributed to building a framework that will provide a comprehensive PdM 4.0 program for the whole CWS in commercial buildings. On a related note, this framework was implemented at another site for external validity purposes. The site is a hotel that is related to the same foundation that manages the university. The DT's prediction accuracy for each CWS component was similar to those at the university. Although this framework has obtained encouraging results, it has some challenges from a research point of view, as follows:

- The availability of the data source;
- The experience of the team who collect the data;
- The organizational culture at the building, which may not be cooperative;
- The associated costs, such as arranging the reading tools, the CU, and the labor.

5. Conclusions

This research proposed a methodological framework for a PdM 4.0 program for commercial buildings. A framework was made for one of the most important utility systems of the commercial buildings, the CWS, which has four components. These are the chillers, cooling towers, pumps, and terminal units. The framework contains three parts, which are the setup, ML, and quality control. Each part of this framework has multiple managerial stages or steps to build the maintenance program.

The setup part of this framework contained three stages. The first stage allowed efficiently understanding the building through analyzing its as-built drawing. By doing so, it was possible to determine the unit numbers of each CWS component in the building, as well as to know their locations on site. A schematic was made in this regard, to make a simplified view of such drawings. The second stage of the setup part focused on the reading tools for the CWS operational parameters. How to make the reading tools available and the best location for each tool was discussed. The readings of the operational parameters were essential for creating the datasets that are were used in the second part of the framework, ML. The operational parameters chosen in this research were the water temperatures for chillers and cooling towers, the pressures for pumps, and the space temperatures for terminal units. The third and last stage of this part addressed the data collection. It presented the data required and proposed a complete plan for collecting them. Therefore, the main goal of the setup part was to provide the datasets that were required to build a prediction model, which was explained in the second part of this framework, ML. As this research was intended to implement a PdM 4.0 program, the second part of the framework utilized ML. The DT technique was chosen to build a model for predicting the CWS faults, as recommended by the SLR study. Two DT algorithms, C4.5 and CART, were proposed to build, train, and test the model. The last part of the framework, which was quality control, proposed a control plan to evaluate the prediction model. The control plan required two actions, which were monitoring and response. Both actions proposed executing the DT model while operating the CWS, and then to control the system via many aspects, such as solving the faults predicted and documenting the outcomes of the predictive model from an engineering management point of view.

This research implemented the proposed framework in a university in Riyadh city, Kingdom of Saudi Arabia. The DT model produced encouraging results, where the prediction accuracy was 98.50 percent for chillers, 99.60 percent for cooling towers, 99.80 percent for pumps, and 99.20 percent for terminal units. Furthermore, the DT model was evaluated over an empirical period. The model gave outstanding performance in predicting the faults of all CWS components, especially when it was compared to the BMS, which was the existing control system at the university. During the said period, the DT made a 27 percent improvement in predicting the faults for chillers, 22 percent for cooling towers, 23 percent for pumps, and 31 percent for terminal units. On a separate note, refrigeration leaks, malfunctioning blowdown systems, noisy non-return valves, and low static pressure faults occurred often during this period in chillers, cooling towers, pumps, and terminal units, respectively. This confirmed the information provided by the IS study with regard to the most common faults.

Though this research, along with the SLR and IS studies, provided significant outcomes towards implementing PdM 4.0 for CWS in commercial buildings, future research agendas could explore further insights about this topic, as follows:

- To discuss how to integrate the ML models with the building automation and management systems such as BMS, for a more efficient prediction model;
- To propose an intelligent system for updating the datasets, which are required to build the prediction model, in order to rise the control efficiency of commercial buildings;
- To investigate and give more focus to the repeated occurrence of faults, especially the aforementioned four faults, which are refrigeration leaks in chillers, malfunctioning blowdown systems in cooling towers, noisy non-return valves in pumps, and low static pressure in terminal units;
- To use the ideas of this research, which built the framework, and extend them to other HVAC systems such as heating systems, as well as for other utility systems, such as the electrical system.

Author Contributions: Conceptualization, M.A. and K.M.; methodology, M.A.; investigation and analysis, M.A. and A.A.; writing—original draft, M.A.; writing—review and editing, A.A.; data curation, M.A.; visualization, M.A.; supervision, K.M. All authors have read and agreed to the published version of the manuscript.

Funding: This research received no external funding.

Institutional Review Board Statement: Not applicable.

Informed Consent Statement: Not applicable.

Data Availability Statement: Data available on request due to privacy restrictions.

Acknowledgments: Many thanks to the DMEM department at the University of Strathclyde, especially Anja Maier, Jorn Mehnen, Xichun Luo, Ian Whitfield, Laura Hay, and Gillian Eadie for their guidance, advice, and support. Many thanks also to Alfaisal University President (Mohammed Alhayaza), VP External relation and advancement (H.R.H. Maha Bint Mishari Bin Abdulaziz AlSaud), VP Admin and Finance (Khaled AlKattan), and Acting Dean of College of Engineering (Muhammad Anan) for their unlimited support. A special thanks goes to the Associate Dean of Alfaisal's College of Engineering for Academic and Student Affairs (Sobhi Mejjaouli) for his perpetual cooperation and support. In addition, thanks go to the operation and maintenance manager at Alfaisal University (Kaleemoddin Ahmed) and his team for their great assistance during this research.

Conflicts of Interest: The authors declare no conflict of interest.

Appendix A

Check Sheet (Building Name: ....)

Component: Terminal Unit #.......

Day & Date: ....

Time	Space Temperature (°C)	Fault Free (0) Fault (1)	Time	Space Temperature (°C)	Fault Free (0) Fault (1)
6:30			14:45		
7:15			15:30		
8:00			16:15		
8:45			17:00		
9:30			17:45		
10:15			18:30		
11:00			19:15		
11:45			20:00		
12:30			20:45		
13:15			21:30		
14:00			22:15		
			23:00		
Inspector Name:			Inspector Name:		
Signature:			Signature:		

Appendix B

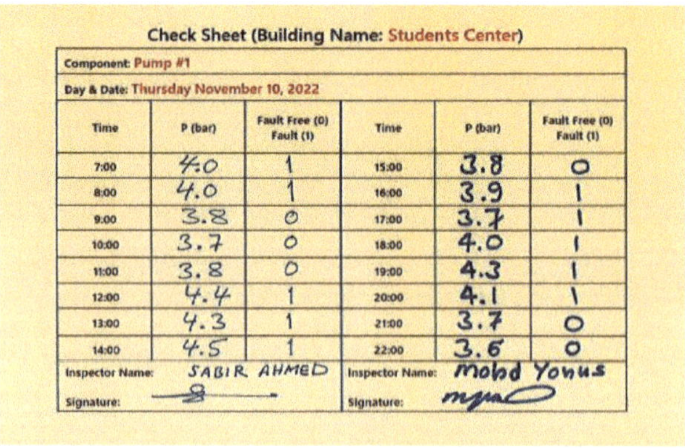

Check Sheet (Building Name: Students Center)

Component: Pump #1

Day & Date: Thursday November 10, 2022

Time	P (bar)	Fault Free (0) Fault (1)	Time	P (bar)	Fault Free (0) Fault (1)
7:00	4.0	1	15:00	3.8	0
8:00	4.0	1	16:00	3.9	1
9:00	3.8	0	17:00	3.7	1
10:00	3.7	0	18:00	4.0	1
11:00	3.8	0	19:00	4.3	1
12:00	4.4	1	20:00	4.1	1
13:00	4.3	1	21:00	3.7	0
14:00	4.5	1	22:00	3.6	0
Inspector Name:	SABIR AHMED		Inspector Name:	Mohd Younus	
Signature:			Signature:		

Appendix C

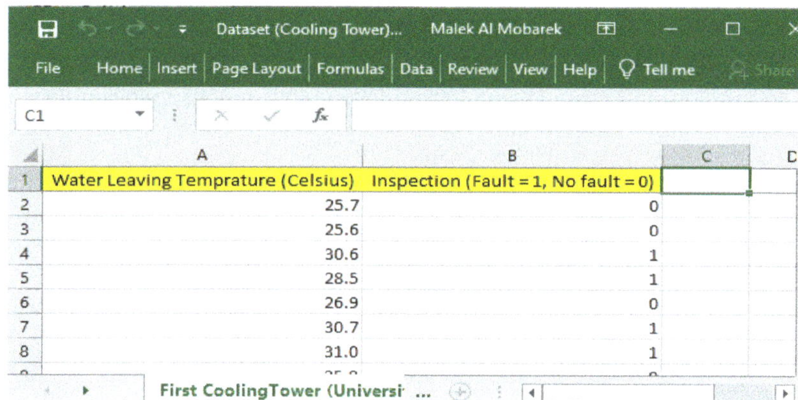

References

1. Kullu, O.; Cinar, E. A Deep-Learning-Based Multi-Modal Sensor Fusion Approach for Detection of Equipment Faults. *Machines* **2022**, *10*, 1105. [CrossRef]
2. Sahal, R.; Breslin, J.G.; Ali, M.I. Big data and stream processing platforms for Industry 4.0 requirements mapping for a predictive maintenance use case. *J. Manuf. Syst.* **2020**, *54*, 138–151. [CrossRef]
3. Bousdekis, A.; Apostolou, D.; Mentzas, G. Predictive maintenance in the 4th industrial revolution: Benefits, business opportunities, and managerial implications. *IEEE Eng. Manag. Rev.* **2019**, *48*, 57–62. [CrossRef]
4. Cotrufo, N.; Saloux, E.; Hardy, J.M.; Candanedo, J.A.; Platon, R. A practical artificial intelligence-based approach for predictive control in commercial and institutional buildings. *Energy Build.* **2020**, *206*, 109563. [CrossRef]
5. Simon, L.; Rauffet, P.; Guérin, C.; Seguin, C. Trust in an autonomous agent for predictive maintenance: How agent transparency could impact compliance. In Proceedings of the 13th AHFE Conference (Applied Human Factors and Ergonomics), New York, NY, USA, 24–28 July 2022; HAL Open Science: Lyon, France, 2022; pp. 61–67.
6. Villa, V.; Bruno, G.; Aliev, K.; Piantanida, P.; Corneli, A.; Antonelli, D. Machine Learning Framework for the Sustainable Maintenance of Building Facilities. *Sustainability* **2022**, *14*, 681. [CrossRef]
7. Achouch, M.; Dimitrova, M.; Ziane, K.; Sattarpanah Karganroudi, S.; Dhouib, R.; Ibrahim, H.; Adda, M. On Predictive Maintenance in Industry 4.0: Overview, Models, and Challenges. *Appl. Sci.* **2022**, *12*, 8081. [CrossRef]
8. Almobarek, M.; Mendibil, K.; Alrashdan, A. Faults handling in chilled water system maintenance program. In Proceedings of the 12th International Conference on Industrial Engineering and Operations Management, Istanbul, Turkey, 7–10 March 2022; IEOM Society International: Southfield, MI, USA, 2022; pp. 1616–1625.
9. Almobarek, M.; Mendibil, K.; Alrshdan, A. Study of Factors Influencing Room Ambient Temperature Using Design of Experiments. In Proceedings of the 11th International Conference on Industrial Engineering and Operations Management, Singapore, Singapore, 7–11 March 2021; IEOM Society International: Southfield, MI, USA, 2021; pp. 1509–1516.
10. Li, D.; Zhou, Y.; Hu, G.; Spanos, C.J. Fusing system configuration information for building cooling plant Fault Detection and severity level identification. In Proceedings of the 2016 IEEE International Conference on Automation Science and Engineering (CASE), Fort Worth, TX, USA, 21–25 August 2016; IEEE: Piscataway, NJ, USA, 2016; pp. 1319–1325.
11. Yan, R.; Ma, Z.; Kokogiannakis, G.; Zhao, Y. A sensor fault detection strategy for air handling units using cluster analysis. *Autom. Constr.* **2016**, *70*, 77–88. [CrossRef]
12. Tehrani, M.M.; Beauregard, Y.; Rioux, M.; Kenne, J.P.; Ouellet, R. A Predictive Preference Model for Maintenance of a Heating Ventilating and Air Conditioning System. *IFAC-Pap* **2015**, *48*, 130–135. [CrossRef]
13. Rueda, E.; Tassou, S.A.; Grace, I.N. Fault detection and diagnosis in liquid chillers. *Proc. Inst. Mech. Eng. Part E J. Process. Mech. Eng.* **2005**, *219*, 117–125. [CrossRef]
14. Zhou, Q.; Wang, S.; Xiao, F. A novel strategy for the fault detection and diagnosis of centrifugal chiller systems. *HVACR Res.* **2009**, *15*, 57–75. [CrossRef]
15. Dudzik, M.; Romanska-Zapala, A.; Bomberg, M. A neural network for monitoring and characterization of buildings with Environmental Quality Management, Part 1: Verification under steady state conditions. *Energies* **2020**, *13*, 3469. [CrossRef]
16. Montazeri, A.; Kargar, S.M. Fault detection and diagnosis in air handling using data-driven methods. *J. Build. Eng.* **2020**, *31*, 101388. [CrossRef]
17. Liang, J.; Du, R. Model-based fault detection and diagnosis of HVAC systems using support vector machine method. *Int. J. Refrig.* **2007**, *30*, 1104–1114. [CrossRef]

18. Hu, R.; Granderson, J.; Auslander, D.; Agogino, A. Design of machine learning models with domain experts for automated sensor selection for energy fault detection. *Appl. Energy* **2018**, *235*, 117–128. [CrossRef]
19. Almobarek, M.; Mendibil, K.; Alrashdan, A. Predictive Maintenance 4.0 for Chilled Water System at Commercial Buildings: A Systematic Literature Review. *Buildings* **2022**, *12*, 1229. [CrossRef]
20. Almobarek, M.; Mendibil, K.; Alrashdan, A.; Mejjaouli, S. Fault Types and Frequencies in Predictive Maintenance 4.0 for Chilled Water System at Commercial Buildings: An Industry Survey. *Buildings* **2022**, *12*, 1995. [CrossRef]
21. Jebreen, I. Using inductive approach as research strategy in requirements engineering. *Int. J. Comput. Inf. Technol.* **2012**, *1*, 162–173.
22. Malachowski, A. *The New Pragmatism*, 1st ed.; Routledge: London, UK, 2014.
23. Bacon, M. *Pragmatism: An Introduction*; Polity Press: Cambridge, UK, 2012.
24. Sakib, N.; Wuest, T. Challenges and opportunities of condition-based predictive maintenance: A review. *Procedia CIRP* **2018**, *78*, 267–272. [CrossRef]
25. Sharma, H.; Kumar, S. A survey on decision tree algorithms of classification in data mining. *Int. J. Sci. Res.* **2016**, *5*, 2094–2097.
26. Fletcher, S.; Islam, M.Z. Decision tree classification with differential privacy: A survey. *ACM Comput. Surv. (CSUR)* **2019**, *52*, 1–33. [CrossRef]
27. Wildenauer, A.; Mbabu, A.; Underwood, J.; Basl, J. Building-as-a-Service: Theoretical Foundations and Conceptual Framework. *Buildings* **2022**, *12*, 1594. [CrossRef]
28. Ellis, G. What Are as Built Drawings? *Digital Builder Blog, Autodesk Construction Cloud.* 2021. Available online: https://constructionblog.autodesk.com/as-built-drawings/ (accessed on 2 December 2022).
29. ASHRAE Handbook. Available online: www.ashrae.org (accessed on 5 December 2022).
30. Lam, H.F.; Yang, J.H.; Hu, Q. How to install sensors for structural model updating? *Procedia Eng.* **2011**, *14*, 450–459. [CrossRef]
31. Beckmann, C.; Consolvo, S.; LaMarca, A. Some Assembly Required: Supporting End-User Sensor Installation in Domestic Ubiquitous Computing Environments. In *Ubiquitous Computing*; Davies, N., Mynatt, E.D., Siio, I., Eds.; Springer: Berlin/Heidelberg, Germany, 2004.
32. Kayastha, N.; Niyato, D.; Hossain, E.; Han, Z. Smart grid sensor data collection, communication, and networking: A tutorial. *Wirel. Commun. Mob. Comput.* **2014**, *14*, 1055–1087. [CrossRef]
33. Javed Mehedi Shamrat, F.M.; Ranjan, R.; Hasib, K.M.; Yadav, A.; Siddique, A.H. Performance Evaluation Among ID3, C4.5, and CART Decision Tree Algorithm. In *Pervasive Computing and Social Networking*; Ranganathan, G., Bestak, R., Palanisamy, R., Rocha, Á., Eds.; Springer: Singapore, 2022.
34. Grąbczewski, K. *Meta-Learning in Decision Tree Induction*; Springer: New York, NY, USA, 2014.
35. Guttag, J. *Introduction to Computation and Programming Using Python: With Applications to Understanding Data*, 2nd ed.; The MIT Press: Cambridge, MA, USA, 2017.
36. Almobarek, M.; Alrashdan, A. Water budget control using DMAIC in commercial buildings. *Int. J. Six Sigma Compet. Advant.* **2022**, *14*, 86–208. [CrossRef]

MDPI

St. Alban-Anlage 66

4052 Basel

Switzerland

www.mdpi.com

Buildings Editorial Office

E-mail: buildings@mdpi.com

www.mdpi.com/journal/buildings